# ATMOSPHERIC POLLUTION

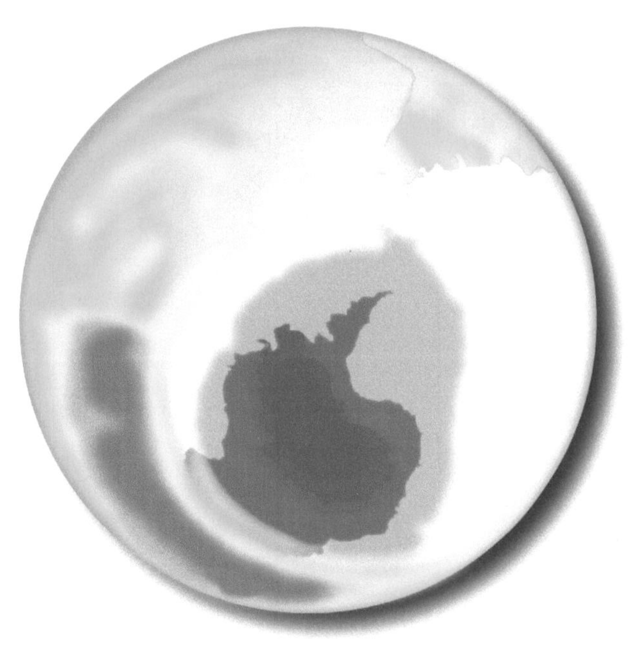

**REVISED EDITION**

**EDITED BY RANJEET S SOKHI**

ANTHEM PRESS
LONDON · NEW YORK · DELHI

In association with the International Union of Air Pollution Prevention and Environmental Protection Associations (IUAPPA) and the Global Atmospheric Pollution Forum

Anthem Press
An imprint of Wimbledon Publishing Company
*www.anthempress.com*

This edition first published in UK and USA 2011
by ANTHEM PRESS
75-76 Blackfriars Road, London SE1 8HA, UK
or PO Box 9779, London SW19 7ZG, UK
and
244 Madison Ave. #116, New York, NY 10016, USA



*British Library Cataloguing in Publication Data*
A catalogue record for this book is available from the British Library.

*Library of Congress Cataloging in Publication Data*
The Library of Congress has cataloged the hardcover edition as follows:
World atlas of atmospheric pollution / edited by Ranjeet S Sokhi.
p. cm.
Includes bibliographical references.
ISBN 978-1-84331-289-5 (hardback)
1. Air pollution. 2. Air quality management. 3. Atlases.
G1021.N852 2007
363.739'20223—dc22
2007047667

ISBN-13: 978 1 84331 891 0 (Pbk)
ISBN-10: 1 84331 891 1 (Pbk)

# CONTENTS

# CONTRIBUTORS

## LEAD AUTHORS AND CO-AUTHORS


### Lead Authors

**MARKUS AMANN** International Institute for Applied Systems Analysis (IIASA), A-2361 Laxenburg, Austria.

**MIKE ASHMORE** Environment Department, University of York, Heslingon, York YO10 5DD, UK.

**PETER BRIMBLECOMBE** School of Environmental Sciences, University of East Anglia, Norwich NR4 7TJ, UK.

**S TRIVIKRAMA RAO** NOAA Atmospheric Sciences Modelling Division, US Environmental Protection Agency, Research Triangle Park, NC 27711, USA.

**RANJEET S SOKHI** Centre for Atmospheric and Instrumentation Research (CAIR), University of Hertfordshire, Hatfield, AL10 9AB, UK.

**RICHARD S STOLARSKI** NASA Goddard Space Flight Center, Atmospheric Chemistry and Dynamics Branch, Code 916, Greenbelt, MD 20771, USA.

**DING YIHUI** National Climate Centre, China Meteorological Administration, No. 46 Zhongguancun Nan Da Jie, Haidian District, Beijing 100081, PRC.

### Co-Authors

**JANUSZ COFALA** International Institute for Applied Systems Analysis (IIASA), A-2361 Laxenburg, Austria.

**WIM DE VRIES** Alterra Green World Research, Droevendaalsesteeg 3, PO Box 47, NL-6700 AA Wageningen, The Netherlands.

**FRANK DENTENER** Institute for Environment and Sustainability, Joint Research Centre, Ispra, Italy.

**JEAN-PAUL HETTELINGH** Netherlands Environmental Assessment Agency (MNP), Coordination Center for Effects, PO Box 303, NL-3720 AH Bilthoven, The Netherlands.

**KEVIN HICKS** Stockholm Environment Institute at York, University of York, York YO10 5DD, UK.

**CHRISTIAN HOGREFE** Atmospheric Sciences Research Center, University at Albany, State University of New York, Albany, NY 12203, USA.

**TRACEY HOLLOWAY** Center for Sustainability and the Global Environment, Nelson Institute for Environmental Studies, University of Wisconsin-Madison, Madison, WI, USA.

**GEORGE KALLOS** Atmospheric Modelling and Weather Forecasting Group, School of Physics, University of Athens, Building PHYS-V, 15784 Athens, Greece.

**NUTTHIDA KITWIROON** Centre for Atmospheric and Instrumentation Research (CAIR), University of Hertfordshire, Hatfield, AL10 9AB, UK.

**MAXIMILIAN POSCH** Netherlands Environmental Assessment Agency (MNP), Coordination Center for Effects, PO Box 303, NL-3720 AH Bilthoven, The Netherlands.

**GERT JAN REINDS** Alterra Green World Research, Droevendaalsesteeg 3, PO Box 47, NL-6700 AA Wageningen, The Netherlands.

**WOLFGANG SCHÖPP** International Institute for Applied Systems Analysis (IIASA), A-2361 Laxenburg, Austria.

**FRED TONNEIJCK** Wageningen University and Research Centre, Plant Research International, PO Box 16, 6700AA Wageningen, The Netherlands.

**LEENDERT VAN BREE** Netherlands Environmental Assessment Agency (MNP), Coordination Center for Effects, PO Box 303, NL-3720 AH Bilthoven, The Netherlands.

**HAN VAN DOBBEN** Alterra Green World Research, Droevendaalsesteeg 3, PO Box 47, NL-6700 AA Wageningen, The Netherlands.

# FOREWORD

We are living in an increasingly shrinking world. Instant communication and the internet have seemingly dissolved time, space and cultural boundaries. The international movement of peoples has reached levels previously unheard of. Globalization has become a catchphrase for our times.

The atmospheric sciences have partly led and partly responded to this process. The extent of continental, hemispheric – and even global – transport of air pollution has become an issue of increasing scientific and policy concern; and nothing emphasises the fragile unity of the planet more graphically than the increasing evidence of climate change – and the portentous implications that emerge as we contemplate the possible consequences of the interaction of air pollution and climate change.

Yet appearances can be deceptive, and are only part of the story. Even for scientists, the 'Big Picture' is never easy, and, for the most part, the pressures of professional life mean that we must concentrate on the particular and limit ourselves to our own field. We are able only from time to time to look outside our own boxes, and this task paradoxically becomes more difficult as the totality of knowledge in our separate specialties increases.

For the ordinary citizen, with no specialist training in the atmospheric sciences, there is a similar problem. The unity of the atmosphere, and of the atmospheric sciences, is less easy to grasp than the variety of seemingly separate problems, such as climate change, ozone depletion, urban pollution, and industrial and vehicle emissions, which at different times rightly command separate and urgent attention.

It is therefore perhaps not surprising that, apart from academic textbooks and one or two international journals, few initiatives have sought to bring the variety of atmospheric issues into a single perspective, still less to highlight it in terms which excite the interest and understanding of the ordinary citizen. Yet action in this area is now urgently needed.

It is for this reason that the International Union of Air Pollution Prevention and Environmental Protection Associations has joined with the United Nations Environment Programme, the United Nations Economic Commission for Europe's Convention on Long-Range Transboundary Air Pollution, the Stockholm Environment Institute, and other relevant bodies in the Global Atmospheric Pollution Forum to promote understanding and debate on these issues. It is also the reason why the International Union and the Global Forum are now delighted to sponsor and support this volume. In fostering international cooperation in the atmospheric sciences the challenge is not simply to promote progress, at scientific and policy level, in each of the separate specialist areas: it is to allow scientists and lay persons alike to stand outside those specialisms and see the wider context. That task becomes the more urgent, month by month, as we come to recognize the extent of interactions among the various issues and realize that measures to address one problem will help or hinder others and must therefore be seen together.

This Atlas is an essential step in that process. And it is one upon which both the International Union and the Forum hope to build.

I would like to express the appreciation of the International Union and the Forum to the Editor and all the contributors for their support and commitment; and to the publishers for their vision and patience in a challenging project.

*Mario Molina, Nobel Laureate*

# PREFACE

Atmospheric pollution affects us all. It affects our health and our environment. It is our activities and actions, however, which are resulting in the continual pollution of the atmosphere. For example, road transport has emerged as one of the most important sources of air pollution, particularly in urbanised areas such as mega-cities. Although emissions from other sources, such as industrial and domestic have reduced in developed countries, they still make a significant contribution to the overall pollution burden of the atmosphere in many less developed regions.

It is not only the directly emitted pollutants that can be hazardous to our health and the environment. Pollutants can also react with each other in the presence of sunlight to produce harmful photochemical smog, which affects many cities around the world. Once released into the atmosphere, pollutants can be dispersed into buildings and along streets as well as affect whole city areas. As a result of meteorological processes air pollution can also be transported across continents and, depending on the particular chemical species, remain in the global atmosphere for long periods of time. Pollution emitted locally by cars, industrial chimneys or forest fires, therefore, can have an impact on regional and global scales. On the other hand, pollutants such as aerosols and carbon dioxide, which influence the global radiation balance of the atmosphere, can lead to changes in the natural climate of the world with direct impacts on urban and local scales. Given the extent and complexity of processes and the multitude of sources and pollutants, atmospheric pollution leads to a range of impacts, including climate change, ozone depletion, acid deposition and photochemical smog, as well as causing damage to vegetation and human health.

The *World Atlas of Atmospheric Pollution* brings together several key scientists in the field to provide a global overview of air pollution and its impacts. The Atlas begins with a short introduction to enable readers who do not have the relevant scientific background to understand the behaviour of pollutants in the atmosphere. After providing a historical perspective, the Atlas addresses key topics spanning local to global scales, namely air pollution in cities, transboundary air pollution, global air pollution and climate change, ozone depletion, environmental and health impacts, and future trends.

Wherever possible, the approach has been to provide a global perspective of the problems, which inevitably relies on the use of a wide base of sources. In some cases, however, data was not available for all regions of the world, and hence an extensive global overview could not always be presented. Analysis of the pollution trends and distributions, for example for Chapter 2, proved to be challenging as the data quality and associated terminology was not always uniform across all sources. Furthermore, in many cases, only limited information was available on air pollution monitoring stations, such as the description of instrumentation, location, calibration schedules and maintenance procedures. With regard to research investigations in the field, there is a vast amount of published literature on atmospheric pollution and consequently, only selected studies and results could be included. In light of these limitations, the main purpose of the Atlas is to provide a descriptive and, in some cases qualitative, comparison of atmospheric pollution for the different regions of our world. The Atlas does not attempt to make a rigorous analysis or interpretation of the data but more to provide a graphical representation of the state of atmospheric pollution. Wherever possible, supporting references have been cited to help the reader to gain further understanding of the figures.

Despite the challenges, we hope that the Atlas will contribute to the field by stimulating further interest among the wider scientific community and professionals working in the area. It will be a comforting reward if the interest in the Atlas also increases our motivation, albeit in a small way, to safeguard the atmosphere for future generations. A parallel ambition for writing the Atlas has been to stimulate coordinated action in areas such as accessibility and sharing of data globally so as to improve our understanding and knowledge in this field. In order to achieve these goals, every effort has been made to present the information in the Atlas in a way that will make it useful to informed readers and to those who wish to find out more about key environmental issues affecting all of us in one way or another. We view this work as an ongoing process, and we look forward to increasing cooperation and exchange of information to improve the Atlas in the future.

*Ranjeet S Sokhi,*
*Centre for Atmospheric and Instrumentation Research (CAIR), University of Hertfordshire, UK*

# ACKNOWLEDGEMENTS

The members of the International Union of Air Pollution Prevention and Environmental Protection Associations (IUAPPA), its International Advisory Board and the Global Atmospheric Pollution Forum are thanked for their continued support during this project. Special thanks go to Richard Mills for his advice and help in making this work possible. We appreciate very much the interest that Mario Molina has taken in the Atlas and for agreeing to write the Foreword.

A major work, such as this Atlas, would not have been possible if it were not for the generous input from a large number of colleagues and their organizational across the world especially for making data and information available. With this in mind, it is important to acknowledge the efforts and devotion of numerous scientists and other individuals who have brought atmospheric pollution to the forefront of major environmental concerns in modern times. Much care has been taken to acknowledge all sources of data and to respect copyright, but if there is any instance of omission, the publishers will be only be too pleased to rectify such errors.

The authors of the Atlas would particularly like to state their deep appreciation for the help provided by the following colleagues and organizations:

Nutthida Kitwiroon (University of Hertfordshire, UK) for her wizardry with graphical software and for undertaking the daunting task of producing the maps and other figures.

Jane Newbold (University of Hertfordshire, UK) for copy-editing all chapters and references and producing the index.

Mike Ashmore is thanked for contributing to the Introduction in relation to the text on environmental and health impacts of atmospheric pollution.

Lucy Sadler and David Hutchinson from the Greater London Authority (GLA), UK, for making the London Emission Inventory available (Chapter 2).

Samantha Martin, Elizabeth Somervell, Srinivas Srimath, Lakhumal Luhana and Hongjun Mao (University of Hertfordshire, UK) for helping to collate data and prepare some figures.

European Environment Agency for making air quality data available through the AirBase database (Chapter 2).

DEFRA (UK Government Department of Environment, Food and Rural Affairs) and AEA Energy Environment for making data available via the UK National Atmospheric Emissions Inventory and National Air Quality Archive (Chapter 2).

GEIA/EDGAR/ACCENT for making global atmospheric emissions data available for figures in chapters 2, 4 and 6.

Pim Martens for contributing to the section on Ecosystems and Health in Chapter 4 (4.5 Projection of Global Climate Change, its Impacts and Atmospheric Composition) and providing Figures 4.22 and 4.23. Dr Martens is the Director of the International Centre for Integrated Assessment and Sustainable Development (ICIS), University Maastricht, The Netherlands.

The following are thanked for their help in providing specific figures for the Atlas chapters:

NASA (Figures 3.1, 3.7 and 3.8); US EPA (Figure 3.3); J Fishman and A Balok (Figure 3.5); K Civerolo and Huiting Mao (Figure 3.10); E Brankov, R F Henry, K L Civerolo, W Hao, P K Misra, R Bloxam and N Reid (Figure 3.11); B Schichtel and R Husar (Figure 3.12); Evelyn Poole-Kober for obtaining copyright permission for the graphics used in this chapter; Svetlana Tsyro and Wenche Aas (NILU, Norway) for producing Figures 3.2 and 3.13; Nick Sundt, US Global Change Research Program/Climate Change Science Program (Figure 4.1); Jos G J Olivier, Netherlands Environmental Assessment Agency (MNP) for providing data for Figures 4.3, 4.4 and 4.5; Hong Liao and John H Seinfeld, California Institute of Technology (Figure 4.21); John Kennedy and Philip Brohan, Hadley Centre, UK Met Office (Figure 4.13) and Frank Dentener, Joint Research Centre, Italy (Figure 4.20).

The permission of the IPCC Secretariat to reproduce several figures in Chapter 4 is acknowledged.

Rakesh Kumar, National Environmental Engineering Research Institute (NEERI), India for providing images for use in Figure I.3 and chapter opening illustrations.

Jeffrey Clark for locating images of stack plume by courtesy of USEPA (Figure I.5).

We are also grateful to several reviewers who made some very helpful comments to improve the content of the Atlas.

# INTERNATIONAL UNION OF AIR POLLUTION PREVENTION AND ENVIRONMENTAL PROTECTION ASSOCIATIONS (IUAPPA)

Founded in 1964, the aim of the Union is to promote progress in the prevention and control of air pollution, the protection of the environment and the adoption of sustainable development, through the promotion of scientific understanding and the development of relevant and effective policies at national and international level.

## INTERNATIONAL ADVISORY BOARD

**MARIO MOLINA, CHAIRMAN** Professor, University of California, San Diego
Center for Atmospheric Sciences at Scripps Institution of Oceanography; President, Molina Center for Strategic Studies in Energy and the Environment

**SIR CRISPIN TICKELL, VICE CHAIRMAN** Former Permanent Representative of the United Kingdom to the United Nations

**HAJIME AKIMOTO** Frontier Research Center for Global Change, Japan

**MEINRAT ANDREAE** Director, Biogeochemistry, Max Planck Institute, Germany

**PAULO ARTAXO** University of São Paulo, Brazil

**PETER BRIMBLECOMBE** Editor, *Atmospheric Environment*

**YUAN TSE LEE** Director, Academia Sinica, Taiwan

**ALAN LLOYD** Secretary, California Environmental Protection Agency

**WON HOON PARK** Chairman, Engineering Research Council, Korea

**V (RAM) RAMANATHAN** Director, Center for Clouds, Chemistry and Climate, University of California, San Diego

**MARTIN WILLIAMS** Chairman, Executive Body, Convention on Long-Range Transboundary Air Pollution

**ALAN GERTLER** President, IUAPPA (United States)

**GAVIN FISHER** Past President, IUAPPA (New Zealand)

**MENACHAM LURIA** Past President, IUAPPA (Israel)

**RICHARD MILLS** Director General, IUAPPA



## NATIONAL AND INTERNATIONAL MEMBER ORGANIZATIONS OF THE UNION

Air & Waste Management Association

Asian Society for Environmental Protection

European Federation for Clean Air

Clean Air Society of Australia & New Zealand

Austrian Society for Air & Soil Pollution

Ecological Society, Azerbaijan

Royal Flemish Chemical Society – Environment Safety Section, Belgium

Brazilian Association for Ecology and Water & Air Pollution Prevention

Chinese Society of Environmental Sciences

Croatian Air Pollution Prevention Association

Czech Association of IUAPPA

Finnish Air Pollution Prevention Society

Association for the Prevention of Atmospheric Pollution, France

Committee on Air Pollution Prevention VDI & DIN – Standards Committee KRdL, Germany

Green Earth Organization, Ghana

Indian Association for Air Pollution Control

Israel Society for Ecology & Environmental Quality Sciences

Air Pollution Study Committee, Italy

Japanese Union of Air Pollution Prevention Associations

Korean Society for Atmospheric Environment

Environment Public Authority, Kuwait

Environmental Protection Society, Malaysia

National Council of Industry Environmentalists, Mexico

Ecological Society, Nepal

VVM-Section for Clean Air in The Netherlands

Institute of Ecological Feasibility Studies, Peru

Peruvian Society for Clean Air & Environmental Management

Meteorology & Environmental Protection Administration, Saudi Arabia

Environmental Engineering Society of Singapore (Clean Air Section)

National Association for Clean Air, South Africa

Swedish Clean Air Society

Cercl'Air, Switzerland

Environmental Protection Society, Taiwan

Tunisian NGO for Sustainable Development

Turkish National Committee for Air Pollution Research & Control

National Society for Clean Air & Environmental Protection, UK

# GLOBAL ATMOSPHERIC POLLUTION FORUM

In 2004 IUAPPA joined with the Stockholm Environment Institute to create the Global Atmospheric Pollution Forum. The Forum links together existing regional air pollution control networks from around the globe so that they can better share information, experience and expertise and, in so doing, more effectively tackle air pollution and climate change. The Global Forum supports the development of solutions to air pollution-related problems by promoting effective cooperation among nations at the regional, hemispheric and global scales. It also supports regional networks in their efforts to find cost-effective solutions that promote economic development and help alleviate poverty. The international and regional networks and organizations participating in the Global Forum include:

United Nations Environment Programme (UNEP);

United Nations Economic Commission for Europe/Convention on Long-Range Transboundary Air Pollution (UNECE/LRTAP);

Clean Air Initiative (CAI), including CAI-Asia; CAI-Latin America and CAI-Sub-Saharan Africa;

International Union of Air Pollution Prevention and Environmental Protection Associations (IUAPPA);

Stockholm Environment Institute (SEI);

Air Pollution Information Network for Africa (APINA); and

The Inter-American Network for Atmospheric/Biospheric Studies (IANABIS).

# ACRONYMS AND ABBREVIATIONS

| | |
|---|---|
| ABC | atmospheric brown cloud (sometimes referred to as Asian brown cloud), also called atmospheric brown haze (ABH) |
| ABL | atmospheric boundary layer |
| ANC | acid-neutralising capacity |
| APHEIS | Air Pollution and Health: A European Information System (http://www.apheis.net) |
| AQG | World Health Organization (WHO) Air Quality Guidelines |
| AR4 | fourth assessment report of the Intergovernmental Panel on Climate Change |
| BC | black carbon |
| BC | base cations, taken as the sum of calcium (Ca), magnesium (Mg) and potassium (K) |
| CAFE | Clean Air for Europe (http://ec.europa.eu/environment/air/cafe/index.htm) |
| CAIR | Centre for Atmospheric and Instrumentation Research, University of Hertfordshire, UK |
| CFC | chlorofluorocarbons |
| CfIT | Commission for Integrated Transport, UK |
| CIESIN | Center for International Earth Science Information Network, Columbia University, USA |
| CL | critical load |
| CLAES | Cryogenic Limb Array Etalon Spectrometer |
| CLRTAP | Convention on Long-Range Transboundary Air Pollution (http://www.unece.org/env/lrtap) |
| CPCB | Central Pollutant Control Board, Delhi, India |
| DEFRA | Department for Environment, Food and Rural Affairs, UK |
| DMS | dimethylsulphide |
| DNA | deoxyribonucleic acid |
| DU | Dobson Unit |
| EC | European Commission (http://ec.europa.eu/index_en.htm) |
| EEA | European Environment Agency (http://www.eea.europa.eu) |
| EMEP | Cooperative Programme for Monitoring and Evaluation of the Long-Range Transport of Air Pollutants in Europe (http://www.emep.int) |
| ENSO | El Niño-Southern Oscillation |
| EU | European Union |
| EUNIS | European Nature Information System |
| FLUXNET | a global collection of micrometeorological flux measurement sites, which measure the exchanges of carbon dioxide, water vapour and energy between the biosphere and atmosphere (http://www.fluxnet.ornl.gov/fluxnet/ index.cfm) |
| GDP | gross domestic product |
| GHG | greenhouse gases |
| GNP | gross national product |
| GWP | global warming potential |
| HFC | hydrofluorocarbons |
| IIASA | International Institute for Applied Systems Analysis, Austria |
| IPCC | Intergovernmental Panel on Climate Change (http://www.ipcc.ch) |
| IR | infrared radiation |
| IS92 | IPCC set of six emission scenarios developed in 1992 |
| ITCZ | Intertropical Convergence Zone |
| JRC | Joint Research Centre |
| LAEI | London Atmospheric Emissions Inventory |

| | |
|---|---|
| LRT | long-range transport |
| LRTAP | Long-Range Transboundary Air Pollution (http://www.unece.org/env/lrtap) |
| MLS | microwave limb sounder (http://mls.jpl.nasa.gov) |
| MSC-W | Meteorological Synthesizing Centre-West (http://www.emep.int/index_mscw.html) |
| NAO | North Atlantic Oscillation |
| NAPAP | National Acid Precipitation Assessment Program (http://gcmd.nasa.gov/records/GCMD_EPA0141.html) |
| NASA | National Aeronautics and Space Administration (http://www.nasa.gov) |
| NOAA | National Oceanic & Atmospheric Administration (http://www.noaa.gov) |
| OC | organic carbon |
| OECD | Organization for Economic Co-operation and Development (http://www.oecd.org) |
| PAN | peroxyacetyl nitrate |
| PFC | perfluorocarbons |
| $PM_{10}$ | particulate matter of aerodynamic size equal to or less than 10 $\mu$m |
| $PM_{2.5}$ | particulate matter of aerodynamic size equal to or less than 2.5 $\mu$m |
| POC | particulate organic carbon |
| PSC | polar stratospheric clouds |
| QA/QC | quality assurance and quality control |
| QBO | quasi-biennial oscillation |
| RAINS | Regional Air Pollution Information and Simulation (http://www.iiasa.ac.at/rains/index.html) |
| RIVM | National Institute for Public Health and the Environment, The Netherlands |
| SAGE | Stratospheric Aerosol and Gas Experiment (http://www-sage3.larc.nasa.gov) |
| SAR | IPPC's Second Assessment Report |
| SECAP | South European Cycles of Air Pollution, European Commission-funded project (1992–95) |
| SEI | Stockholm Environment Institute, Sweden |
| SMB | simple mass balance |
| SRES | Special Report on Emission Scenarios by IPCC |
| SRLULUCF | IPCC's Special Report on Land Use, Land Use Change and Forestry |
| SRRF | IPCC's Special Report on Radiative Forcing |
| SST | sea surface temperature |
| STOCHEM | UK Meteorological Office global three-dimensional Lagrangian tropospheric chemistry model |
| TAR | IPCC's Third Assessment Report |
| TOMS | Total Ozone Mapping Spectrometer |
| TSP | total suspended particles |
| T-TRAPEM | transport and transformation of air pollutants from Europe to the Mediterranean region, European Commission-funded project, 1993–95 |
| UAM-V | Urban Airshed Model-Variable grid version |
| UAQ | urban air quality |
| UARS | Upper Atmosphere Research Satellite |
| UDI PHAHA | Institute of Transportation Engineering of the City of Prague (http://www.udi-praha.cz) |
| UHI | urban heat island |
| UN | United Nations (http://www.un.org) |
| UNECE | United Nations Economic Commission for Europe (http://www.unece.org) |
| UNEP | United Nations Environment Programme (http://www.unep.org) |
| UNEP/DEWA/GRID | United Nations Environment Program, Division of Early Warning and Assessment, Global Resource Information Database |
| UNFCCC | United Nations Framework Convention on Climate Change (http://unfccc.int/2860.php) |
| USEPA | United Nations Environment Programme (http://www.unep.org) |
| UV | ultraviolet radiation |
| UVB | ultraviolet radiation with wavelengths from 280 to 320 nm |
| VOC | volatile organic compounds |
| WHO | World Health Organization (http://www.who.int) |
| WMO | World Meteorological Organization (http://www.wmo.ch) |
| WRI | World Resources Institute (http://www.wri.org) |

# SELECTED UNITS USED IN ATMOSPHERIC POLLUTION SCIENCE

| | |
|---|---|
| $CO_2$ eq | equivalence of the global warming potential (usually over a 100 year period) of greenhouse gases relative for carbon dioxide ($CO_2$) |
| DU | Dobson Units are used in atmospheric ozone research (1 DU is 0.01 mm thickness at standard temperature and pressure) |
| Gt | giga-tonne ($10^9$ or billion tonnes or $10^{15}$ grams) e.g. Gt C means giga-tonnes of carbon |
| ha | hectare, used for measuring land area and equivalent to 10,000 $m^2$ |
| mol | measure of the amount of substance where 1 mol is the amount of substance that contains the same number of its elementary entities (e.g. atoms or molecules) as there are atoms in 0.012 kg of carbon-12 ($6.022 \times 10^{23}$, a number known as the Avogadro's constant) |
| Mt | mega-tonnes ($10^6$ tonnes or $10^{12}$ grams) |
| $\mu g/m^3$ or $\mu gm^{-3}$ | air pollution concentration unit (micro- or $10^{-6}$ grams of a particular specie per cubic metre of air) |
| Pg | peta grams ($10^{15}$ grams) |
| ppm | parts per million (1 in $10^6$) |
| ppb | parts per billion (1 in $10^9$) |
| ppbv | parts per billion as a volume to volume ratio |
| Tg | tera grams ($10^{12}$ grams) |
| Tonne | one thousand kilograms ($10^3$ kg or $10^6$ grams) |
| $W/m^2$ or $Wm^{-2}$ | unit to measure amount of radiation incident on a surface such as that of the earth (Watts per square metre) |

# INTRODUCTION

*Ranjeet S Sokhi*



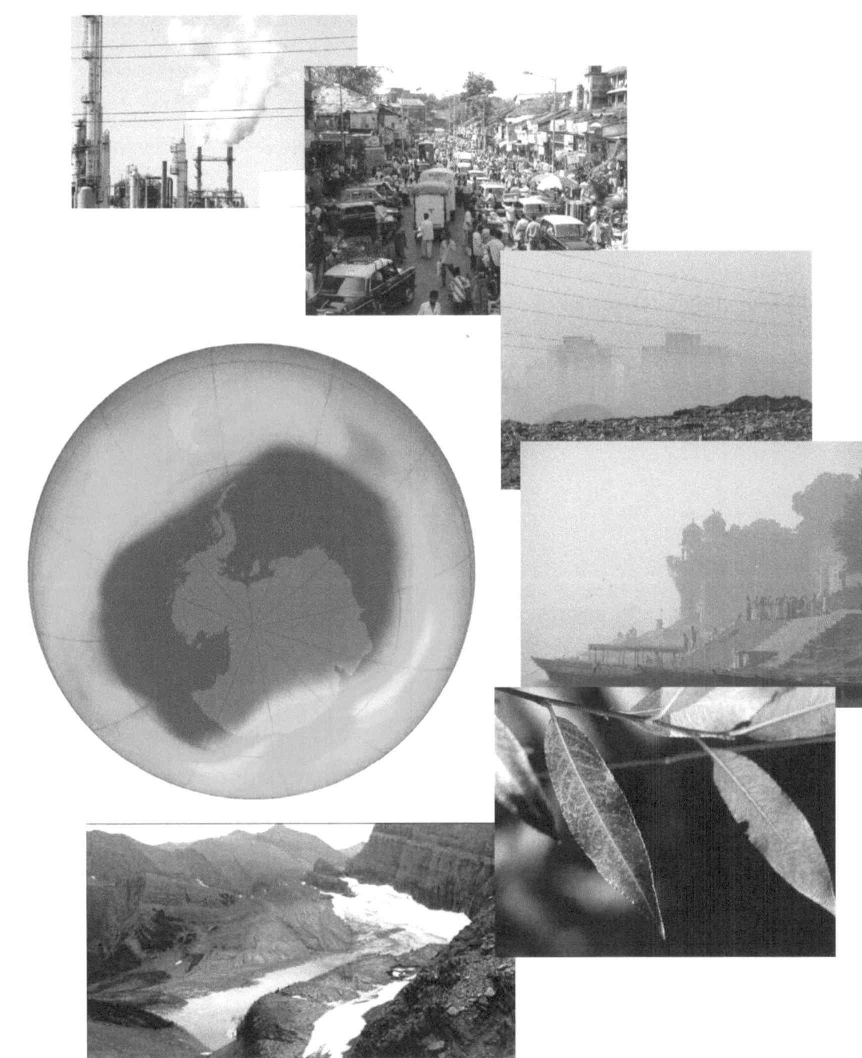

Images show air pollution sources of industry and road traffic and examples of smog. Air pollution can damage vegetation and lead to global impacts of climate change and ozone depletion (globe).

This introduction brings together the separate, but interlinked, themes of the chapters. Some of the key concepts of atmospheric pollution are explained to help those readers who are not familiar with the subject area but have a science background. This section also explains the overall structure of the Atlas and the approach adopted by the contributors. A list of useful reading material is provided at the end of the Atlas for those who are interested in an in-depth treatment of atmospheric science and pollution.

## I.1 The Earth's Atmosphere

### Radiation Balance of the Atmosphere

The overall atmospheric dynamics and the climate system are driven by the energy from the sun. Much of the incoming solar radiation is transmitted through the atmosphere and absorbed by the earth's surface (see Figure I.1). Some of this short-wave radiation is reflected by ground surfaces such as snow and deserts, and by clouds. About 30 per cent of the solar radiation incident on the earth is reflected back into space by clouds, the atmosphere and the earth's surface (Kiehl and Trenberth 1997). Approximately 20 per cent of the incident radiation is absorbed by the atmosphere, and the remaining proportion (about 50 per cent) warms the earth's surface (land and the oceans).

The proportion of incoming radiation that is reflected back by a surface depends on its reflectivity, and is termed 'albedo'. For example, albedo of fresh snow is typically 0.8 and that of the earth and the atmosphere, 0.3. Some of the long-wave infrared radiation emitted by the earth's surface, is absorbed by atmospheric constituents such as gases (e.g. carbon dioxide) and water vapour. The absorbed energy is re-radiated back as infra-red radiation towards the earth's surface, making the atmosphere warmer than it would be otherwise. This is the 'natural greenhouse effect', sometimes also called the 'atmospheric effect' or the 'atmospheric greenhouse effect' (e.g. Ahrens 2003; Le Treut et al. 2007). Clouds can have both a warming and a cooling effect, and hence play a crucial role in determining the radiation balance of the earth. As a result of the greenhouse effect the earth's average surface temperature is about 15 °C, which is 33 °C warmer than it would be if it had no atmosphere. Some of the main greenhouse gases (GHGs) include water vapour, carbon dioxide, methane, nitrous oxide and ozone. The role of GHGs in affecting the earth's climate is discussed further in Chapter 4.

Over long periods radiative components such as incoming and outgoing radiation are in balance for the earth as a whole. The energy is redistributed across the earth's surface through the transference of sensible heat flux, latent heat flux and surface heat flux into oceans. Sensible heat flux is the direct energy which is transferred from the Earth's surface to the atmosphere by conduction and convection and by advection from the tropics to the poles, leading to large-scale atmospheric circulations. Latent heat flux is the energy that is stored in water vapour as it evaporates. The atmospheric circulation transports this vapour vertically and horizontally to cooler locations where it is condensed as rain or is deposited as snow, releasing the stored heat energy. A large amount of radiation energy is also transferred to the tropical oceans where conduction and convection processes cause the movement of warm surface waters deeper into the water column. The ocean currents transfer this heat energy horizontally from the equator to the poles. A Schematic of the earth's radiation balance and the greenhouse effect is shown in Figure I.1.

### Structure of the Atmosphere

The structure of the atmosphere can be divided into four layers, according to the variation of air

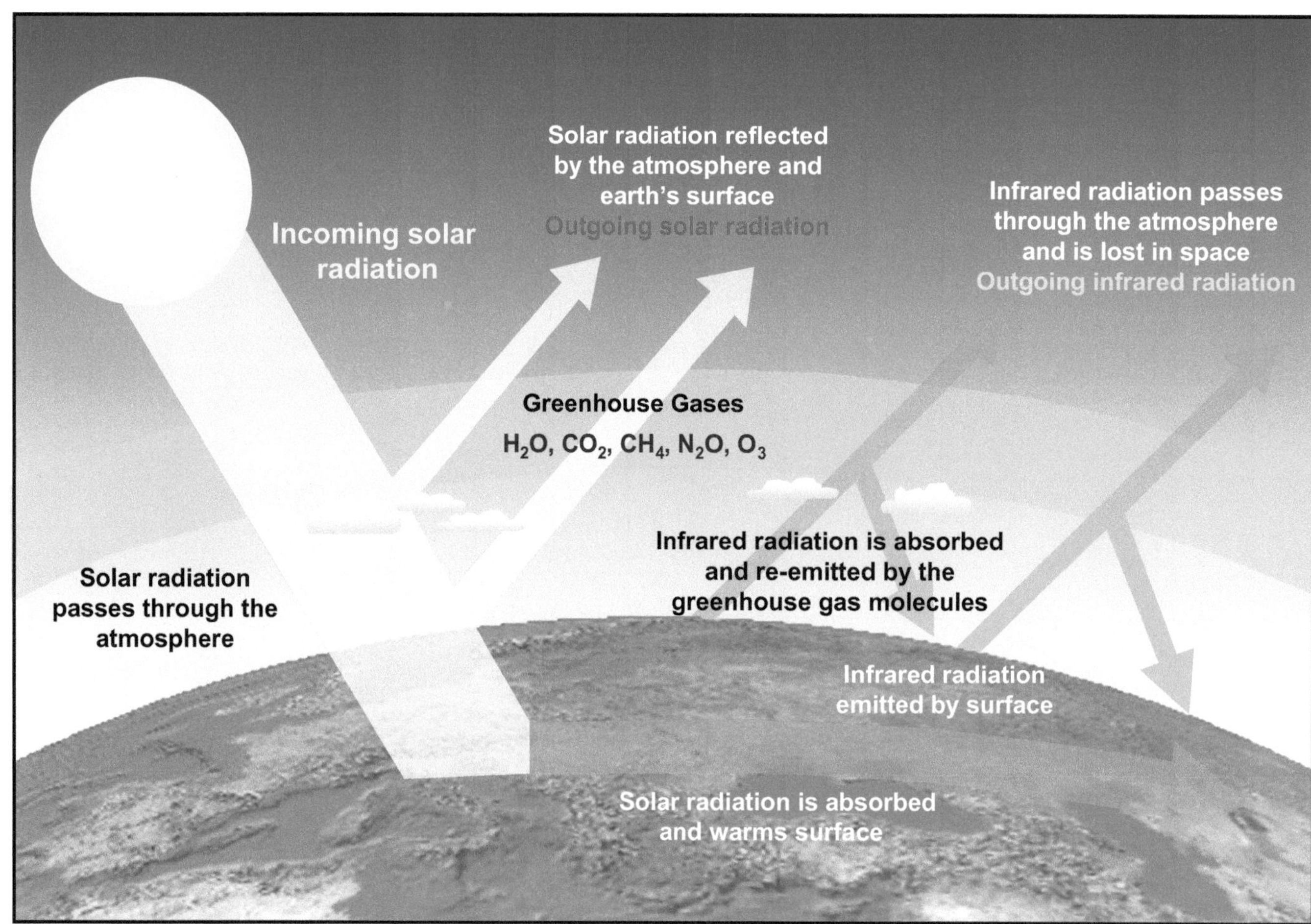


Figure I.1 Radiation balance of the earth's atmosphere, showing the main radiation interactions leading to the greenhouse effect (adapted from UNEP/WMO 1996).

temperature with height (as shown in Figure I.2). Most of the mass of the atmosphere is within the troposphere, which extends up to about 10–15 km above sea level. It is in this region that much of the weather and atmospheric circulation patterns occur. The lower part of the troposphere consists of the atmospheric boundary layer (ABL), which is heavily influenced by the earth's surface characteristics over short timescales (about an hour), including temperature and roughness. This is the region where we live and where most of the pollution is emitted, and hence is of particular interest. This layer can extend up to a few thousand kilometres in height and has a complicated structure that changes with time of day and season (see for example, Stull 1998, Jacobson 2002 and Piringer and Joffre 2005).

Pollutants that are released within the ABL become subject to vertical and horizontal dispersion processes, resulting in their mixing and dilution. Under certain situations, the temperature of the air within this layer and its sub-layers can increase with height (known as a thermal inversion), causing the atmosphere to be stable, with little mixing. With restricted mixing at times of increased emissions, pollution can accumulate, leading to excessively high concentrations (air pollution episodes). Under the action of sunlight, the cocktail of pollutants present in the atmosphere can lead to what is termed photochemical smog, which is experienced in many cities across the world. Originally, the term 'smog' was coined to describe both smoke and fog (as a portmanteau of the two words). These processes and interactions within the ABL are particularly important for understanding local and urban pollution problems, which are addressed in Chapter 2. Figures I.3 and I.4 shows examples of air pollution incidents affecting urban and nearby locations.

The layer between the top of the ABL and the tropopause is known as the free troposphere. Mean temperatures in this layer decrease with altitude as rising air expands and cools. Above the troposphere is the stratospheric layer, where the average temperature remains approximately constant at first, before increasing with height. Much of the aircraft traffic takes place in the lower part of the stratosphere. Most of the ozone is found in the stratosphere which extends from about 15 km to approximately 50 km in altitude. The increase in temperature in this layer is caused by the presence of ozone, which absorbs the ultraviolet (UV) radiation and re-emits infrared radiation, causing the stratosphere to warm. Importantly for human health, the stratosphere protects us from harmful solar radiation by absorbing the UVB component of the radiation. A major environmental concern on global scales is the destruction of this 'protective' ozone layer, which is discussed in Chapter 5.

The layer above the stratosphere is known as the mesosphere. As the density of ozone is very low in the mesosphere, the ozone does not lead to any significant absorption of UV radiation to offset the decrease in temperatures with increasing altitude. Above the mesopause, the temperatures increase again in the thermosphere, as the oxygen and nitrogen molecules absorb the large amount of very short wavelength radiation from the sun.

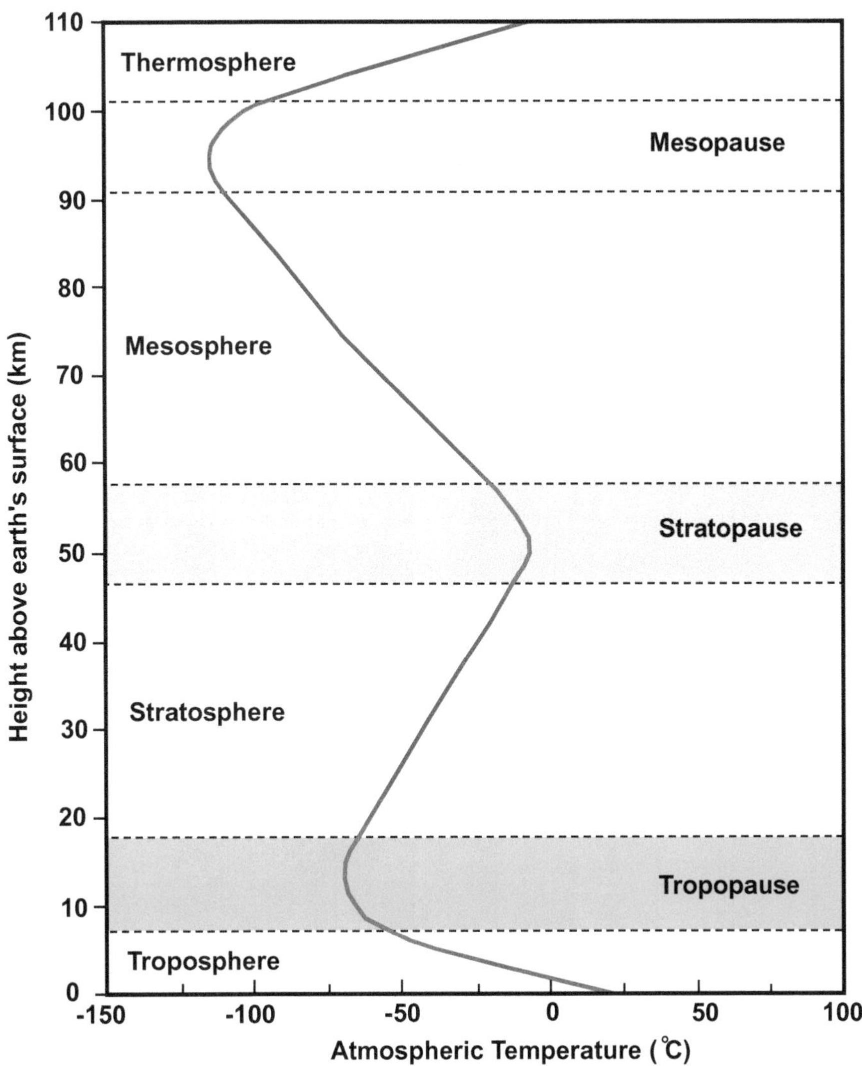


Figure I.2 Major layers of the atmosphere according to the mean variation of air temperature with height (solid line).

## I.2 Transport of Pollutants within the Atmosphere

Transport processes within the atmosphere can take place on a range of scales. As a general guide, the spatial scales can be categorised as: micro and local scales (a few metres to hundreds of metres), urban scales (a few kilometres to about one hundred kilometres), regional scales (hundreds to thousands of kilometres), continental and hemispheric scales (several thousands to about twenty thousand kilometres) and, finally, global scales. Emissions that result from localized industrial plants or from road traffic are transported and transformed on all spatial scales, and on temporal scales ranging from seconds and minutes to years, depending on a range of factors, including the type of substance released, emission characteristics, its chemical reactivity and meteorological parameters.

### Large-Scale Transport

Large-scale wind patterns can affect transport behaviour of pollutants on regional to global scales. Low pressure systems are associated with upward air motion, high or varying surface wind conditions and often cloudy skies. Under these conditions, pollution is dispersed and transported horizontally and vertically, reducing ground-level concentrations. Strong winds, however, can cause significant suspension and re-suspension of dust, especially when the ground is dry. High pressure systems, on the other hand, are associated with downward air motion, lower surface wind speeds, cloud-free skies and often sunny conditions. Under these conditions, dispersion is limited and pollution can accumulate, leading to air pollution episodes. Complex photochemical reactions can also occur in clear, sunny skies, leading to the formation of urban smog.

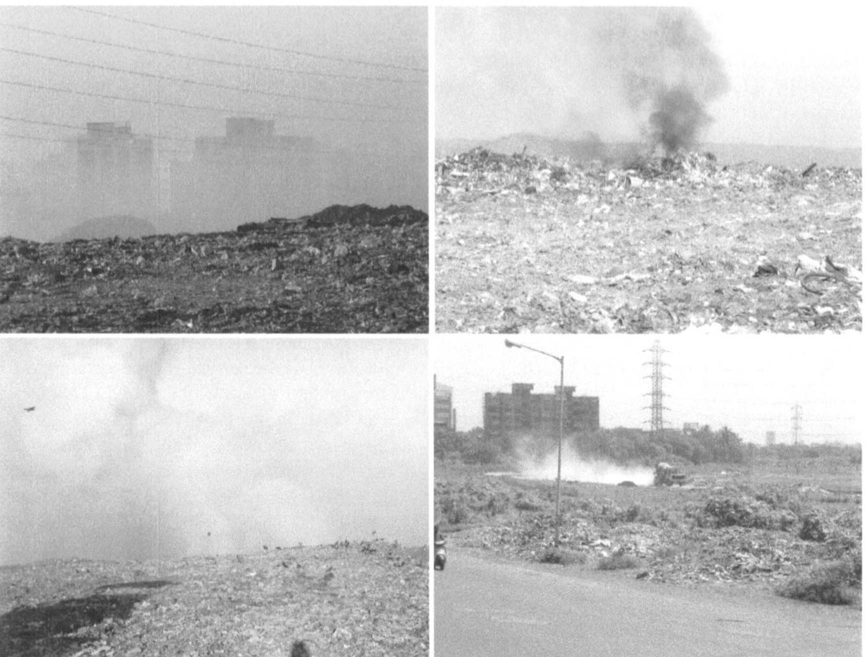

Figure I.3 A variety of air pollution incidents can affect a city and nearby locations. The examples shown here are photochemical smog over Mumbai (India) (top left), smoke from waste burning (top right and bottom left) and dust from construction activities (bottom right). (Source: Kumar 2007 personal communication.)

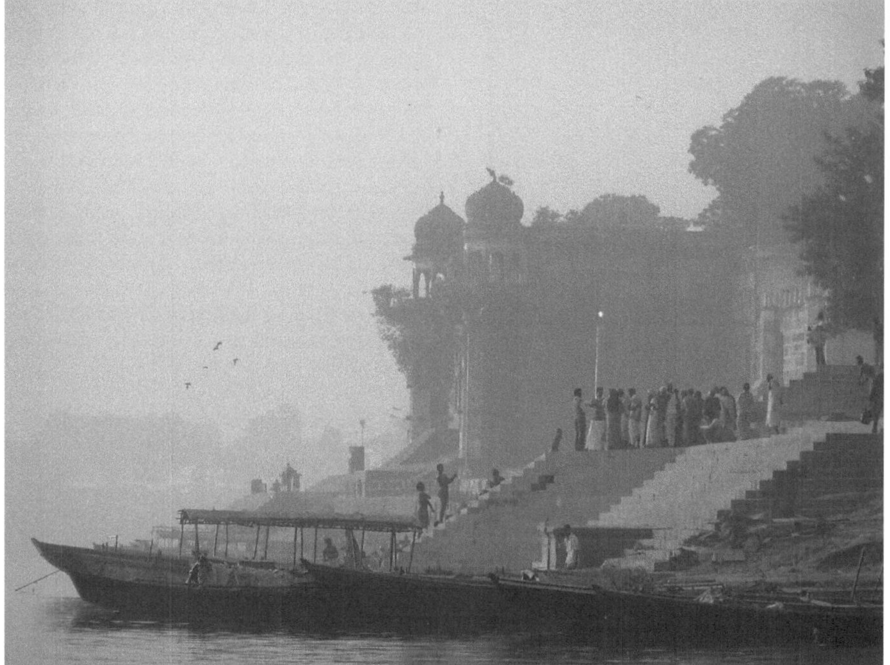

Figure I.4 Build up of air pollution haze over Varanasi (India). The image has been taken on the shores of River Ganges. (Source: Emberson 2007 personal communication.)

A number of meteorological processes cause the horizontal and vertical movement of air, with advection and convection being most important. Advection is largely responsible for the horizontal transport of wind and properties of the atmosphere, such as temperature and moisture, from one location to another. The pollution that is mixed within the air is then also transported from the point of emission to the receptor location (point at which the impact of the pollution is measured or observed). In particular, the direction of wind, along with its strength (wind speed), will largely determine the horizontal spatial distribution of the pollution, that is, where the pollution is transported and how far it has travelled. The vertical distribution of pollution is mainly determined by the convection, which can be induced thermally (e.g. warm air rising and cool air falling) or mechanically (e.g. air passing over mountainous terrain), within the boundary layer. Vertical movement of pollution is discussed further below.

Pollutants that reach higher altitudes, above built-up areas, for example, or are emitted from a high chimney stack, can be transported horizontally by winds aloft. The pollution can be transported over very large distances (up to thousands of kilometres) away from the source, taking the pollutants beyond geographical and political boundaries. Examples of long-range transport of pollution include acid deposition and transport and formation of ozone and particles on regional scales (see Chapter 3).

## Local and Urban Scale Transport

On smaller local scales, meteorological processes still play a crucial role in determining transport characteristics of air pollution. For example, urban areas exhibit higher air temperatures than in the surrounding regions. This is known as the urban heat island (UHI) effect. Higher temperatures in urban areas can lead to enhanced convection and surface wind speeds, higher mixing depths and, hence, lower pollution levels. However, this is counterbalanced by the high level of pollution resulting from sources such as road traffic. Large-scale wind patterns can be modified by local pressure gradients caused by a range of factors. These include variable topography (either natural terrain or built-up areas), ground heating and turbulence. Examples of local wind patterns include sea breeze, mountain and valley flows, and wind around and within urban areas. Wind patterns in urban areas can be complex, for example, near buildings or within canyons (city streets bordered on both sides by lofty buildings), leading to high localized pollution concentrations.

Although the processes discussed above will lead to larger-scale movement of air, smaller-scale, unsteady and irregular flows, termed turbulence, also play a key role in determining the transport characteristics and distribution of atmospheric properties such as momentum, heat and matter (e.g. pollutants). Turbulence processes operate in three dimensions and lead to random and irregular flows, termed eddies, over scales of about 100 metres down to a few millimetres. One obvious example of the effect of turbulence is that of the near chaotic meandering of stack plumes.

Turbulence can be generated when there is a disturbance to the flow of wind. This can be caused by thermal gradients, uneven surfaces and obstacles (e.g. buildings). In areas of variable topography (e.g. urban areas and mountainous regions), wind speed will fall rapidly near the surface, leading to wind-shear effects (where layers of air are travelling at different speeds). Wind shear is a very important mechanism for disturbing the wind and hence generating turbulent flows. Turbulence processes then generate eddies, for example, around buildings, which can also trap the pollution in a localized area.

### Vertical Stability of the Atmosphere

The vertical stability of the atmosphere can have a significant influence on pollution levels near the surface. Under summer conditions, the surface heats up and causes the air to warm and rise, leading to a thermal low pressure system. When the near-surface air is sufficiently warm, thermal convective currents are generated and remove the pollution upwards. Such unstable conditions normally lead to lower levels of ground-level pollution. During winter seasons, the surface cools, causing the cold air to descend, forming a thermal high pressure system. The cold air near the surface becomes stagnant as it is unable to rise above its original position. Under these stable conditions, pollution levels tend to build up. If the air near the surface is colder than the air above, a ground level thermal inversion forms, which restricts any dispersion of pollution vertically. Inversions can also form aloft, acting as a 'cap' and limiting the mixing volume available to disperse low-level emissions.

Stability of the atmosphere also exhibits diurnal variations. For example, a ground-level inversion will eventually dissipate as solar radiation increases during the day and warms up the surface. At night-time, as the surface cools, an inversion can again be formed. Some of the plume shapes that can arise as a result of variations in atmospheric stability are shown in Figure I.5.

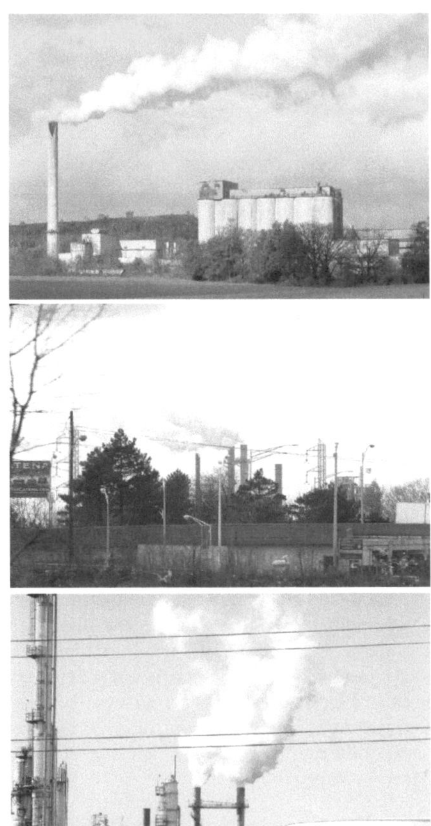

**Figure I.5** Typical shapes of plumes observed from industrial stacks. The shapes are caused by the different stability conditions that can occur in the atmosphere. (Source: USEPA 2007.)

## I.3 Transformation of Pollutants within the Atmosphere

### Global-Scale Transformation

On a global scale, the chemistry of compounds with long life becomes important. Chlorofluorocarbons (CFCs), for example, typically have atmospheric lifetimes of tens to hundreds of years and are responsible for the destruction of stratospheric ozone. CFCs are normally chemically unreactive and, once mixed in the troposphere, they penetrate into the stratosphere where they are broken down by the action of far-UV radiation, with chlorine as a product. The chlorine then reacts with ozone, causing its depletion in the stratosphere. There are other substances in addition to chlorine that lead to the destruction of ozone (see Chapter 5).

Other long-lived compounds, such as carbon dioxide ($CO_2$) and methane ($CH_4$), are associated with global warming and climate change. The anthropogenic increase in these gases in the atmosphere is leading to the 'enhanced greenhouse effect'. As mentioned above, such gaseous species are generally referred to as greenhouse gases (GHG). Although these gases transmit solar radiation, they absorb certain wavelengths of thermal infrared (IR) radiation from the earth, increasing the temperature of the atmosphere. Aerosols tend to have shorter lifetimes than greenhouse gases, but still have a significant effect on global temperatures. Whereas black carbon is efficient at absorbing solar radiation as well as IR radiation (Jacobson 2001), sulphate and nitrate aerosols scatter the radiation away from the earth's surface and hence cause cooling of the atmosphere. Global pollution and climate change are discussed in Chapter 4.

### Photo-Oxidant Reactions

As pollutants enter the atmosphere they are subjected to a range of transport processes, as mentioned above. However, these primary pollutants can also be subjected to many other complex processes that can transform them into harmful secondary products. For example, nitric oxide (NO) will be oxidised quickly by ozone ($O_3$) in the atmosphere to form nitrogen dioxide ($NO_2$), which is harmful to susceptible groups such as asthmatics and children. In urban areas, where emissions of NO can be high (e.g. from road traffic), this reaction has important implications because of the health effects of $NO_2$. The $NO_2$ molecule will photodissociate under the action of sunlight, to form nitric oxide and an oxygen (O) atom, which in turn reacts with an oxygen molecule ($O_2$) to form ozone. Therefore, although NO depletes ozone through oxidation, it is also a precursor specie for the formation of ozone. This is not the only route for ozone formation – there are a large number of complex chemical reaction pathways by which it can be formed (see for example, Jacobson 2002).

During the night, photolysis of $NO_2$ does not occur and hence no more ozone is produced through this route. Any emission of NO will lead to further destruction of ozone. However, $NO_2$ which now is not being photodissociated is oxidised further by ozone, to form the nitrate radical and then $N_2O_5$ (dinitrogen pentoxide). $N_2O_5$ reacts with water ($H_2O$) to form nitric acid. During the daytime, $NO_2$ can react with the hydroxyl radical (OH) to form nitric acid.

Photochemical smog, which is associated with pollution over many cities, results from chemical reactions under the action of sunlight, involving oxides of nitrogen ($NO_x$=NO + $NO_2$) and reactive volatile organic compounds (VOC), leading to a cocktail of products, including ozone. Various radicals (e.g. hydroxyl and peroxy radicals) are involved in the formation of ozone, creating photochemical smog conditions. The reaction of the hydroxyl radical (OH), which is a key species in the formation of ozone, is influenced by the relative proportions of VOC and $NO_x$ present in the atmosphere as the reactions take place competitively. The ratio of VOC to $NO_x$ has proved to be a useful parameter when estimating the production of ozone. In urban environments $NO_x$ is high (for example, due to emissions from road vehicles), yielding a low VOC to $NO_x$ ratio; and in rural environments $NO_x$ is generally lower, giving a higher ratio. Overall, high VOCs will lead to higher ozone production, but increasing $NO_x$ for a given concentration of VOCs can lead to higher as well as lower ozone, depending on the level of $NO_x$. (Further discussion on this can be found in Seinfeld and Pandis 1998).

### Other Transformation Processes

In addition to gas phase chemistry, there are also a host of heterogeneous (gas-particle and gas-liquid) reactions occurring in the atmosphere. Gas phase acids, such as sulphuric and nitric, can condense or dissolve into raindrops, creating acid rain. Furthermore, removal of gases and aerosols can take place through dry or wet deposition. Sulphur dioxide ($SO_2$), for example, is an important gaseous industrial pollutant and is emitted normally from high chimney stacks. It can be transformed into aqueous sulphuric acid and then deposited as acid rain over large distances. Pollutants in the atmosphere are also subject to various physical changes, for example, through growth processes in the case of ultra-fine aerosols.

## I.4 Major Atmospheric Pollution Problems

Historically, air pollution was closely associated with industrial activity, but with the advent of the motor vehicle and the relocation of large industrial plants outside urban areas, road traffic-related pollution has become the dominant concern within towns and cities. This is particularly the case for large urbanised areas or mega-cities (e.g. Molina and Molina 2004). Air pollution across the globe, however, is not only caused by road traffic emissions. In many cities, particularly those in developing regions, emissions from other sources, such as industrial and domestic, still play a crucial role in determining the quality of air.

Climate change has probably become the most important global challenge facing humankind. On a global scale, there is increasing evidence that the mean temperature of the world is increasing, leading to a range of impacts (IPCC 2001a; IPCC 2007). These changes are influenced and caused by internal variability within the climate system, as well as by natural and anthropogenic emissions. Chapters 4 and 5 both illustrate the extent and impact of global-scale pollution in terms of climate change and ozone depletion respectively. The cause of ozone depletion has been confirmed to be human induced (e.g. WMO 2003b) and now there is stronger evidence that much of global warming observed over the last 50 years is also attributable to human activities. These are pertinent examples of how locally generated emissions can have major environmental consequences on a global scale. Atmospheric pollution, therefore, has to be viewed in terms of a variety of anthropogenic sources, including traffic, industrial and domestic. These in turn have impacts on local scales (e.g. street-level pollution), urban scales (e.g. photochemical smog), regional scales (e.g. acid rain and long-range transport of precursors of particles and ozone) and global scales (e.g. ozone depletion and global warming).

It is not only the physical and chemical impacts of atmospheric pollution that are important. In fact, the intense interest in atmospheric pollution has been motivated by its direct effects on human health and environmental quality, which are discussed in Chapter 6. Health effects may be related to both individual smog episodes and long-term exposure to pollutants such as sulphur dioxide, particles and ozone. The proximity of people to pollution sources, such as factories and major roads, may be associated with a greater disease burden. Such sources of atmospheric pollution can also have local impacts on crop yield and forest health, but of greater concern is the much larger-scale damage to sensitive ecosystems, including soil and water quality, that is caused by long-range transport of secondary pollutants such as acid rain and ozone. Since both health and ecological impacts can result from cumulative exposure to atmospheric pollution over several decades, an important challenge, which is discussed in Chapter 6 and 7, is to assess both the short-term and long-term benefits of different policies to reduce emissions to the atmosphere. As recognized in Chapter 4, climate change is also likely to lead to major and irreversible impacts on human health and the wider environment over the twenty-first century.

A range of socio-economic factors also affect the extent to which we pollute the atmosphere. These include population growth and migration, economic development, industrial development, resource management, energy consumption patterns and wealth creation and distribution. In response, numerous measures have been initiated on various levels of government to limit pollution and curtail its impacts. These can be classified in terms of emission abatement technologies, pollution control and management strategies and policy frameworks on local (e.g. local air quality management action plans), regional (e.g. UNECE Convention on Long Range Transboundary Air Pollution, CLRTAP) and global dimensions (e.g. Kyoto and Montreal Protocols). At the same time, however, population, especially in urban areas, is projected to increase, along with the amount of road traffic. How, then, will air pollution levels change in the future across the world? How will this influence the impact of air pollution? Such questions are examined in Chapter 4 and 6 and especially in Chapter 7 in relation to air pollution emissions, air quality trends and changes in impacts.

As our understanding of the atmosphere has improved, it has been recognized that air pollution problems that have hitherto been treated separately are closely interlinked. A good example is that of interactions and feedbacks between air quality and climate change (e.g. AQEG 2007; Wilson et al. 2007) which is now an emerging environmental issue. Increasingly there is the requirement for science and policy to adopt more integrated approaches with a global outlook to investigate and develop solutions to solve current and future atmospheric pollution problems.

## I.5 Scope and Structure of the Atlas

### Scope of the Atlas

The Atlas addresses a range of key problems associated with atmospheric pollution. Over the past few decades there has been increasing recognition of interactions, not only between spatial and temporal scales affecting pollution, but also between the atmospheric, biospheric and hydrospheric systems. The focus of this Atlas, however, is on the state of the atmosphere and the impact of anthropogenic (caused by human activity) pollution on local, urban, regional and global scales.

Traditionally, the term 'air pollution' has been associated with local emissions of contaminants, such as those arising from industrial stacks or from road traffic. The term 'air quality' is also routinely used, but mainly to describe the state of the air in relation to limit values or standards. As the Atlas attempts to provide a comprehensive coverage of key air pollution problems, the term 'atmospheric pollution' is also employed to provide a more global perspective. Within individual chapters, however, air pollution and atmospheric pollution terms are used interchangeably.

Each chapter of the Atlas has been written to be self-contained, so that the reader can select a particular issue of interest without first having to read previous sections. Although the topics are based on complex concepts, the text and the associated tables and illustrations have been presented to be understandable to any informed reader who has a relevant science background equivalent to first-year level at university. This introduction should help those with limited background knowledge in the field.

It is also important to state what we have not done or attempted. We have not adopted a theoretical approach. Instead the emphasis is on providing a description of the state of atmospheric pollution along with explanations to aid the understanding of the problems in question. Given that the overall purpose of the Atlas is to provide a graphical representation of the state of our atmosphere, the approach has been to present an overview of the topics rather than to undertake an in-depth scientific critique or analysis of the wider literature. An extensive set of references to support the illustrations, however, has been provided along with a useful reading list for those readers who wish to pursue the subject in more detail.

### Structure of the Atlas

The Atlas begins with an introduction to the earth's atmosphere and a brief description of the main transport and transformation processes that influence atmospheric pollution on local to global scales. It is hoped that the Introduction will also serve to highlight the importance of treating atmospheric pollution in a more integrated manner and not just in the narrow terms of spatial scales. A historical perspective on atmospheric pollution is provided in Chapter 1. It illustrates the key milestones along the path to the current understanding of the state of the atmosphere. The population and traffic trends across the main regions of the world are discussed in Chapter 2, before examining air pollution in different cities around the world. Chapter 3 considers pollution transported over regional and hemispheric scales, and how it influences air quality on continental scales. Global-scale problems are addressed in the next two chapters. Atmospheric pollution and the resulting climate change is the focus of Chapter 4 and ozone depletion is discussed in Chapter 5. As stated earlier, the primary motivation to understand atmospheric pollution stems from the recognition of the serious environmental and health impacts. These are considered in Chapter 6. The final chapter provides a window into the future, by estimating how atmospheric pollution and its impacts are likely to change over the coming decades.

Although every effort has been made to ensure a comprehensive coverage of the literature, especially when producing the illustrations and maps, there will be many sources that we have not been able to include or cite. We do hope, however, that the Atlas as a whole, provides insight into some of the most pressing environmental concerns of this century.

# AIR POLLUTION HISTORY

*Peter Brimblecombe*



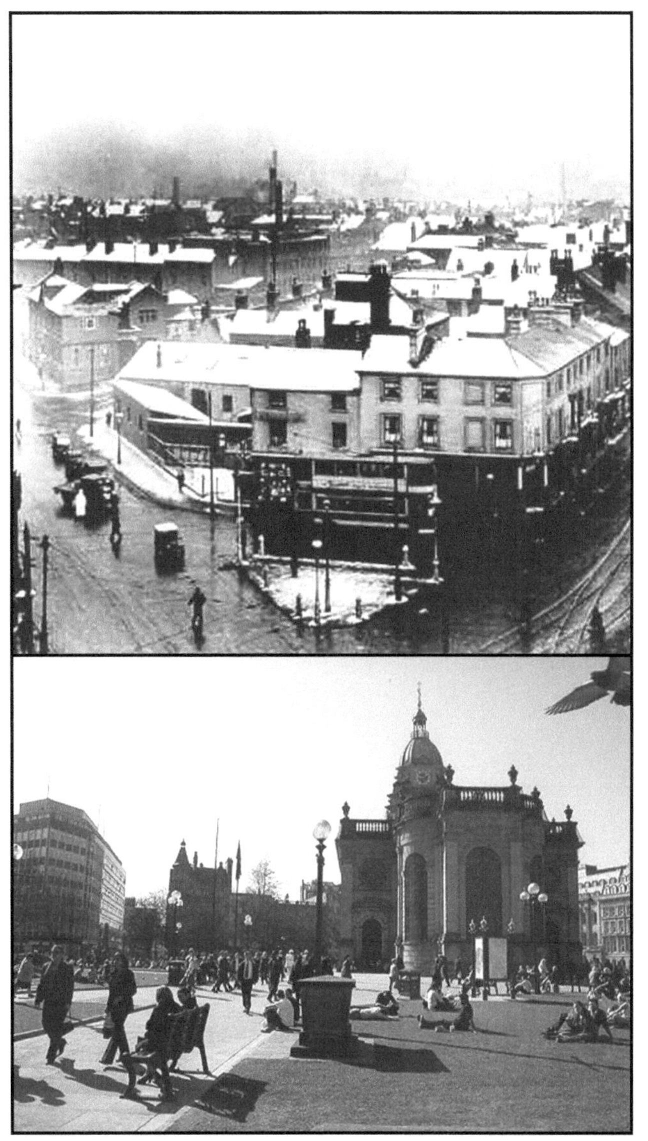

Upper: Lancaster Place, Birmingham in the early 20th c. showing winter pollution from extensive coal use (source: Brimblecombe 2007 personal communication). Lower: Modern day Birmingham, St. Philip's Cathedral and churchyard showing the improvement in the air quality of the city (source: Birmingham City Council, UK).

Although interest in indoor air pollution seems relatively recent, our earliest evidence of air pollutants often comes from indoor environments, such as dwellings filled with smoke and associated pollutants from poorly ventilated fires. When cities developed, these also became associated with pollution problems. The development of air pollution over the last 700–800 years seems to follow consistent patterns. Air pollution has often been related to the history of fuel use and the perceptible change in air pollution that arises from the fuels. Increasing energy demands and the adoption of new fuels (sequentially: coal, petrol, diesel) have caused air pollution problems. Mieck (1990) has argued that the numerous pollution decrees from the Middle Ages are essentially a response to single sources of what he terms *pollution artisanale*. These were usually just one particular type of pollution and distinct from the later and broader *pollution industrielle* that characterised an industrialising world.

Air pollution has often been visible as smoke, photochemical smog and diesel smoke. The concentration of air pollutants from a given source, such as coal, seems to increase for a long period and undergo a decrease due to declining emission strength. The pollution from one source is often replaced by another (e.g. coal smoke by petrol-derived pollution).The patterns of changing air pollution, although similar from one country to another, can take place over very different timescales. The changes, which took almost 800 years in Britain, all seem to have occurred in about 50 years in China, as it has moved from wood, to coal, to oil and then to gas.

Air pollution problems have not been easy to solve and the slow rate of improvement has often interested historians. The obvious cause is the reluctance of industry to expend money on abatement and limit technological progress. It is also possible that citizens in polluted cities have come to accept the state of the air where they live and work. The cosiness of the open coal fire and the fear of job losses (Mosley 2001) may have limited the strength of public protest. More recently, the implications of air pollution control on personal freedom (i.e. not having access to a car) seems an additional source of resistance to change.

From the second half of the twentieth century, air pollution problems have also been more global. There is a wide social awareness of the enhanced greenhouse effect, acid rain, the ozone hole and Asian brown haze.

The history of air pollution shows that our atmospheric environment is in a state of continual change. Problems emerge, reach some kind of crisis and then decline, only to be overtaken by others. The scales involved have become ever larger. The ability to detect pollutants and their effects has led to increasing instrumentation rather than influencing human perception. As people often interpret air pollution from local perceptions (Bickerstaff and Walker 2001), it may be increasingly difficult to maintain interest on larger temporal and spatial scales involving other pollution problems that are ever more subtle.

## 1.1 Europe and the Near East: Early History and Legislation

Our understanding of the first few thousand years of air pollution history is clearest for Europe and the Near East, where there are the most numerous written records, see Figure 1.1.

### *Sinusitis in Anglo-Saxon England*

Examination of skulls from burial sites can be used to establish the frequency of sinusitis (Figure 1.1a). An increased incidence of sinusitis in the Anglo-Saxon period has suggested smoky interiors to huts which lacked chimneys. In earlier periods there may have been a greater tendency to cook outside, so interiors may have been less smoky (Brimblecombe 1987a).

### *Anthracosis in mummies*

Soot deposits in desiccated lung tissue from mummies, most particularly in Egypt, suggest long exposure to smoke (Figure 1.1b).

### *Air pollution in dwellings, Sweden*

Studies of indoor air pollution in reconstructed houses, shown in Figure 1.1c, from the Scandinavian Iron Age attest to pollutant concentrations sufficient to affect health (Edgren and Herschend 1982; Skov et al. 2000).

### *Babylon*

Babylonian and Assyrian law included clauses that affected neighbours' property. Although the earliest laws, those of Hammurabi (twenty-third century BC) relate mostly to water (Driver and Miles 1952), smoke was typically treated in the same way in ancient law (Brimblecombe 1987b). Around AD 200, the Hebrew Mishnah, and its interpretation through the Jerusalem and Babylonian Talmud, details pollution issues (Mamane 1987).

### *Hermopolis, Egypt*

The Victory Stela of King Piye tells of the Nubian king's campaigns in Egypt, and that stench and a lack of air caused the city of Hermopolis to surrender *c*.734 BC (Lichtheim 1980).

### *Greece*

Cities of the ancient world were often small, but the inhabitants lived in high density, which led to pollutants becoming concentrated. Policy decisions regarding pollution in classical Greece were made by the *astynomoi* (controllers of the town), who were to ensure that pollution sources were well beyond the city walls; fortunately, industrial processes often took place in forests where fuel was abundant.

### *Rome*

Sextus Julius Frontinus (*c*. AD 30–100) oversaw water supply to imperial Rome (recorded in his book, *De Aquaeductu Urbis Romae*) and believed his actions also improved Rome's air. Civil claims over smoke pollution were brought before Roman courts almost 2000 years ago (Brimblecombe 1987b).

### *Indoor air pollution at Herculaneum*

Well-preserved skeletons from Herculaneum show lesions on the ribs that suggest a high frequency of pleurisy, see for example Figure 1.1d. Such lung infections have been seen as the result of indoor pollution from oil lamps and cooking (Capasso 2000).

### *Mining in Spain*

The geographer Strabo described (*c*.7 BC) the high chimneys required to disperse the air pollutants from silver production in Spain.

### *Justinian Code*

In AD 535 the *Institutes* issued under the Roman emperor Justinian were used as a text in law schools. Under the section Law of Things, our right to the air is clear: 'By the law of nature these things are common to mankind – the air, running water, the sea, and consequently the shores of the sea.'

### *Coal, industry and urban pollution in medieval London*

Wood was in such short supply in thirteenth-century London that coal brought by ships from England's north began to be used, especially to produce lime as a cement. The strange-smelling coal smoke was thought unhealthy, so by the 1280s there were attempts to prevent its use. As Sea Coal Lane and Limeburner's Lane lay to the west of the city (Figure 1.1e), prevailing winds carried smoke across the city towards busy St Paul's Cathedral. The area was further troubled by odours from the River Fleet. These were said to affect the health of the White Friars. The Knights Templar were accused of blocking the river in 1306, perhaps unfairly, as a commission of 1307 found tanning and butchers' waste from Smithfield market in the river. Domestic smoke also created problems and there were complaints that chimneys were not high enough to disperse it (Brimblecombe 1987a).

## 1.2 Early Ideas about Air and Its Pollution

Key ideas and discoveries which significantly influenced our understanding of air pollution are shown in Figure 1.2.

### *Miasmatic theories of disease and Hippocrates (c.460–377 BC)*

Ancient writings of the classical world (e.g. *Air, Water and Places* in the *Hippocratic Corpus*)

Coal, industry and urban pollution in medieval London
Wood was in such short supply in thirteenth-century London that coal brought by ships from England's north began to be used, especially to produce lime as a cement. The strange-smelling coal smoke was thought unhealthy, so by the 1280s there were attempts to prevent its use.
Air pollution in dwellings, Sweden
Concentrations of indoor pollutants were sufficient to affect the health of Iron Age people.
Rome
Civil claims over smoke pollution were brought before Roman courts almost 2000 years ago.
Sinusitis in Anglo-Saxon England
An increased incidence of sinusitis in the Anglo-Saxon period has suggested smoky interiors to huts which lacked chimneys.
Greece
Cities of the ancient world were often small, but the inhabitants lived in high density, which led to pollutants becoming concentrated.
Justinian Code
Under the Law of Things, our right to the air is clear: 'By the law of nature these things are common to mankind – the air, running water, the sea, and consequently the shores of the sea'.
Indoor air pollution at Herculaneum
Well-preserved skeletons show lesions on the ribs that suggest a high frequency of pleurisy.
Anthracosis in mummies
Soot deposits in desiccated lung tissue from mummies, most particularly in Egypt, suggest long exposure to smoke.
Mining in Spain
The geographer Strabo described (c.7 BC) the high chimneys required to disperse the air pollutants from silver production in Spain.
Hermopolis, Egypt
The Victory Stela of King Piye tells of the Nubian king's campaigns in Egypt, and that the stench and a lack of air caused the city of Hermopolis to surrender c.734 BC.
Babylon
Babylonian and Assyrian law included clauses that affected neighbours' property. These related to water and air.

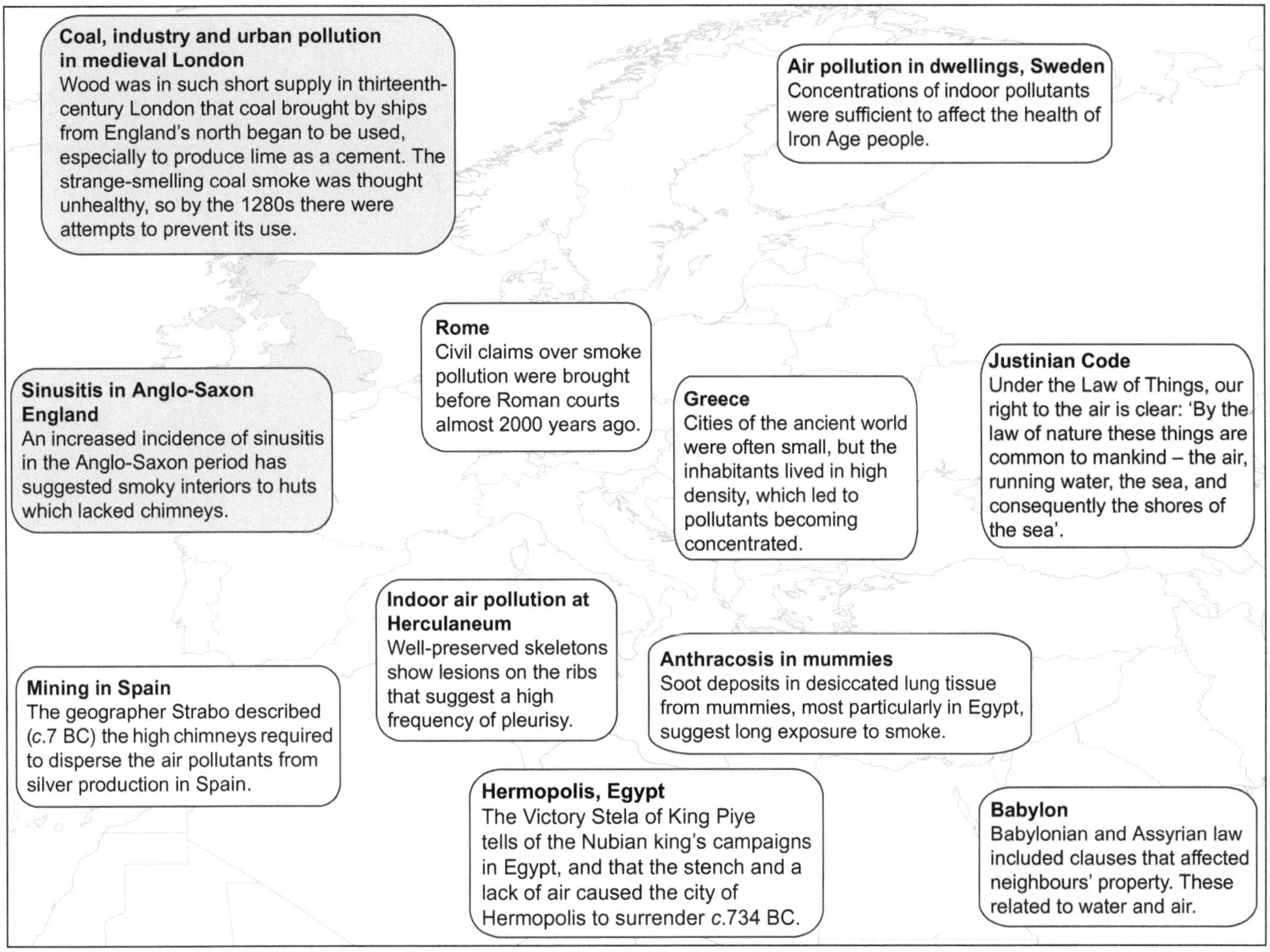

**Figure 1.1** Europe and the Near East: early history and legislation of air pollution.

describe the importance of climate and the properties of air relevant to health. Such environmental factors were seen as important in the treatment of disease.

### *Imperial Rome*

In Imperial Rome, Nero's tutor, Lucius Annaeus Seneca (*c*.4 BC–AD 65), was often in poor health and suffered from asthma, so his doctor ordered him to leave Rome; he found that no sooner had he escaped its oppressive atmosphere and awful culinary stenches, his health improved.

### *Pliny*

Pliny the Elder (AD 23–79) observed that saline rain damaged crops.

### *Arabic sources*

With the loss of understanding of classical writings in Europe, the *Hippocratic Corpus* became better known in the Arab world, so miasmatic theories of disease made it easy for air pollution there to be linked to health (Gari 1987). There were many important treatises, such as that on avoiding epidemics by at-Tamīmi (AD 932–1000), a great physician who grew up in Jerusalem.

### *Hildegard von Bingen*

The German mystic Hildegard von Bingen (1098–1179) thought that the dust of the atmosphere was harmful for plants.

### *Spontaneous generation*

In the Middle Ages, it was generally accepted that some life forms arose spontaneously from non-living matter, which could explain the minute organisms and small animals found in rainwater. Scientists gradually began to doubt this and experiments by the Italian physician Francesco Redi (1626–97) suggested that spontaneous generation was unlikely.

### *Agricola*

Georgius Agricola (1494–1555) wrote *De Re Metallica* and drew attention to the dangers of mining and the exposure of miners and metalworkers to diseases of air that caused damage to the lungs.

### *Theophrastus Bombastus von Hohenheim ('Paracelsus')*

Paracelsus (1493–1541) wrote the first monograph dedicated to diseases of miners and smelter workers, beginning a long interest in the toxicity of metals.

### *Margaret Cavendish and Kenelme Digby: atoms and air pollution*

Margaret Cavendish (1623–73) wrote much about atoms, and in her book *Poems and Fancies* (1653) she speculated on atoms from burning coal: 'Why that a coale should set a house on fire/Is, Atomes sharpe are in that coale entire/ Being strong armed with Points, do quite pierce through;/Those flat dull Atoms, and their forms undo.'

Sir Kenelme Digby (1603–65), who admired Cavendish, wrote in *A Discourse on Sympathetic Powder* (1658) that the corrosiveness of coal smoke arose when it dissipated to atoms that were claimed to be a 'volatile salt very sharp …', suggesting that the smoke was acidic.

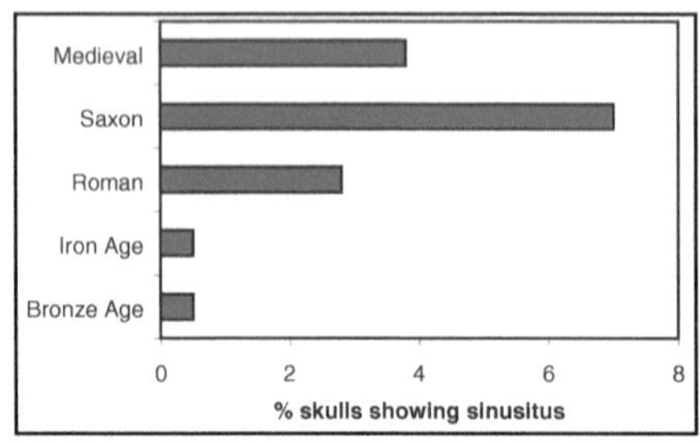


Figure 1.1a Sinusitis frequency in Britain over the ages.

Figure 1.1c Reconstructed Swedish huts used for indoor air pollution studies.

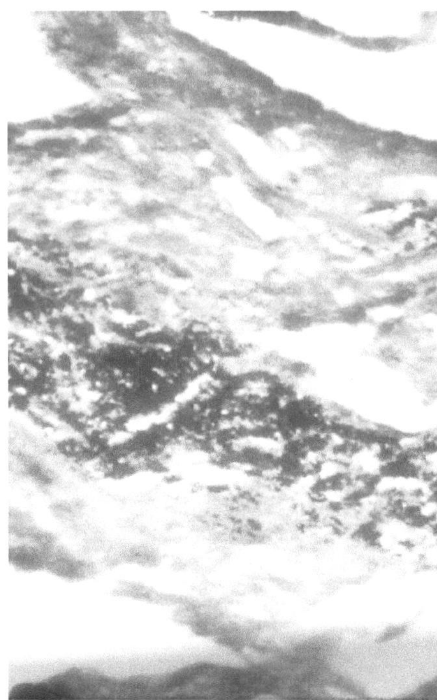

Figure 1.1b Lung tissue from mummies showing soot deposits.

Figure 1.1d Lesions on ribs from skeletons found in the archaeological area of Herculaneum.

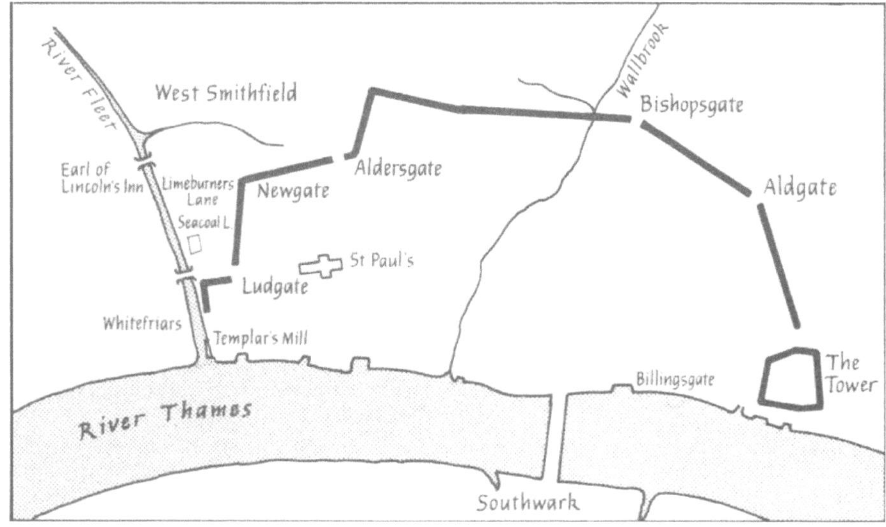


Figure 1.1e Map of medieval London showing locations described in the text.

### *John Evelyn*

John Evelyn (1620–1706), in the earliest book on air pollution, *Fumifugium* (1661), sought a broad explanation for the corrosive effects of coal-burning, which damaged plants, materials and health. He observed long-range transport of pollutants from the Great Fire of London and pressed for laws about air pollution that never got on to the statute books.

### *John Graunt*

John Graunt (1620–74), an early demographer, wrote *Natural and Political Observations Made Upon the Bills of Mortality* (1662), which suggested that the high death rate in London could be attributed partly to coal smoke.

### *Robert Boyle*

Robert Boyle (1627–91), who we remember for Boyle's Law, was interested in the corrosiveness of trace components of air in his book *A General History of the Air* (1692).

### *Bernardo Ramazzini*

Ramazzini (1633–1714) is often considered the father of occupational medicine. *De Morbis Artificum Diatriba* described the diseases of particular trades, including leather-tanning, wrestling and grave-digging. Ramazzini says that with a general improvement in diet and less arduous work, people would be better able to resist attacks on their health.

### *Joseph Black*

Joseph Black (1728–99) wrote *Experiments upon Magnesia Alba, Quick-Lime, and some other Alkaline Substances* (Edinburgh, 1756), which describes carbon dioxide.

### *Lavoisier, Scheele and Priestley*

Lavoisier (1743–94), Scheele (1742–86) and Priestley (1733–1804) are often linked with the discovery of oxygen. In *Réflexions sur le Phlogistique* (1783), Lavoisier showed the phlogiston theory to be inconsistent, so the modern ideas of atmospheric composition developed.

### *Henry Cavendish*

Cavendish (1731–1810) perfected the technique of collecting gases above water, publishing *On Fractious Airs* (1766). He investigated 'fixed air' and isolated 'inflammable air' (hydrogen) in 1766 and studied its properties. He showed that it produced a dew, which appeared to be water, upon being burnt. He also investigated the concentrations of oxygen above England using a balloon flight (Brimblecombe 1977).

### *Humphrey Davy*

Sir Humphrey Davy (1778–1829) investigated firedamp (methane) in mines and developed the safety lamp to detect it.

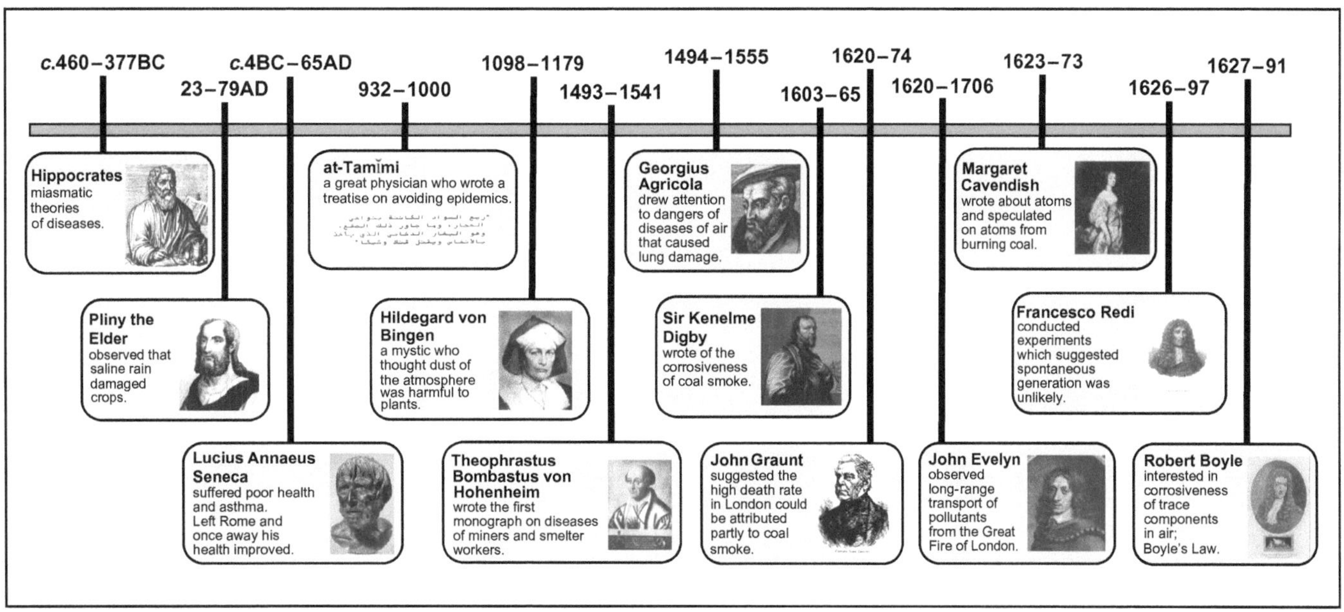


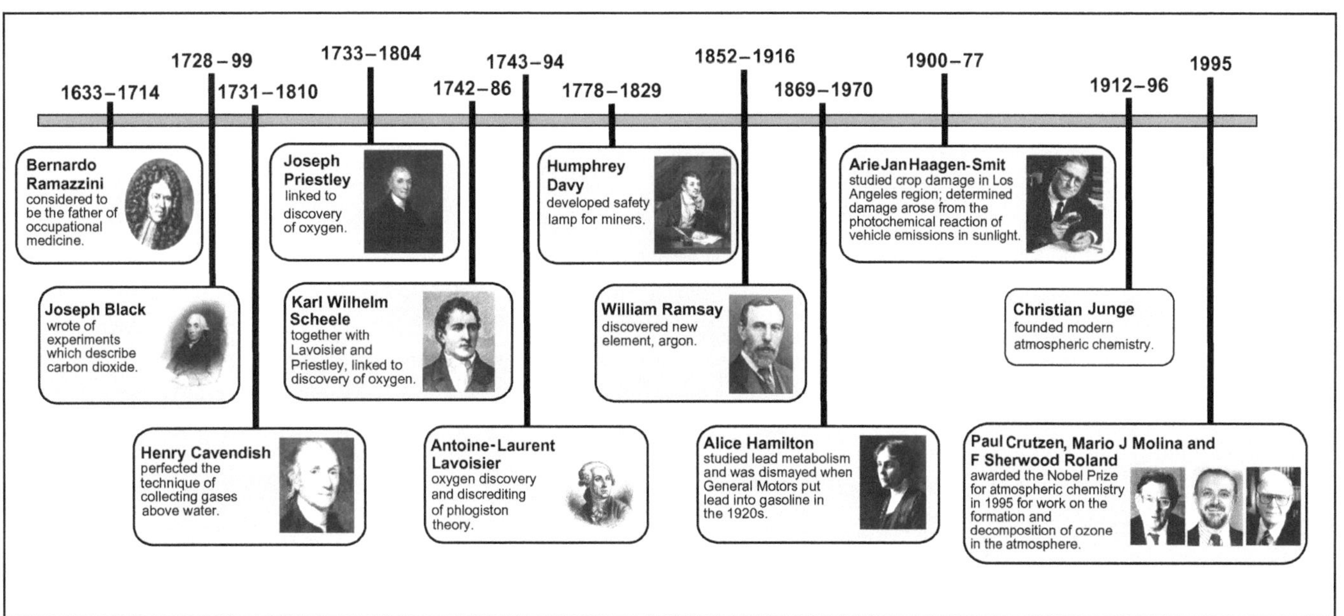


Figure 1.2 Early ideas about air and its pollution. The quote in Arabic script is from at-Tamīmi and translates to 'the black storm which is located in Hijaz and surroundings, is the smoky vapour which asphyxiates and kills'.

### *William Ramsay*

Henry Cavendish had noticed that a small volume of air could not be combined with nitrogen using electrical sparks. The experiment was ignored until Ramsay (1852–1916) examined it using spectroscopy, recognising it as a new element, which he termed argon (from a Greek word for inert).

### *Alice Hamilton*

Alice Hamilton (1869–1970) of Harvard University studied lead metabolism in the human body and was particularly dismayed when General Motors began to put lead into gasoline in the 1920s.

### *Arie Jan Haagen-Smit*

Dr Arie Haagen-Smit (1900–77) was interested in crop damage in the Los Angeles region. He realized that the damage arose from the photochemical reaction of vehicle emissions in sunlight. He saw the need for emissions controls on automobiles.

### *Christian Junge*

Christian Junge (1912–96) made many valuable contributions to atmospheric research, including his realisation in 1952 of the continuous level of distribution of atmospheric aerosols, and the first direct observation (1961) of the stratospheric sulphate aerosol layer (often called the 'Junge layer'). His book *Atmospheric Chemistry and Radioactivity* (1963) gave a sense of unity to a new research field, so important to our understanding of the environment today.

### *Nobel Prize for Atmospheric Chemistry, 1995*

Paul Crutzen, Mario J Molina and F Sherwood Roland were awarded the Nobel prize in chemistry for their work concerning the formation and decomposition of ozone in the atmosphere.

## 1.3 Urban Histories of Air Pollution

As economic activity moved from the countryside to the city, so did the use of fuels which generate pollutants. During the Industrial Revolution especially, cities experienced profound changes in air pollution, see Figure 1.3.

### *Athens*

Athens remained a small city in the millennia that followed its classical glory. However, growth in the mid-twentieth century led to increased air pollution and the development of a brown cloud, known locally as the *nephos*. Automotive emissions have proved to be a major factor in the development of the cloud and give photochemical conditions akin to those of Los Angeles. Sulphur content of diesel and fuel oils were strictly controlled, so an ageing vehicle fleet required replacement with newer, better-performing cars (Valaoras et al. 1988).

### *Auckland/Christchurch*

Auckland, New Zealand, is a mid-oceanic city situated on an isthmus, where it is exposed to marine aerosols, complex sea breezes, shallow harbours and emissions from mudflats. Ultimately motor vehicles have become an important source, but for a long time the city suffered odour problems or fume attacks (Sparrow 1968; Sparrow et al. 1969):

- 1840s: odour problems from putrefying waste at Port of Auckland.
- 1950s: fume attacks in South Auckland before sewage works.
- 1970s: civil emergency in Parnell-Merphos.
- 1997, 16 October: fume alert at Nelson Street, Onehunga.

Christchurch, by contrast, remained a city where solid fuel (wood and coal) burnt to heat homes caused severe pollution in winter until the end of the twentieth century.

### *Kolkata (formerly Calcutta)*

Kolkata, the second city of the British Empire, experienced environmental problems from the eighteenth century, and adopted smoke pollution legislation in 1863. In the early twentieth century, a number of UK experts, including Fredrick Grover of Leeds and William Nicholson of Sheffield (Nicholson 1907–8), aided in the development of a stringent smoke inspection policy and pressed for furnaces to be carefully stoked. Although the policy had some measure of success (as seen in the Figure 1.3a), smoke observation was not always reliable (Brimblecombe 2004a) and the action placed unreasonable demands on poorly trained stokers, who bore the brunt of penalties for producing smoke (Anderson 1997). Kolkata developed into an overcrowded city with a serious pollution problem, but low sulphur coal meant that smoke and increased pollution from traffic became more problematic than sulphur dioxide (WHO/UNEP 1992).

### *Japan*

In Japan, cottage industry traditionally relied on wood in rural areas. Early copper-smelting operations to cast enormous bronze statues caused sulphur dioxide pollution from the eighth century, see Figure 1.3b (Satake 2001). The urban history of air pollution in Japan began with concerns over dust from industry in the last years of the nineteenth century. After the Second World War this involved stronger concerns about the smoke that resulted from the growing utilisation of energy. Protests by women in Tobata City who suffered the effects of the air pollution began in 1950. As elsewhere in the world, concerns evolved from smoke to worries about trace gases, particularly sulphur dioxide, which began to show reductions from 1967 (Sawa 1997). Tokyo has often been seen as an example of how a large city can successfully control its air pollution (WHO/UNEP 1992).

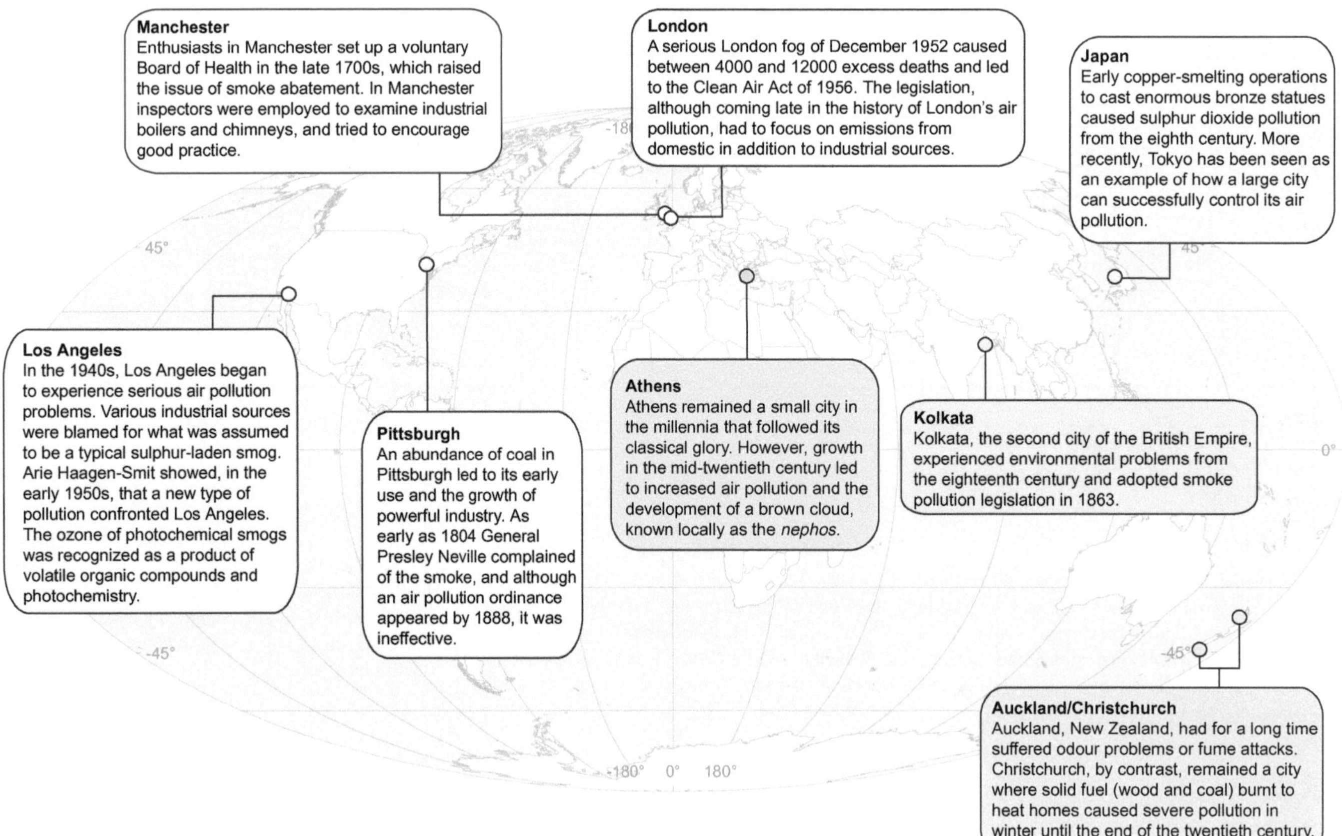
Manchester
Enthusiasts in Manchester set up a voluntary Board of Health in the late 1700s, which raised the issue of smoke abatement. In Manchester inspectors were employed to examine industrial boilers and chimneys, and tried to encourage good practice.
London
A serious London fog of December 1952 caused between 4000 and 12000 excess deaths and led to the Clean Air Act of 1956. The legislation, although coming late in the history of London's air pollution, had to focus on emissions from domestic in addition to industrial sources.
Japan
Early copper-smelting operations to cast enormous bronze statues caused sulphur dioxide pollution from the eighth century. More recently, Tokyo has been seen as an example of how a large city can successfully control its air pollution.
Los Angeles
In the 1940s, Los Angeles began to experience serious air pollution problems. Various industrial sources were blamed for what was assumed to be a typical sulphur-laden smog. Arie Haagen-Smit showed, in the early 1950s, that a new type of pollution confronted Los Angeles. The ozone of photochemical smogs was recognized as a product of volatile organic compounds and photochemistry.
Pittsburgh
An abundance of coal in Pittsburgh led to its early use and the growth of powerful industry. As early as 1804 General Presley Neville complained of the smoke, and although an air pollution ordinance appeared by 1888, it was ineffective.
Athens
Athens remained a small city in the millennia that followed its classical glory. However, growth in the mid-twentieth century led to increased air pollution and the development of a brown cloud, known locally as the *nephos*.
Kolkata
Kolkata, the second city of the British Empire, experienced environmental problems from the eighteenth century and adopted smoke pollution legislation in 1863.
Auckland/Christchurch
Auckland, New Zealand, had for a long time suffered odour problems or fume attacks. Christchurch, by contrast, remained a city where solid fuel (wood and coal) burnt to heat homes caused severe pollution in winter until the end of the twentieth century.
45°
-45°
0°
-180°
180°

**Figure 1.3** Urban histories of air pollution.

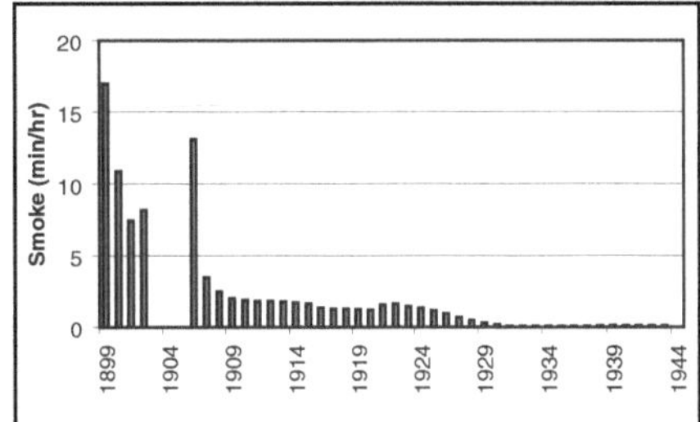


Figure 1.3a Reduction in smoke levels in Kolkata through visual observation of chimney plumes. Units of min/hour refer to the number of minutes for which the plumes were observed during an hour (Brimblecombe 2004a).

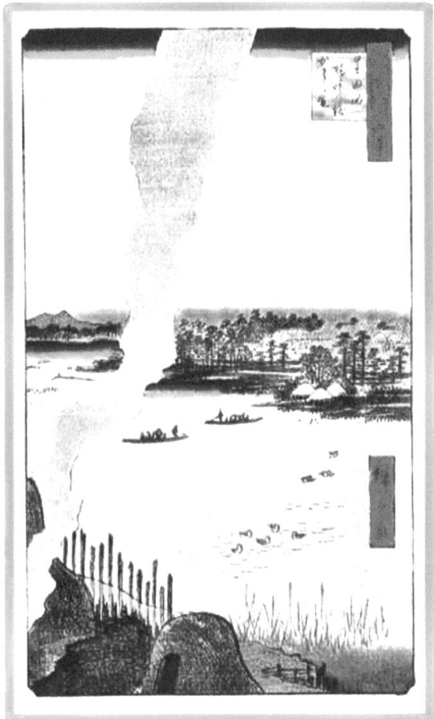

Figure 1.3b Smoke plume from some kilns from the picture by Hiroshige: Lime kilns at Hasiba Ferry, Sumida River.

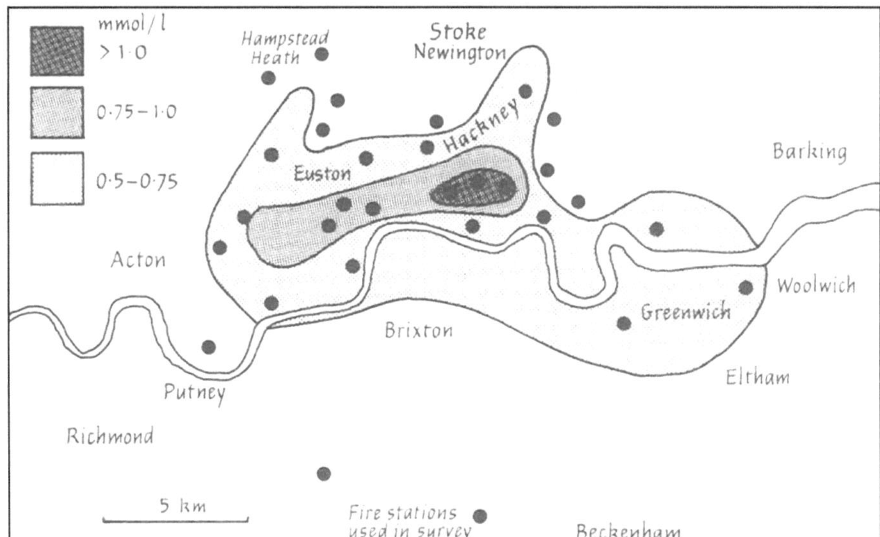


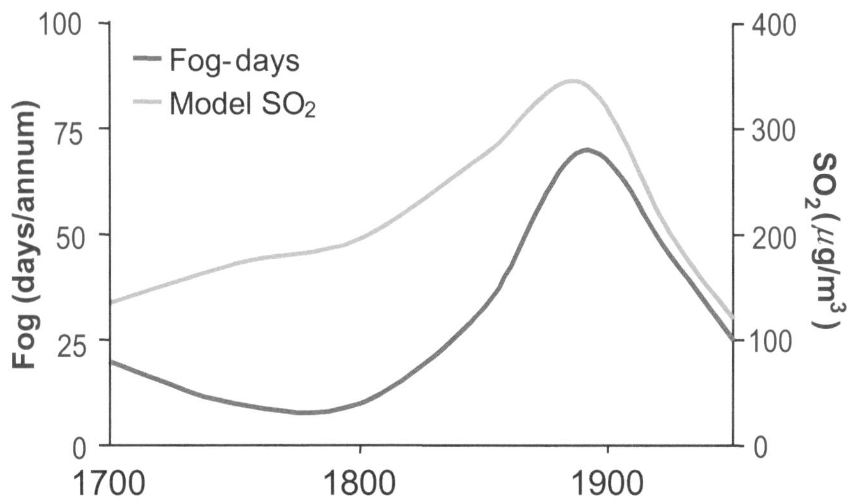


Figure 1.3c Two figures for London showing concentrations of sulphate in rain (upper) and comparison of frequency of fog with the predicted concentrations of sulphur dioxide (lower).

### London

From the late eighteenth century steam engines began to find increasing use and added to the noise and smoke of London, becoming a focus of concern. The nineteenth century saw laws to improve urban health, but smoke abatement did not come about in an age which emphasised economic progress. In 1870, measurements of sulphate in rain showed a broad spread along the river that peaked where the poor lived in the East End (as shown in Figure 1.3c (upper)). The predicted concentrations of sulphur dioxide from coal in London and the frequency of smoky fog both appear to peak at the end of the nineteenth century (as can be seen in Figure 1.3c (lower)). Declines came from urban expansion, which decreased the density of emissions (Brimblecombe 2004a). However, a serious London fog of December 1952 caused between 4000 and 12000 excess deaths and led to the Clean Air Act 1956. The legislation, although coming late in the history of London's air pollution, had to focus on emissions from domestic in addition to industrial sources. This was an important recognition that personal freedom might have to be reduced in the face of environmental pressure (Brimblecombe 2002; Brimblecombe 2006). In the 1990s, attention shifted to particles in the atmosphere, an area which had been neglected since the 1960s. The fine particles, often attributed to the use of diesel engines, play a critical role in health effects of air pollution.

### Los Angeles

In the 1940s, Los Angeles began to experience serious air pollution problems. Although these were called smogs, it was some time before their true nature was understood. Various industrial sources were blamed for what was assumed to be a typical sulphur-laden smog. Arie Haagen-Smit showed, in the early 1950s, that a new type of pollution confronted Los Angeles. The ozone of photochemical smogs was recognized as a product of volatile organic compounds and photochemistry. The automobile proved to be the major contributor to the problem and control required the reduction in emissions from cars. Figure 1.3d shows improvements in ozone at Crestline California (Lee et al. 2003). Los Angeles was the archetype of a major transition of the twentieth century, where primary pollutants from stationary sources were to become less important than photochemical precursors from mobile sources.

### Manchester

From medieval times smoke and other nuisance in England were dealt with by local courts (e.g. the Court Leet), but in cities where steam engines were used, more modern approaches of control were required. Enthusiasts in Manchester set up a voluntary Board of Health in the late 1700s, which raised the issue of smoke abatement (Bowler and Brimblecombe 2000). Subsequent developments in

Manchester anticipated the sanitary reforms that became more general in Britain, Europe and North America by the mid-nineteenth century. In Manchester inspectors were employed to examine industrial boilers and chimneys and tried to encourage good practice. The problems of the early nineteenth century were significant because population had shifted from the countryside to the polluted cities. Early attempts to control pollutants were largely overwhelmed by a lack of effective technology, industrial pressure and insufficient political will (Bowler and Brimblecombe 1990; Mosley 2001).

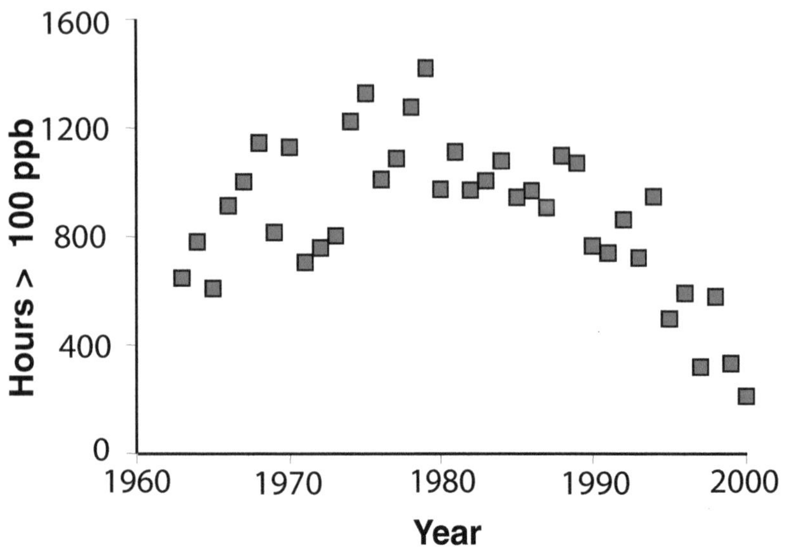


Figure 1.3d Variation of ozone concentrations from 1950 to 2000 measured in Crestline, California.

### *Pittsburgh*

An abundance of coal in Pittsburgh led to its early use and the growth of powerful industry. As early as 1804 General Presley Neville complained of the smoke, and although an air pollution ordinance appeared by 1888, it was ineffective. For a short time in the 1880s, natural gas was cheap and readily available, which lowered the amount of smoke. The situation was so bad again by 1912 that the Mellon Institute undertook an in-depth study of the smoke problem. The early twentieth century saw faltering local steps to control the smoke, which finally took effect with the major changes in fuel use of the 1950s. Improvements continued, although increasingly driven by national legislation (Davidson 1979).

## 1.4 Air Pollution Disasters and Episodes

Air pollution episodes and disasters may be brief, but their social and legislative impacts can be exceedingly important. Some major air pollution disasters are described below and highlighted in Figure 1.4.

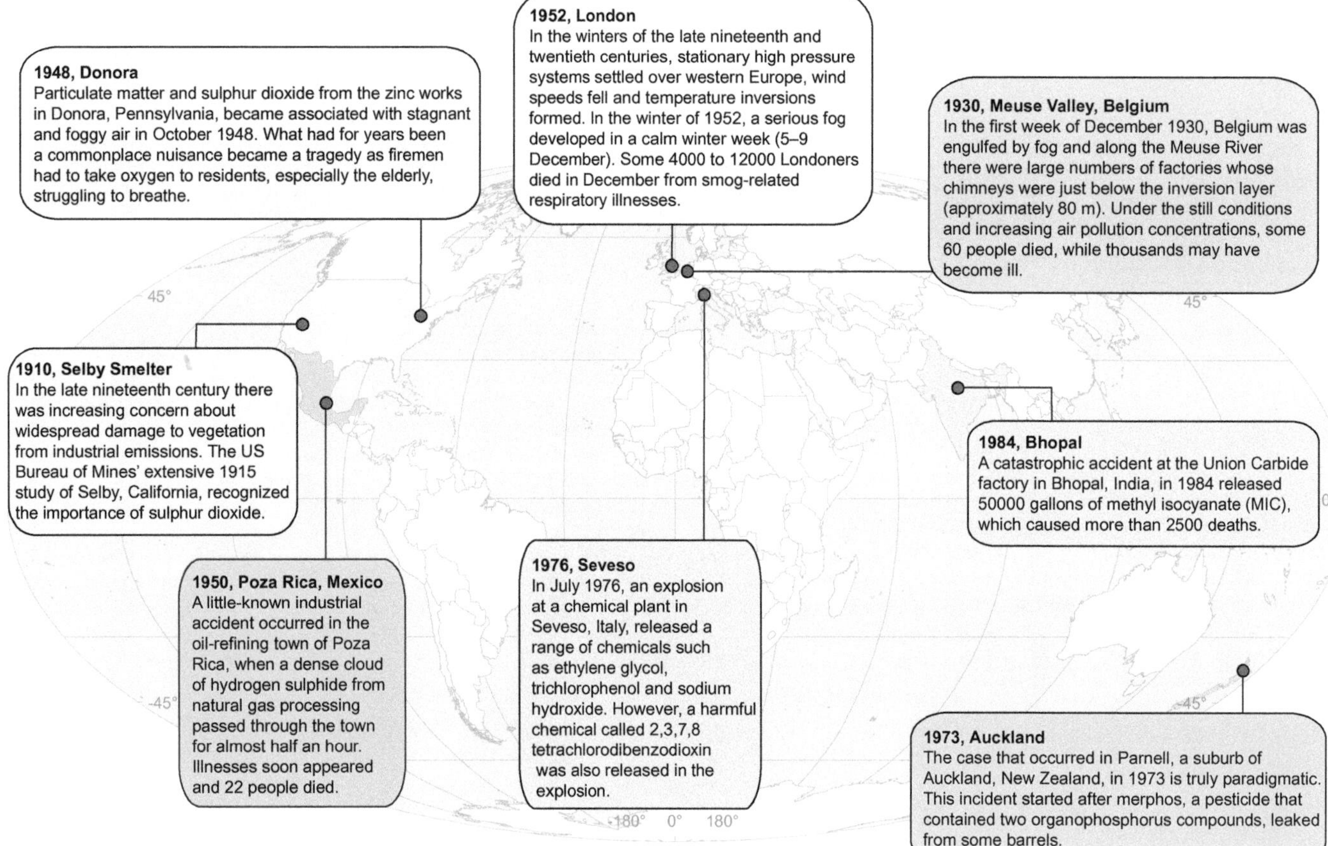


Figure 1.4 World map of air pollution disasters and episodes.

### *1910, Selby Smelter*

In the late nineteenth century there was increasing concern about widespread damage to vegetation from industrial emissions. This was first handled by the Alkali Acts (1863), England (MacLeod 1965), and later scientific studies of smelter smoke in Germany (Schramm 1990) and the United States (Holmes et al. 1915). One of the best known may be the US Bureau of Mines' extensive 1915 study of Selby, California, which recognized the importance of sulphur dioxide. The smelter emissions attracted much attention and were mentioned by Jack London in his book *John Barleycorn*: 'Out of the Oakland Estuary and the Carquinez Straits off the Selby Smelter were smoking.'

### *1930, Meuse Valley, Belgium*

In the first week of December 1930, Belgium was engulfed by fog, and along the Meuse River there were a large number of factories whose chimneys were just below the inversion layer (approximately 80 m). Under the still conditions and increasing air pollution concentrations, some 60 people died while thousands may have become ill. Although these were respiratory illnesses, the cause was difficult to establish. At the time sulphur oxides were seen as the main culprit, but fine particles (perhaps correctly) and fluorides were also considered as playing a role (Nemery et al. 2001). Although an important incident, it led to little immediate change in air pollution control.

### *1948, Donora*

Particulate matter and sulphur dioxide from the zinc works in Donora, Pennsylvania, became associated with stagnant and foggy air in October 1948. What had for years been a commonplace nuisance became a tragedy as firemen had to take oxygen to residents, especially the elderly, struggling to breathe. Some 20 died during this episode, which became subject to an investigation from the US Public Health Service. Like so many investigations of the time it was difficult to attribute cause of death to a specific agent in the air. However, like many episodes it influenced the development of local, regional, state and national laws to reduce and control factory smoke, culminating with the US Clean Air Act-1970 (Kiester 1999; Helfand et al. 2001).

### *1950, Poza Rica, Mexico*

A little-known industrial accident occurred in the oil-refining town of Poza Rica, when a dense cloud of hydrogen sulphide from natural gas processing passed through the town for almost half an hour. Illnesses soon appeared and 22 people died.

### *1952, London*

In the winters of the late nineteenth and twentieth centuries, stationary high pressure systems settled over western Europe, wind speeds fell and temperature inversions formed. Pollutant concentrations increased and fog became widespread in Britain, with London severely affected by these conditions (Brimblecombe 2002). Its fogs were a backdrop to all that was magical or evil about the city. For almost a century it had been widely known that the death rate increased in these smogs. London's smogs became central to books such as Sherlock Holmes, and an inspiration to painters such as Claude Monet, André Derain and Yoshio Markino.

In the winter of 1952, a serious fog developed in a calm winter week (5–9 December):

- Transport came to a standstill, buses could not see the kerb nor trains the signals, so people became housebound.
- Fog was so thick that people became lost, and sometimes blind people were found leading them home.
- Some 4000 to 12000 Londoners died in December from smog-related respiratory illnesses.
- Smog was so bad that cattle died at Smithfield market.
- A performance of *La Traviata* at the Sadler's Wells theatre had to be abandoned.

Widespread concern over this smog period and its impacts on people led to the Clean Air Act 1956.

### *1973, Auckland*

The case that occurred in Parnell, a suburb of Auckland, New Zealand, in 1973 is truly paradigmatic. This incident started after merphos, a pesticide that contained two organophosphorus compounds, leaked from some barrels. Inquires concerning potential threats to health were conducted immediately and the authorities were wrongly informed that the compound was extremely toxic. Just after the announcement of its danger, nearly 400 workers and nearby residents started exhibiting symptoms: breathing difficulty, eye irritation, headache and nausea. Although merphos was finally judged of low toxicity, an inquiry blamed butyl mercaptan for the effects. However, a number of writers have seen the symptoms as a result of mass hysteria in the face of the failure of medical authorities to take a firm stand on the issue (Christophers 1982).

### *1976, Seveso*

In July 1976, an explosion at a chemical plant in Seveso, Italy, released a range of chemicals such as ethylene glycol, trichlorophenol and sodium hydroxide. However, 2, 3, 7, 8 tetrachlorodibenzodioxin, better known simply as 'dioxin', was also released in the explosion. This only became evident more than a week after the accident. Small animals died and children began to show symptoms where they had been exposed to the emissions. Concern over dioxin exposure led to difficult decisions over the area to evacuate and some 50 cases of chloracne were observed, mostly among children. These gradually subsided over the next year or two. Fortunately, pathological data gathered in subsequent years did not show differences between those exposed and the controls. In general, the early recognition of the disaster and the recovery of the region are seen as successes (Mocarelli 2001). The accident led to the 'Seveso Directive' in 1982 to prevent major-accident hazards from industry, and amendments have continued to develop the European approach to the regulation of dangerous activities.

### *1984, Bhopal*

'The accident at Union Carbide's pesticide plant in Bhopal in 1984 killed 8000 people immediately and injured at least 150,000. It remains the worst industrial disaster on record, and the victims are still dying' (Source: NewScientist.com, December 2002).

A catastrophic accident at the Union Carbide factory in Bhopal, India, in 1984 released 50000 gallons of methyl isocyanate (MIC), which caused more than 2500 deaths. The final death toll related to the incident will ultimately be many times higher. Water used for washing the lines entered a tank through leaking valves. An exothermic reaction caused the release of a lethal gas mixture. Hundreds of thousands were injured and a bitter debate about causes and responsibility ensued, which often left those affected without proper medical attention or just compensation. The US Congress passed the Emergency Planning and Community Right-to-Know Act in 1986, so that citizens can gain access to information about an increasing list of industrial emissions. Even though Union Carbide accepted moral responsibility, the issue remains unresolved, as many victims are still seeking redress and there are concerns over continued contamination.

## 1.5 Environmental Damage by Acid Rain

Acid rain is typically the result of long-range transport of pollutants, such as sulphur dioxide and nitrogen oxides, which yield sulphuric and nitric acid. Acid rain might seem to be a twentieth-century phenomenon, but the idea of pollutant contamination of rain has a long history, see Figure 1.5. For example, the late 1860s witnessed interest in rainfall composition in terms of drinking water quality and urban pollution from Franklin and Angus Smith (who used the term 'acid rain'). The second half of the twentieth century and the early years of the twenty-first, however, have been characterised by an increasing worry about damage to forests, aquatic life and building materials from acid deposition.

Work by Hans Egner and Erik Erikkson from the late 1940s gave ample material for Svante Odén to introduce the problem of acid rain; with pH values less than 4.0 and its thin soil cover, Sweden was seen as particularly vulnerable. The government realized it would have to encourage other nations to recognize this transnational problem. Sweden hosted the 1972 UN Conference on the Human Environment in Stockholm, which encouraged a cooperative programme and convention on the long-range transport of air pollutants (LRTAP) in 1979. Major coal user, Britain, had to be pressured to reduce sulphur emissions, which led to improvements,

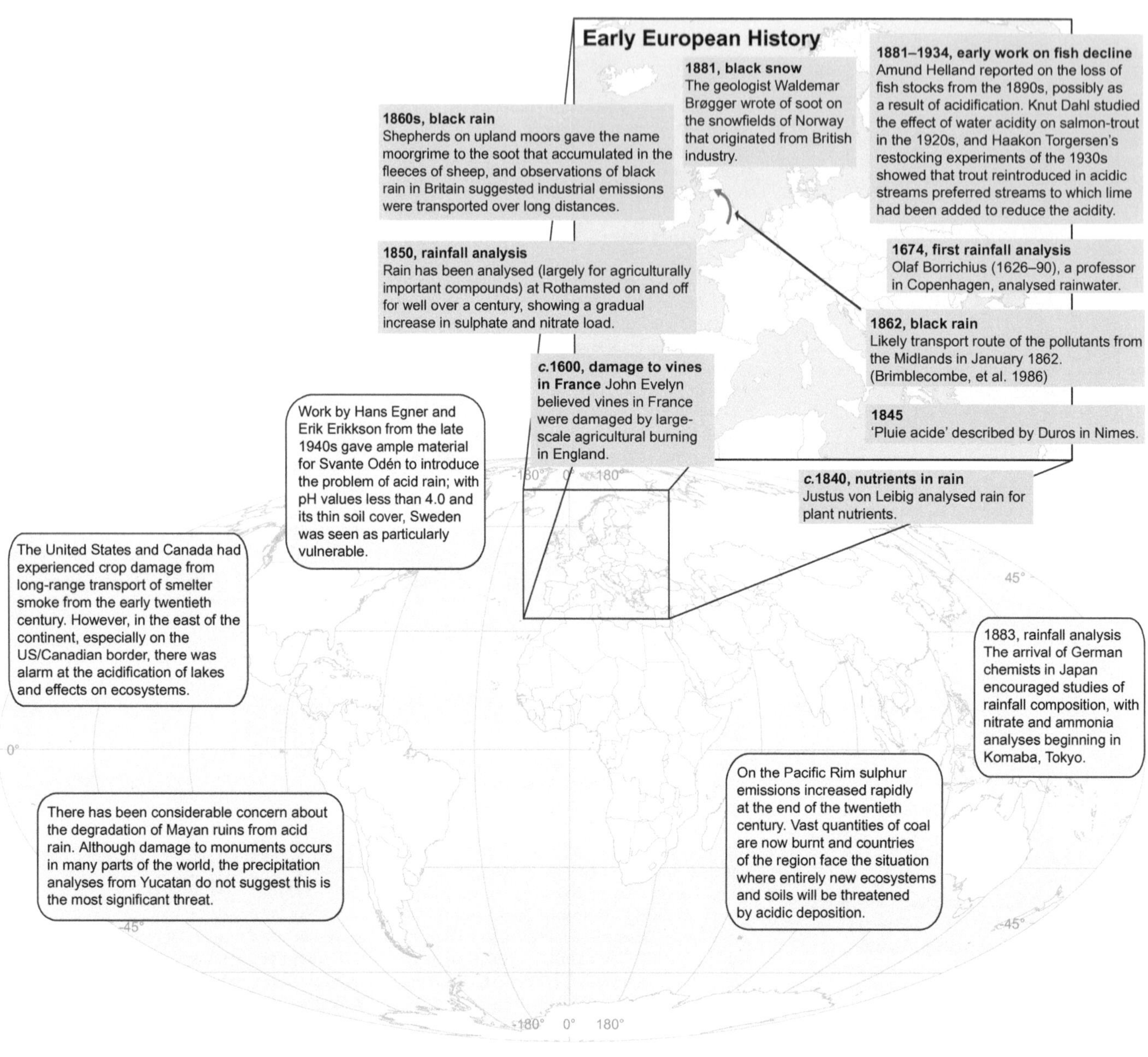

Early European History
1881, black snow
The geologist Waldemar Brøgger wrote of soot on the snowfields of Norway that originated from British industry.
1881–1934, early work on fish decline
Amund Helland reported on the loss of fish stocks from the 1890s, possibly as a result of acidification. Knut Dahl studied the effect of water acidity on salmon-trout in the 1920s, and Haakon Torgersen's restocking experiments of the 1930s showed that trout reintroduced in acidic streams preferred streams to which lime had been added to reduce the acidity.
1860s, black rain
Shepherds on upland moors gave the name moorgrime to the soot that accumulated in the fleeces of sheep, and observations of black rain in Britain suggested industrial emissions were transported over long distances.
1850, rainfall analysis
Rain has been analysed (largely for agriculturally important compounds) at Rothamsted on and off for well over a century, showing a gradual increase in sulphate and nitrate load.
1674, first rainfall analysis
Olaf Borrichius (1626–90), a professor in Copenhagen, analysed rainwater.
1862, black rain
Likely transport route of the pollutants from the Midlands in January 1862. (Brimblecombe, et al. 1986)
c.1600, damage to vines in France John Evelyn believed vines in France were damaged by large-scale agricultural burning in England.
1845
'Pluie acide' described by Duros in Nimes.
c.1840, nutrients in rain
Justus von Leibig analysed rain for plant nutrients.
Work by Hans Egner and Erik Erikkson from the late 1940s gave ample material for Svante Odén to introduce the problem of acid rain; with pH values less than 4.0 and its thin soil cover, Sweden was seen as particularly vulnerable.
The United States and Canada had experienced crop damage from long-range transport of smelter smoke from the early twentieth century. However, in the east of the continent, especially on the US/Canadian border, there was alarm at the acidification of lakes and effects on ecosystems.
1883, rainfall analysis
The arrival of German chemists in Japan encouraged studies of rainfall composition, with nitrate and ammonia analyses beginning in Komaba, Tokyo.
On the Pacific Rim sulphur emissions increased rapidly at the end of the twentieth century. Vast quantities of coal are now burnt and countries of the region face the situation where entirely new ecosystems and soils will be threatened by acidic deposition.
There has been considerable concern about the degradation of Mayan ruins from acid rain. Although damage to monuments occurs in many parts of the world, the precipitation analyses from Yucatan do not suggest this is the most significant threat.
45°
0°
-45°
-180° 0° 180°


**Figure 1.5** Impacts of acid rain.

although a parallel need to reduce nitrogen emissions soon emerged.

The United States and Canada had experienced crop damage from long-range transport of smelter smoke from the early twentieth century. However, in the east of the continent, especially on the US/Canadian border, there was alarm at the acidification of lakes and effects on ecosystems. The National Acid Precipitation Assessment Program (NAPAP) began in 1980, an inter-agency task force under the auspices of the Council on Environmental Quality. It coordinated long-term monitoring of precipitation. As with Europe, gradual improvements have come as the result of reductions in sulphur emissions, especially from coal-burning power stations (Brimblecombe 2004b).

On the Pacific Rim sulphur emissions increased rapidly at the end of the twentieth century. Vast quantities of coal are now burnt and countries of the region face the situation where entirely new ecosystems and soils will be threatened by acidic deposition. The chemistry of rain in these regions is different because of the presence of alkaline dust and smoke from forest fires. These emerging problems have required new initiatives, such as the development of an acid precipitation monitoring network in East Asia from the Toyama meeting of 1993. Fortunately, China, where sulphur dioxide emissions increased most rapidly, is shifting its fuel base to cleaner liquid and gaseous fuels.

There has been considerable concern about the degradation of Mayan ruins from acid rain. Although damage to monuments occurs in many parts of the world, the precipitation analyses from Yucatan do not suggest this is the most significant threat (Bravo et al. 2000).

## 1.6 Global Air Pollution Issues

A range of global air pollution issues have become increasingly apparent over the last hundred years and necessitated international cooperation to resolve the problems they cause. Some of these are indicated graphically in Figure 1.6.

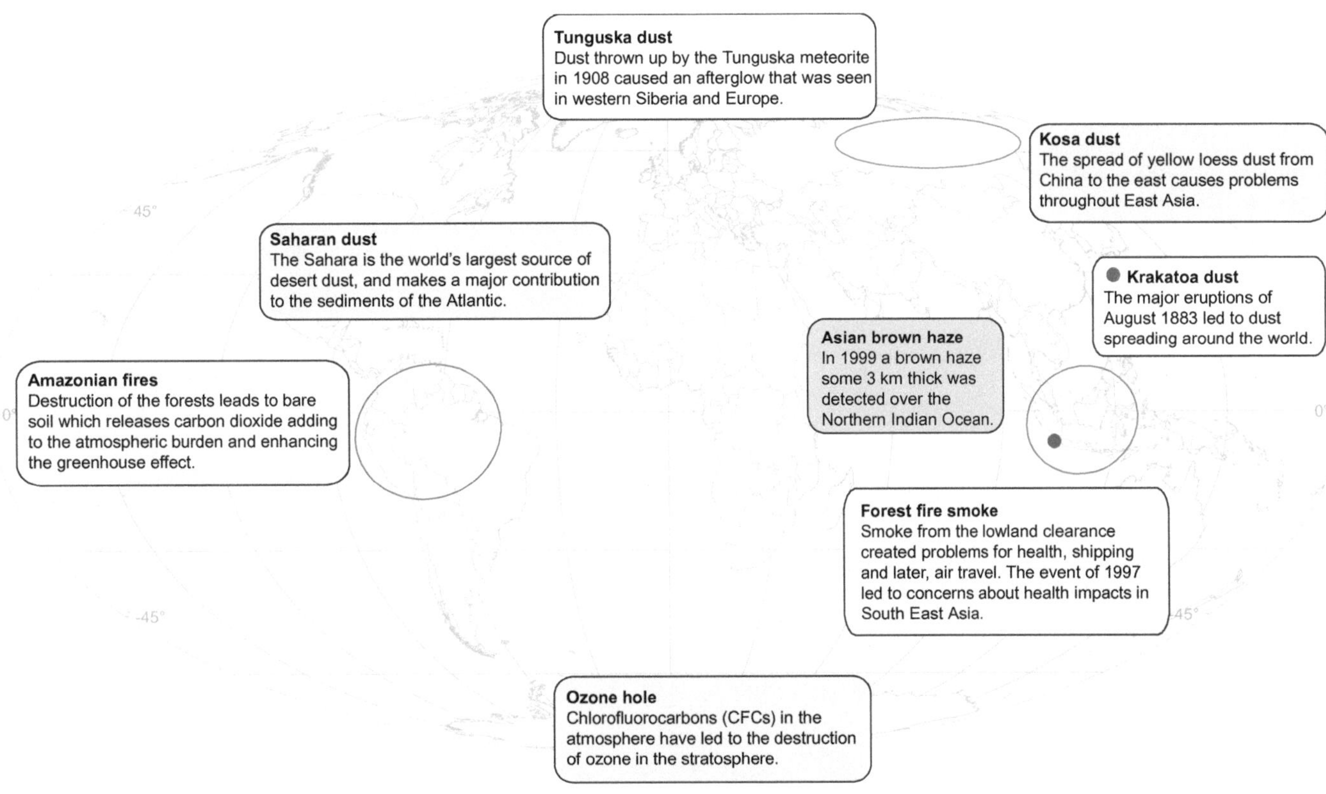


**Figure 1.6** Global air pollution issues.

### *Kosa dust*

The spread of yellow loess dust from China to the east each spring causes problems throughout East Asia. Records of these events have been kept in documents for more than 3000 years. There are also records from lake deposits, such as the one below from Taiwan, which shows a notable increase in dust since the mid-1300s (Chen et al. 2001).

### *Forest fire smoke*

Early Portuguese, Dutch and British merchants exploited the Indonesian archipelago from the 1600s and encouraged lowland clearance to grow spices. Smoke from these activities, especially in times of drought, created problems for health, shipping and later, air travel. Dry periods related to El Niño seem to influence the magnitude of the fires, such as in 1982–83. Striking occurrences come from 1877–78, 1891, 1902–10, 1914, 1972, 1982–83 and 1997, but some (e.g. 1910) were agricultural rather than climatic changes (Potter 2001; Brimblecombe 2005). The event of 1997 received international attention, with great worries about health impacts in South East Asia.

### *Amazonian fires*

The forest clearance within the Amazon has caused extensive smoke palls, but perhaps more importantly, the destruction of the forests leads to bare soil which releases carbon dioxide adding to the atmospheric burden and enhancing the greenhouse effect.

### *Krakatoa dust*

The major eruptions of August 1883 led to dust spreading around the world and gave rise to spectacular sunsets (Austin 1983). The event gave new insight into the global nature of the atmosphere.

### *Tunguska dust*

Dust thrown up by the Tunguska meteorite in 1908 caused an afterglow that was seen in western Siberia and Europe from 30 June to 2 July (Vasil'ev and Fast 1973). There is also some evidence that the meteor caused ozone depletion in the upper atmosphere (Turco et al. 1981).

### *Saharan dust*

The Sahara is the world's largest source of desert dust, and makes a major contribution to the sediments of the Atlantic. The dust can spread across and into the Americas and during extreme events can cause rain to become coloured, leading to the phenomenon of 'blood rain' that has been known for more than 2000 years.

### *Asian brown haze*

In 1999, a brown haze some 3 km thick was detected over the Northern Indian Ocean. The United Nations Environment Programme believes this has a significant impact on the regional and global water budget, agriculture and health. Further studies suggest they are not restricted to Asia, so really should be referred to as atmospheric brown hazes. The haze is a mixture of anthropogenic sulphate, nitrate, organics, black carbon, dust and fly ash particles, and natural aerosols such as sea salt and mineral dust (Ramanathan and Crutzen 2003).

### *Ozone hole*

The destruction of ozone in the stratosphere is enhanced by the presence of chlorofluorocarbons (CFCs), which were used as refrigerants from the early part of the twentieth century. The inventor Thomas Midgley was required to produce something non-toxic, so the CFCs seemed ideal. However, they are very stable and are transferred to the stratosphere, creating the ozone hole which forms over the Antarctic each spring. A range of protocols have increasingly limited their use and improvements seem at hand. Although there has been concern over an illegal trade in CFCs, the quantities are likely to be quite small.

## 1.7 Final Thoughts

Pollution has a long history, yet we can recognize that the air pollution that emerged in the twentieth century is more complex than that in the past. Pollutants are often generated by chemical reactions in the atmosphere and the changes and effects can be detected in a global context.

The problem no longer derives simply from single and obvious sources such as the smoky industrial chimney. It is often invisible and more importantly its source may be far from obvious. In the case of photochemical smog, it derives from a wide range of sources and is transformed by the action of sunlight and air chemistry into a new and resistant form of air pollution. Solving modern air pollution problems can involve detailed study, analysis and modelling, so they seem removed from the satisfyingly direct approaches of the past. The lack of apparent connections and the emergence of new and more subtle pollution problems are often hard for politicians and policy makers to bring before the public. It must at times seem politically expedient to ignore them especially where so they are distant in space or in the future.

We cannot be nostalgic about the old localized forms of air pollution, the smell of coal smoke or the plumes from factory chimneys. However, these problems often took on a local character and hence promoted local action. The globalization of air pollution problems may ultimately weaken the power of local concerns. This favours the actions of distant bureaucracies potentially out of tune with local situations. Thus the complexity is not merely one of air chemistry, as social and political complexities are equally relevant.

# AIR POLLUTION IN URBAN AREAS

*Ranjeet S Sokhi and Nutthida Kitwiroon*



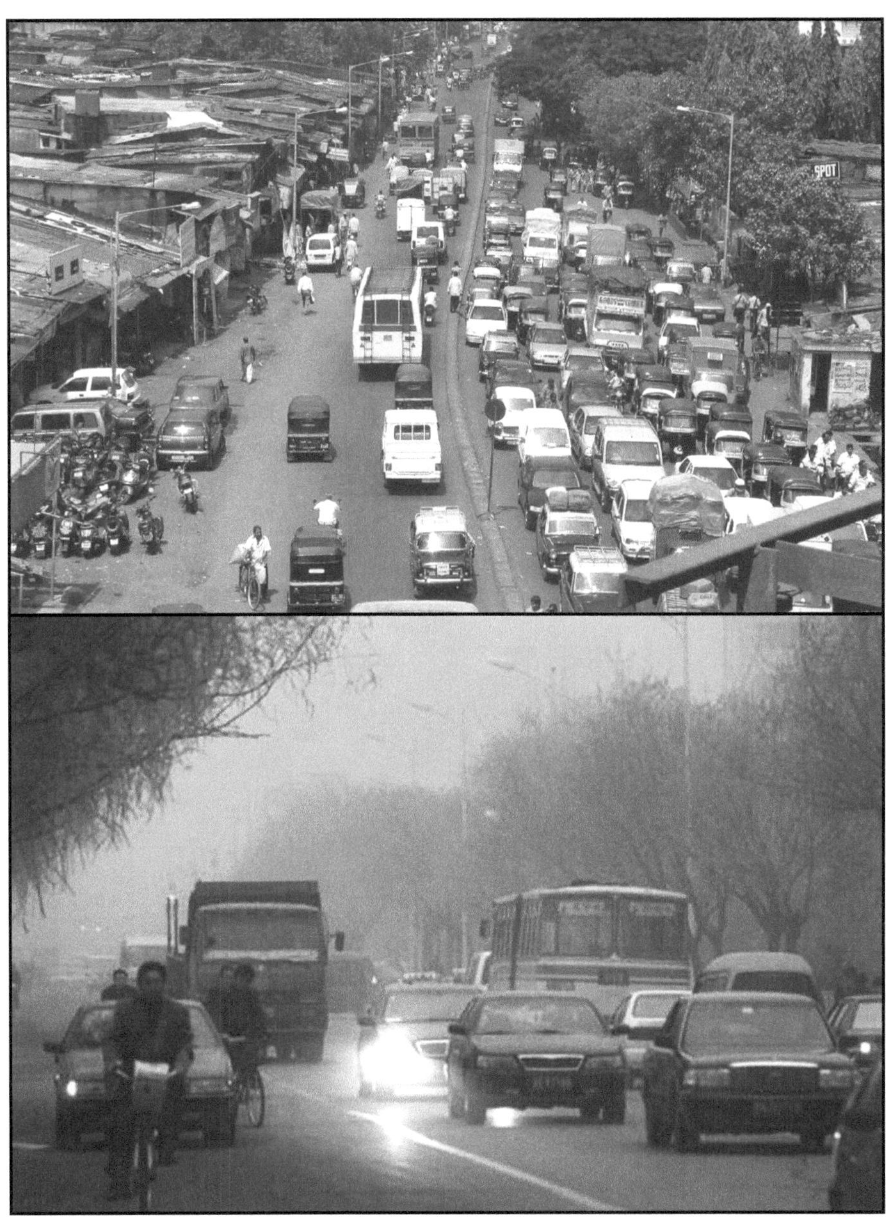

Road transport in an area of Mumbai (upper, courtesy of Rakesh Kumar 2007) and dust episode caused by long-range transport affecting Beijing (lower, courtesy of China Meteorological Administration 2004).

Air pollution is one of the most important environmental concerns. This is particularly the case in urban areas, where the majority of people live in developed countries and, increasingly so, in the developing regions of the world. It is now widely recognized that air pollution can affect our health as well as the environment. Particles and other pollutants adversely affect the quality of life of critical groups such as children and the elderly, and can lead to a significant reduction in life span (Pope et al. 2002; WHO 2003; Anderson, H R et al. 2004).

With rising population, pressure on urban environments is increasing. For example, there is the ever greater demand for travel and the need to increase energy production and consumption. Although other sources, such as industrial pollution, are still a problem in some parts of the world, the greatest threat to clean air is coming from increasing traffic pollution. The link between poor air quality and adverse health conditions is also becoming clearer. Our response to improve air quality in cities at national and local levels, however, is not homogeneous across the globe, with richer nations usually having more stringent and comprehensive pollution management strategies. For example, in the European Union comprehensive legislative frameworks exist to ensure that member states comply with limit values set in the air quality directives and daughter directives (see Directives 96/62/EC, 99/30/EC, 2000/69/EC, 2002/3/EC). Robust strategies are also in place in the USA to control and manage air pollution in cities (e.g. Clean Air Act of 1990). Similarly, urban air quality management strategies and policies to control traffic and implement new, cleaner technologies and fuels are more advanced in industrialized cities than in developing nations.

The general approach of this Chapter is to survey and describe the air pollution within major urban areas from a global perspective. It provides the context of the air pollution challenges facing cities by first considering trends in population and traffic growth in urban areas of the world. Emission and concentration levels of key air pollutants in cities across the world are then compared. Where possible, data is examined for the main continental regions, highlighting the differences in air quality experienced by peoples in different parts of the world.

It is worth noting that during this analysis primary data sources were not always accessible. This was especially difficult for the developing countries where the data is not generally available in the public domain. In addition, information on station types and QA/QC procedures were not available in all instances. Such difficulties have also been noted by Schwela et al. (2006) for cities in developing countries. Wherever possible, it was ensured that the air pollution measurement data sets used for comparison of concentrations in different cities were attributed to stations which were representative of the overall pollution levels. Thus, when comparing urban air pollution levels, data from stations located near major roads or industrial stacks were excluded. Despite these limitations with the data quality, graphical comparisons are intended to provide an overall description and comparison of the state of air pollution in some of the world's major urban areas.

## 2.1 Growing Interest in the Air Quality of Major Cities

In many developed countries, most people live in towns and cities and hence there is considerable scientific interest to understand the processes and mechanisms that influence urban air pollution and its impact (see for example, papers contained in Sokhi and Bartzis 2002; Sokhi 2005, 2006). The history of air pollution has been discussed in Chapter 1 already, but it is worth noting that since the episode of the London smog in 1952 (UK Ministry of Health 1954; Bell and Davis 2001; Bell et al. 2004) and the Clean Air Act that followed in the UK, and the Air Pollution Control Act 1955 in the USA, much effort has been directed in developed nations to establish air pollution control and management infrastructures. As part of these infrastructures extensive monitoring networks have been set up, along with frameworks to collate information on emissions and concentration levels for various pollutants identified to have environmental and health impacts. In some countries, like the UK, this has led to comprehensive, quality-assured and publicly available databases containing detailed data on air quality (DEFRA 2005). Such frameworks provide easy access to air quality information for the public and policy makers to monitor and check compliance with limit values and to assess the impact of pollution management strategies. An extensive database, AIRBASE, now exists for the European Union (AIRBASE 2005). The United States Environmental Protection Agency (EPA) similarly provides a portal to a vast amount of air quality information. While such resources have aided and stimulated research efforts and policy development in this field in the USA and Europe, they are much less commonly available in the developing regions of the world. Interest in air pollution in the major cities of developing nations, however, is now coming to the forefront of scientific attention and a number of studies have been reported in the literature (see for example, Molina and Molina 2002; Baldasano et al. 2003; Gurjar et al. 2004).

This Chapter first considers population changes in cities of the developing and developed regions of the world. Sources and emissions of the major air pollutants and concentration levels across world cities are discussed in the subsequent sections of this Chapter. An overview of recent studies on indoor air pollution is presented as it plays a critical role in terms of our total exposure to air pollution. Possible ways of controlling urban air pollution along with air quality limit values are discussed briefly at the end of the Chapter. The environmental and health effects which result from air pollutants are considered in Chapter 6 and the future changes in air pollution on regional and global scales are the subject of Chapter 7.

## 2.2 Increasing Population in Urban Areas: Rise of Mega-Cities

It is estimated that the population of the world will double over the next 40 years. Overall, nearly half of the world's population lives in towns and cities. Whereas about 70 to 80 per cent of people in developed countries live in urban areas, in the developing countries this proportion is around 20–30 per cent (UN 2004, 2006). Population rises, however, are increasingly occurring in urban regions and are caused by various factors, including economic, security, improved travel links and, in some cases, higher birth rates. The proportion of the total populations living in urban areas, therefore, is also increasing. Figure 2.1 shows the spatial distribution of cities worldwide with a population greater than 2.5 million, while Figure 2.2 shows the proportion of the population living in urban areas around the world. Figure 2.2 clearly shows the contrast between America, Europe and Australia, where most people live in urban areas, and central Africa and Asia, where most live in rural areas.

Although the population increase in developed countries will be around 11 per cent by 2030, in the developing world the population will increase to 4.9 billion people – nearly double that of 2000 levels (UN 2004, 2006).

By 2015 it is estimated that the number of cities with populations greater than 1 million will reach over 480, with two-thirds of these being in developing countries. London was the first major city in terms of population, with 1 million inhabitants by 1800. By the beginning of the last century most of the highly populated cities were in the developed countries, whereas by 2000 most were in the developing regions. By 2000 the list of mega-cities (with populations greater than 10 million) expanded with the inclusion of cities such as Lagos, Dhaka, Cairo, Tianjin, Hyderabad and Lahore. Since 1980 major cities in the industrialized world, such as Milan, Essen and London, have moved out of the list of the 30 largest world cities (UN 2001). Currently, the largest five of the urban conurbations in terms of population are Tokyo, Mexico City, Mumbai (formerly Bombay), São Paulo and New York. It is projected that by 2010 (UN 2001) Mumbai will take second place after Tokyo, and Lagos (sixth in 2000) will be the third largest city, with New York falling to seventh position. Tokyo will remain the largest mega-city in the world, at least until 2010.

The number of mega-cities is expected to increase from 17 to 21, with most of these in the developing world. Although the definition of a mega-city is normally taken to be a 'city' with a population of more than 10 million people, it is not always the case that it refers to a single, large, urbanised area with clearly defined boundaries. A mega-city often consists of a large, central, highly urbanised region, with surrounding towns or zones in close proximity which are associated with the urban area through commercial, economic, housing, transport and security activities or policies (Molina and Molina 2004). Such 'metropolitan areas' are often treated as mega-cities, and Mexico City is a good example of a mega-city in this context (Molina and Molina 2002).

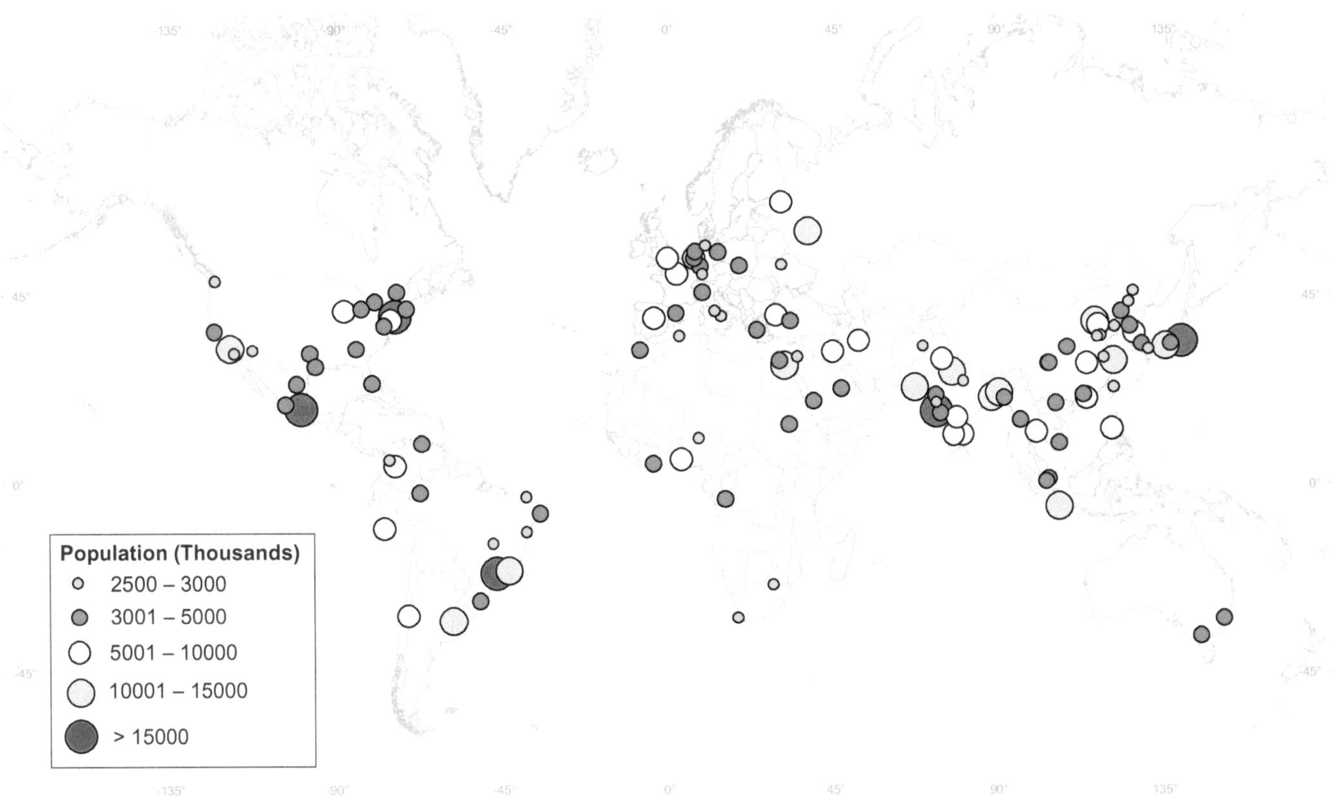


**Figure 2.1** Cities across the world with a population greater than 2.5 million (data from UN 2006).

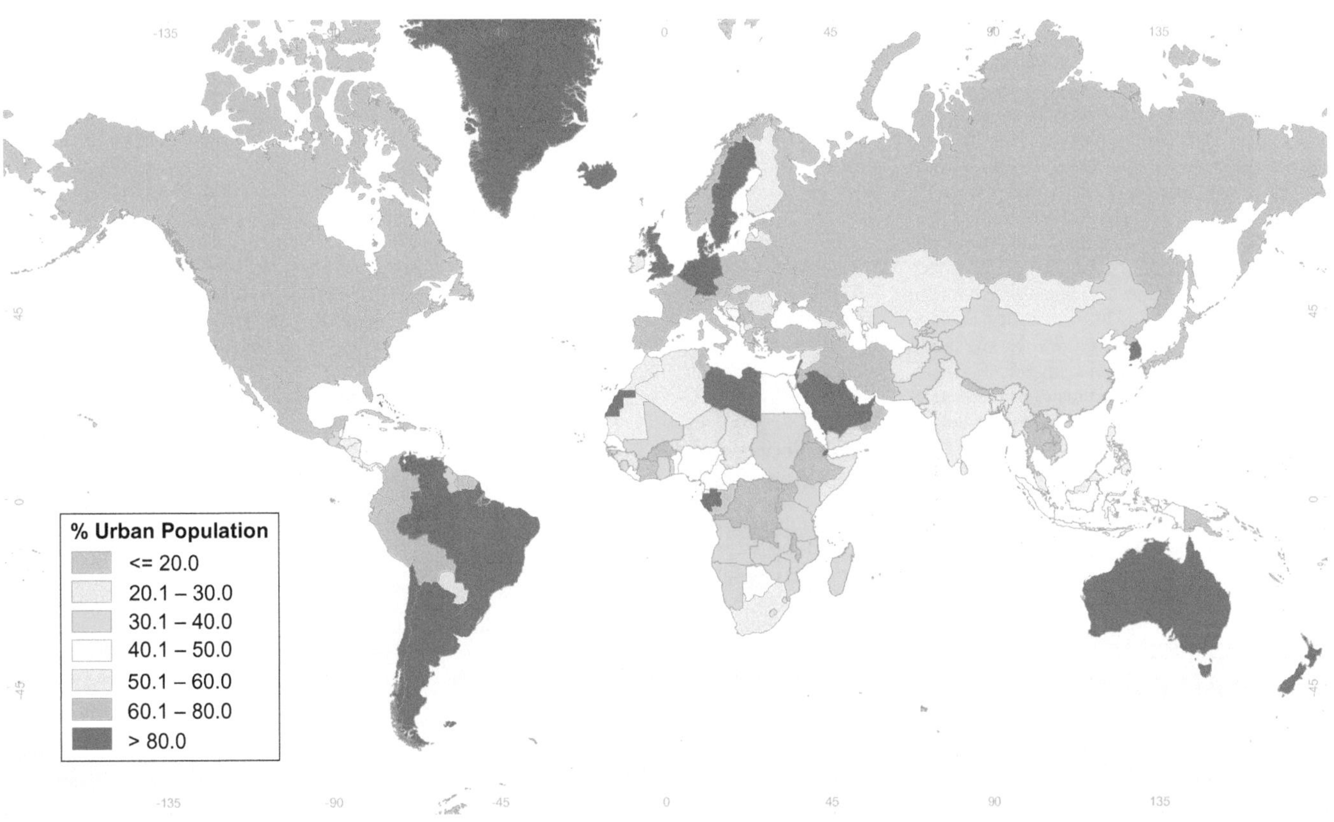


**Figure 2.2** Proportion of the population living in urban areas around the world (data from UN 2006).

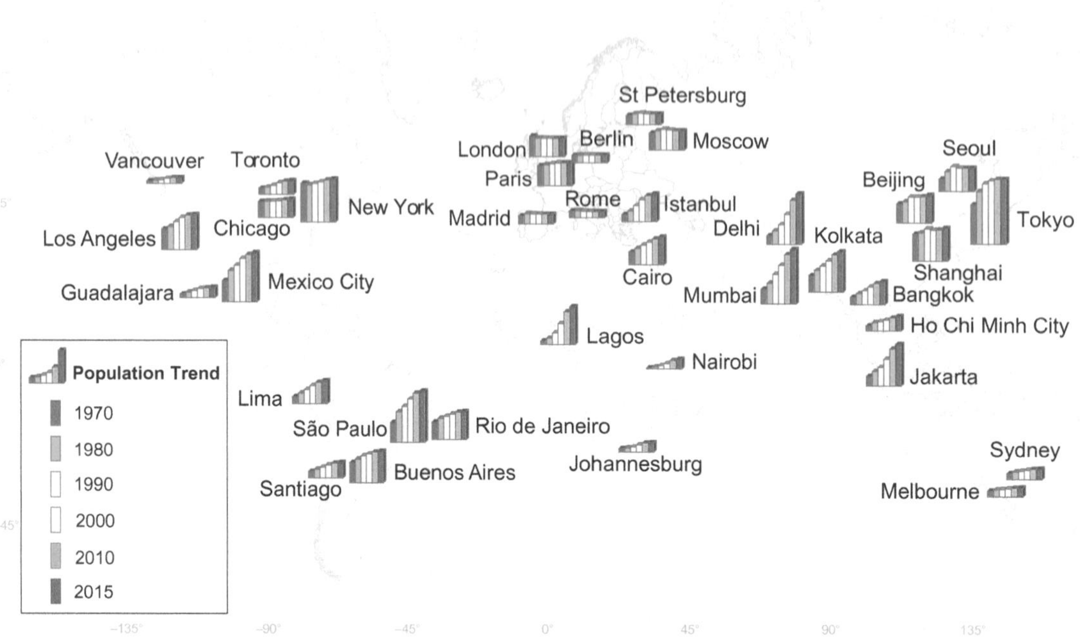


**Figure 2.3** Population growth trends for selected major cities (data from UN 2002, 2006).

The causes of urban growth are complex and interrelated. These include changes in national economy and wealth, resource utilisation, consumption and degradation, national and international migration, conflict, and, most importantly, increase in population, as mentioned above. Trends in population growth in some of the largest cities are shown in Figure 2.3.

Although in some of the cities in the industrialized nations, such as Los Angeles, there has been positive growth in the population, it is the cities in the developing countries that exhibit the fastest increase. This can be seen for Mexico City, São Paulo, Mumbai, Delhi and Tokyo. This is supported by a study by the United Nations (UN 2002), which considered the trends of population increase on a continental scale from 1970 to predicted levels in 2015. Analysis of this data reveals that the fastest rate of growth of urban populations is in Asia. Over the next 30 years, the UN estimates that the rate of global population growth in urban areas will be 1.85 per cent, compared to 1 per cent for the total world population (UN 2002). For the developing countries, over the same period, the rate of population growth in urban areas has been estimated to be 2.35 per cent, compared to just 0.2 per cent in rural areas.

## 2.3 Sources and Emissions of Air Pollution in Urban Areas

### Important Air Pollutants

Some of the key air pollutants that have health and environmental impacts include sulphur dioxide ($SO_2$), nitrogen dioxide ($NO_2$), ozone ($O_3$), carbon monoxide (CO), particulate matter ($PM_{10}$) and volatile organic compounds (VOCs). When estimating emissions of nitrogen oxides, it is usual to consider the sum of nitric oxide (NO) and nitrogen dioxide ($NO_2$), which is denoted by $NO_x$. VOCs are a group of compounds which include pollutants such as benzene, found in unleaded petrol. Particulate matter ($PM_{10}$) represents suspended particles with aerodynamic diameter equal to or less than 10 $\mu$m. As a result of recent studies highlighting the potential health effects of finer particles (WHO 2003), there is now considerable interest in $PM_{2.5}$, which represents the mass concentration of particles with aerodynamic size of 2.5 $\mu$m or less. In many parts of the world, total suspended particles (TSP) is used as an index for air quality. TSP normally consists of suspended particles with size of less than about 40 $\mu$m.

### Major Sources of Air Pollution and Influencing Factors

Given the high concentration of human activity in cities it is not surprising that urban areas are major contributors to global air pollution emissions. Similarly, most major cities are in turn affected directly by high levels of air pollution resulting from human activities. As a consequence of the transport and transformation processes occurring in the atmosphere, locally generated pollution can have an impact on a range of spatial and temporal scales. Pollution from large urban conurbations, for example, will influence the air quality of surrounding areas, but long-range transport (LRT) of pollutants resulting from emissions hundreds or even thousands of kilometres away can also affect cities. A detailed treatment of long-range transport contributions to air pollution is given in chapter 3.

Local emissions of $CO_2$ can lead to global consequences through climate change. Aerosols, including black carbon, sulphates and nitrates, can directly and indirectly influence the radiative balance of our atmosphere. Cities can also cause changes to the meteorology and climate on urban and local scales. For example, the presence of buildings causes increased roughness and hence lowers the wind speed and can create complex wind-flow patterns. Urban materials and buildings can modify the radiation balance in cities, causing higher temperatures within the city domain than in the surrounding rural areas. This is called the urban heat island (UHI) effect, and it can modify the local climate, for example, by altering the local wind flows and precipitation rates, which will then influence the dispersion characteristics of pollutants within the cities.

The major emission sources that contribute to urban air pollution include:

- Road transport (e.g. vehicle exhaust emissions);
- Industrial processes (e.g. chemical processing plants);
- Power generation (e.g. coal- and gas-fired power stations);
- Domestic (e.g. coal heating);
- Construction (e.g. building works);
- Natural (e.g. dust, pollen);
- Long-range transport (e.g. regional transport of particles or ozone precursors).

In relation to what people breathe, it is important to appreciate that both indoor and outdoor sources contribute to the total personal exposure burden. Table 2.1 (below) lists the main sources for some of the key air pollutants found in urban atmospheres.

Table 2.1 Key pollutants and their main sources.

| Pollutant | Main Sources in Urban Areas |
|---|---|
| Carbon monoxide (CO) | Outdoors: mainly road traffic, industrial plants<br>Indoors: cookers, heaters, boilers, environmental tobacco smoke |
| Carbon dioxide ($CO_2$) | Outdoors: industry and road traffic, metabolic activity<br>Indoors: cookers, heaters, boilers, environmental tobacco smoke |
| Nitrogen oxides ($NO_x$) | Outdoors: industry and road traffic<br>Indoors: cookers, heaters, boilers, environmental tobacco smoke |
| Sulphur dioxide ($SO_2$) | Outdoors: mainly power generation plants and smelters<br>Indoors: coal heating |
| Volatile organic compounds (VOCs) | Outdoors: road traffic and industry, evaporation of fuel, solvents, herbicides<br>Indoors: paint, solvents, adhesives, environmental tobacco smoke |
| $PM_{10}$ | Outdoors: road traffic and industry, construction, (re)suspended dust and soil<br>Indoors: house dust, cookers, boilers, heaters |
| $PM_{2.5}$ | Mainly outdoors: road traffic and industry, secondary aerosols through reactions, long-range transport |
| Ozone ($O_3$) | Mainly outdoors: photochemical reactions involving sunlight and chemicals such as $NO_x$ and VOCs |

Road transport emissions play a major role in most urban areas. Figure 2.4 shows the emission contributions from different source sectors for London, with road transport being the main source of CO, $NO_x$ and $PM_{10}$. In the case of $SO_2$, the most significant contributions are from industrial sources.

In many non-European cities the contribution of industry, power generation and the transport sectors is particularly important. Figure 2.5 shows the total emissions (year 2000) for Beijing and Tokyo (Guttikunda et al. 2005). In both cases the contribution of transportation (all main forms) to CO, $NO_x$ and NMVOC emissions is significant, especially for Tokyo. In the case of Beijing, contributions from industry and power generation sectors make up most of the $NO_x$ and $SO_2$ emissions. This is similar for Tokyo for $SO_2$ emissions. Over the past few decades the proportion of industrial contributions has dropped mainly due to the imposition of emission controls and the rise of road traffic. An example of Delhi is given in Figure 2.6 which shows how air pollution emissions from industrial, transport and domestic sectors have changed since 1970.

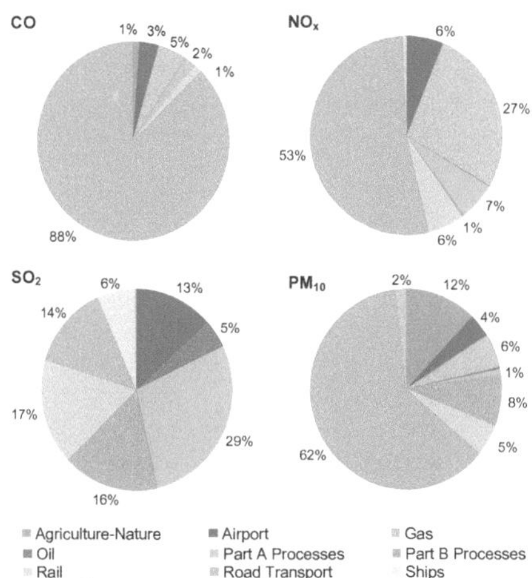


Figure 2.4 Proportion of air pollutant emissions for London (2008) from main source sectors (GLA 2010 and http://data.london.gov.uk/laei-2008). Part A refers to larger scale industrial sources and Part B to smaller scale sources.

The figure illustrates marked increases in emissions from transport, whereas the contribution from the other two sectors has declined. Similar trends are observed in many other urban areas.

In the case of road traffic contributions, the precise level of emissions from vehicles will depend on a range of factors, such as driving behaviour, age and type of vehicle, maintenance history, type of fuel, engine size and technology. In general, however, it is inevitable that if the number of vehicles in cities increases, air pollution will also increase. The number of cars per head of population in major cities across the world is increasing, with a marked contrast between the ownership in Asian cities and those in Europe and North America. Developed cities such as Berlin, Athens, Reykjavik, Rome, Madrid, London, Paris, Ottawa and Toronto have car ownership of around 400–600 per 1,000 people, with the less developed cities such as Bangkok, Delhi, Mumbai, Jakarta and Kuala Lumpur all having significantly lower ownership. Dhaka currently exhibits the lowest ownership of the selected cities (see for example, Barter 1999; City of Reykjavik 2002; UDI PRAHA 2002; WRI 1996, 1998; CfIT 2001; Dunning 2005).

Figure 2.7 shows the number of passenger cars per 1,000 population for selected European and North American countries (UNECE 2005). Most of the countries now have 250–500 cars per 1,000 population, with the USA reaching 765, whereas the newer central and eastern European countries have very few cars per head of population. Vehicle ownership, however, has increased significantly and is projected to grow further in the next two decades, as shown in Figure 2.8. An important observation is the large rate of increase in vehicle ownership in the developing areas compared to other regions (UNEP 2000; APERC 2006).

There are a number of significant differences between developed and less developed countries in relation to emissions and other factors. Some of these are listed in Table 2.2. Whereas cars dominate the road fleet in cities in developed nations, the vehicle split can be very different in developing nations, such as India and Thailand, where two-wheelers can form the largest proportion of the fleet (UNEP 2000). In Delhi, for example, the proportion of two-wheelers is nearly two-thirds of the total fleet (Sharma et al. 2002). In Jakarta the figure is

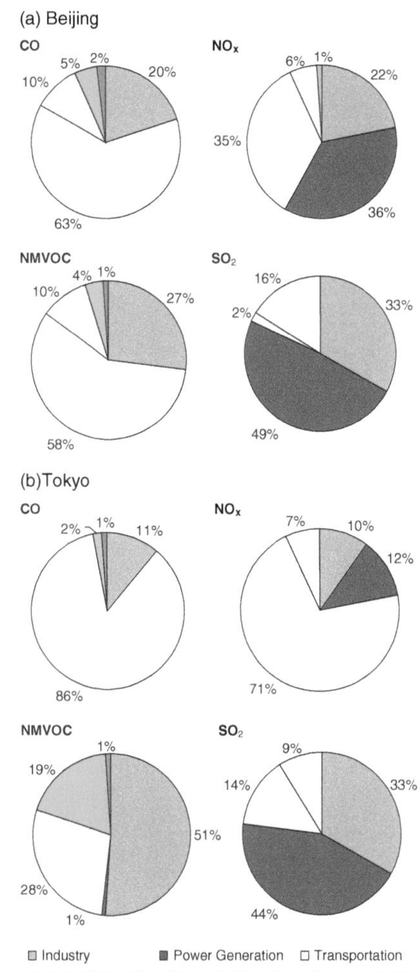


Figure 2.5 Total air pollutant emissions for (a) Beijing and (b) Tokyo for the year 2000 (Guttikunda et al. 2005). Emissions are shown for carbon monoxide (CO), nitrogen oxides ($NO_x$), non-methanic volatile organic compounds (NMVOC) and sulphur dioxide ($SO_2$).

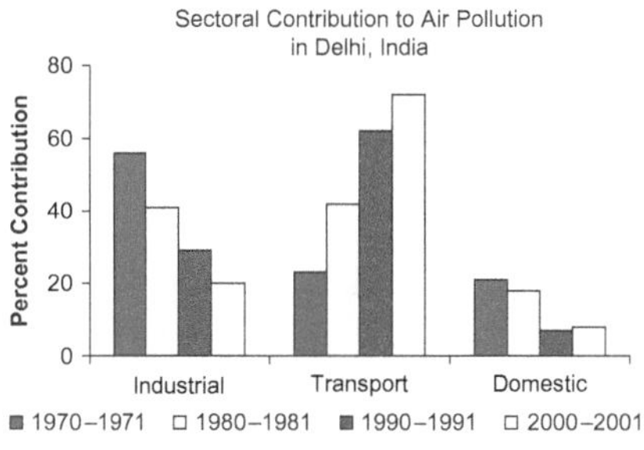


Figure 2.6 Contribution to air pollution emissions in Delhi from industrial, transport and domestic sectors, 1970–2001 (CPCB 2003).

around 50 per cent for 2001 (Wirahadikusumah 2002). It is important to understand these relative proportions of vehicle categories, as two-wheelers (especially two-stroke motorcycles), for example, emit very high levels of particles and hydrocarbons.

In developing countries it is common to find industrial areas in close proximity to residential areas, and hence urban air quality is also affected

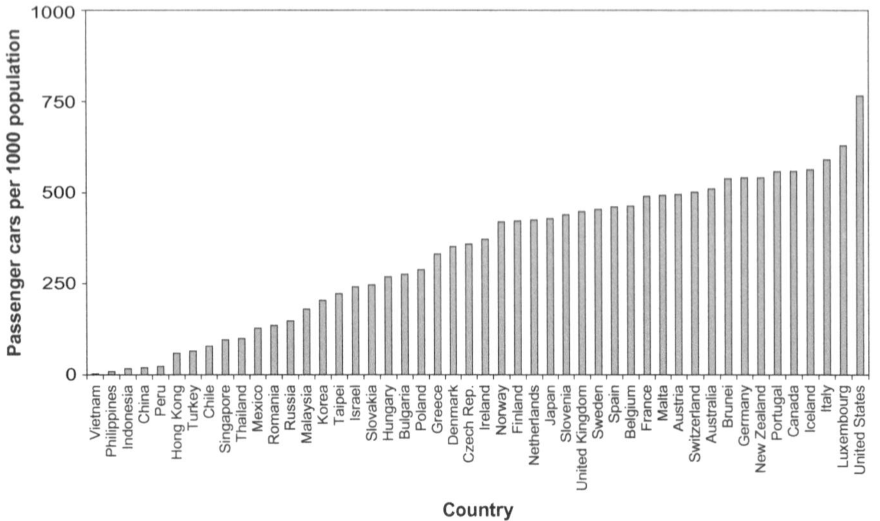


**Figure 2.7** Number of passenger cars per 1,000 population by country for 2002 (UNECE 2005; APERC 2006; EUROSTAT 2006).

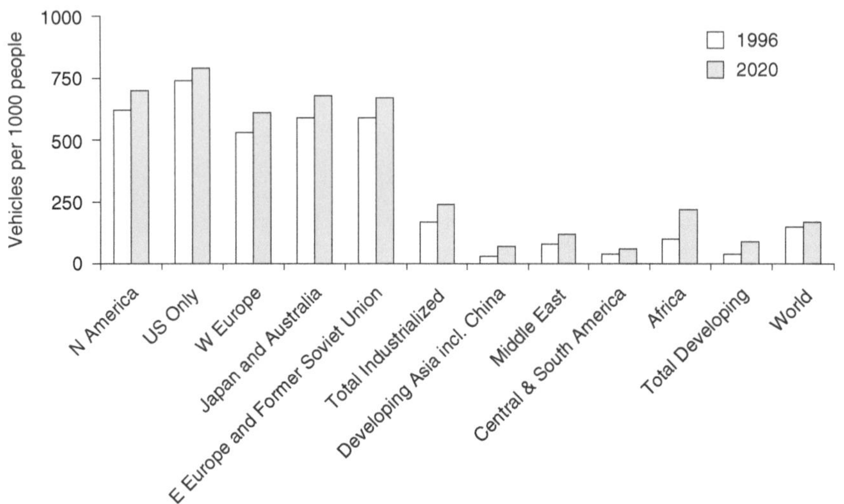


**Figure 2.8** Projected increase in vehicle ownership, 1996–2020 (UNEP 2000; APERC 2006).

by these emissions and not just by road transport. Furthermore, the type of heating and cooking fuels will also affect the quality of the local air. In many developed countries heating of buildings is achieved mainly by gas, but coal, oil or wood are also used. In developing regions, the main fuels tend to be coal, wood or other biofuels.

Although high levels of road traffic are common in most cities, in the developed regions there is higher ownership for cars and higher travel in terms of vehicle kilometres per person. Road surfaces are normally poorly maintained in developing regions, leading to high levels of dust, especially under dry conditions. Changes in urban planning practices over the last couple of decades have encouraged the siting and development of most large industries outside city boundaries. This is especially the case in developed countries. In less developed countries, however, $SO_2$ remains a serious problem where industrial complexes have been located within or close to city regions. Air pollution control infrastructures, as well as legislation and its implementation, tend to be more stringent in developed countries. There is also a higher degree of coordination of air pollution monitoring and management activities in cities of the developed nations as compared to urban areas in the developing countries.

## Emissions of Pollutants in Urban Areas

Carbon dioxide is normally associated with global warming and climate change. Urban areas are a major contributor to the overall $CO_2$ emissions into the atmosphere (see for example, Dhakal et al. 2003; Dhakal 2004; Svirejeva-Hopkins et al. 2004). Figure 2.9 illustrates the $CO_2$ emissions per capita from a range of large cities of the world. It is evident from Figure 2.9 that much of the urban $CO_2$ emissions into the atmosphere result from cities in the USA, Canada, Australia and Europe.

The total atmospheric emissions for major cities are shown in Figure 2.10 for nitrogen oxides ($NO_x$), carbon monoxide (CO) and sulphur dioxide ($SO_2$). It should be noted that data was not generally available for the same year and hence it was not possible to undertake a consistent comparison, although some trends can be highlighted. USA cities are some of the highest emitters of air pollutants,

**Table 2.2** Differences between developed and less developed countries.

| | Developed Countries | Less Developed Countries |
|---|---|---|
| Road transport | Higher ownership<br>Cars dominant<br>Higher vehicle kilometres travelled per person<br>Newer fleet, lower emissions<br>Road surfaces maintained | Significantly less car ownership<br>Mixture of two-/three-wheelers<br>Lower vehicle kilometres travelled per person<br>Older fleet, higher emissions<br>Poorly maintained vehicles<br>Low maintenance of local roads |
| Distribution of sources | Industry normally segregated from residential areas | Industry in close proximity to residential areas |
| Emissions | Reliable emissions in parts | Large uncertainties, greater understanding is needed |
| Heating/cooking fuel | Usually gas, electricity, but others used | Coal, wood, oil, biofuels, gas<br>Reliant on local sources |
| Pollutants of concern | $PM_{10}$, $PM_{2.5}$, $O_3$, $NO_2$ | Additional pollutants: $SO_2$, VOCs and Pb |
| Infrastructure for air pollution control | Well developed in some regions, stringent | Less developed, fragmented, poorly controlled emissions |
| Economic | Loose dependence of economy on polluting industry | Strong coupling (e.g. local jobs depend on polluting industry) |
| Implementation of legislation | Robust | Can be poor |
| Urban planning, governance | Links becoming stronger | Fragmented, patchy |
| Pollution monitoring | Well structured in parts | Requires firmer coordination |

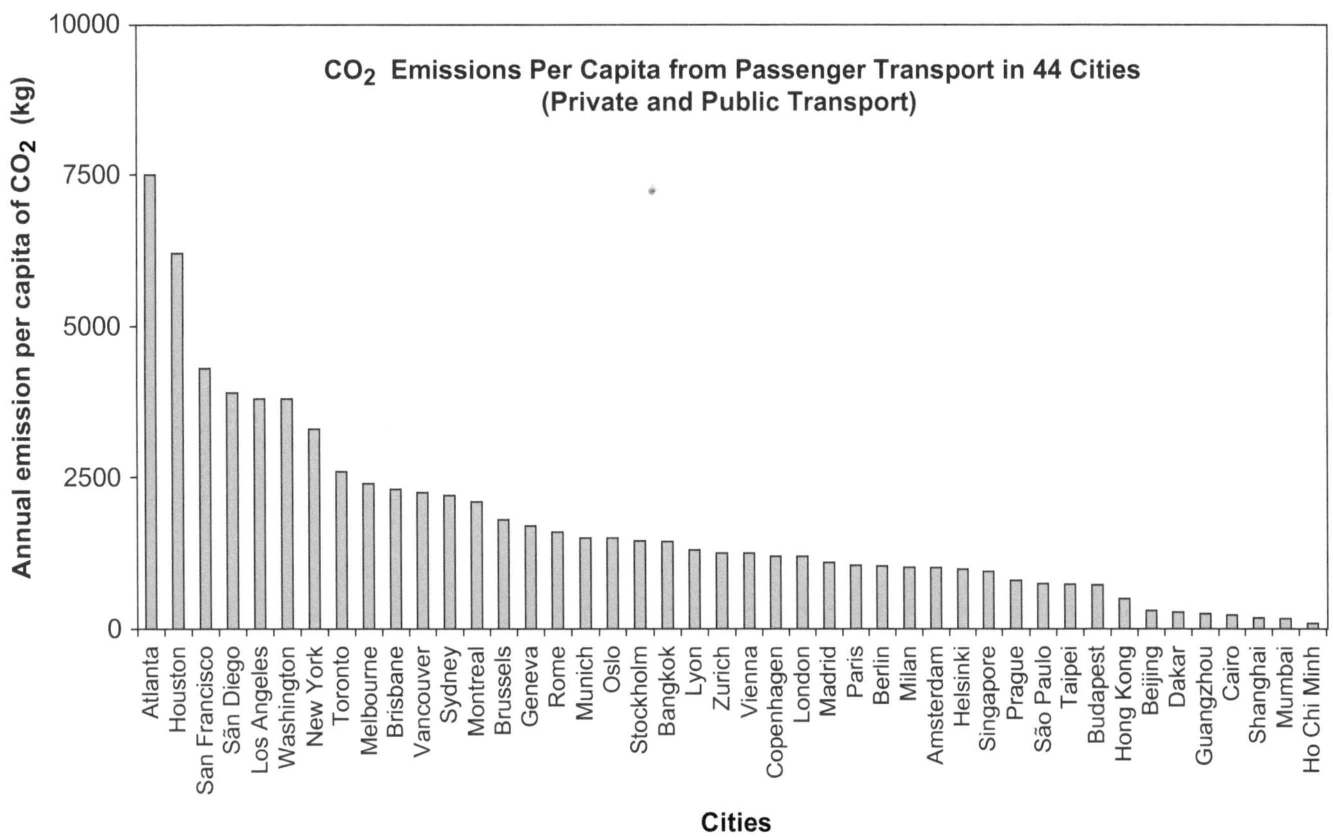


Figure 2.9 Carbon dioxide ($CO_2$) emissions per capita from passenger transport (private and public) from selected large cities (adapted from Kenworthy 2003).

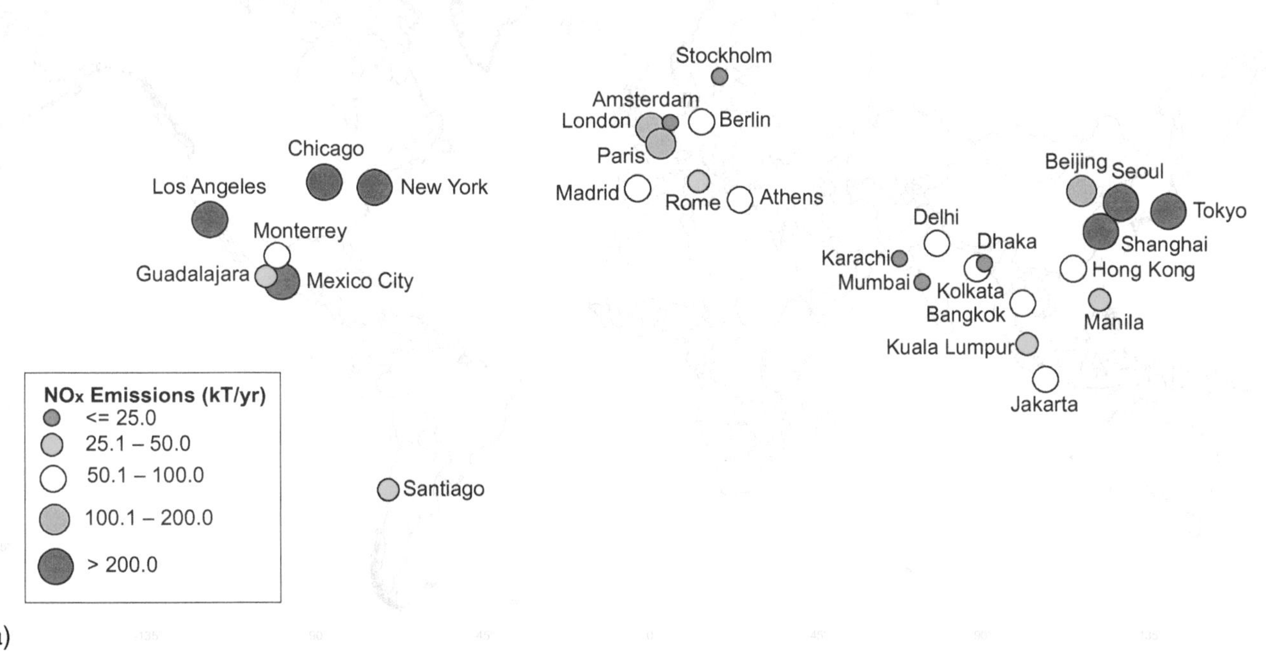


Figure 2.10 Total annual emissions for major cities for (a) nitrogen oxides ($NO_x$) as $NO_2$, (b) carbon monoxide (CO) and (c) sulphur dioxide ($SO_2$) (EEA 2001; ENV ECO 2001; de Leeuw et al. 2001; Montero 2004; Guttikunda et al. 2005; NAEI 2007). Emissions are shown for different years ranging from 1995 to 2000 as data was not available for the same year for all cities.

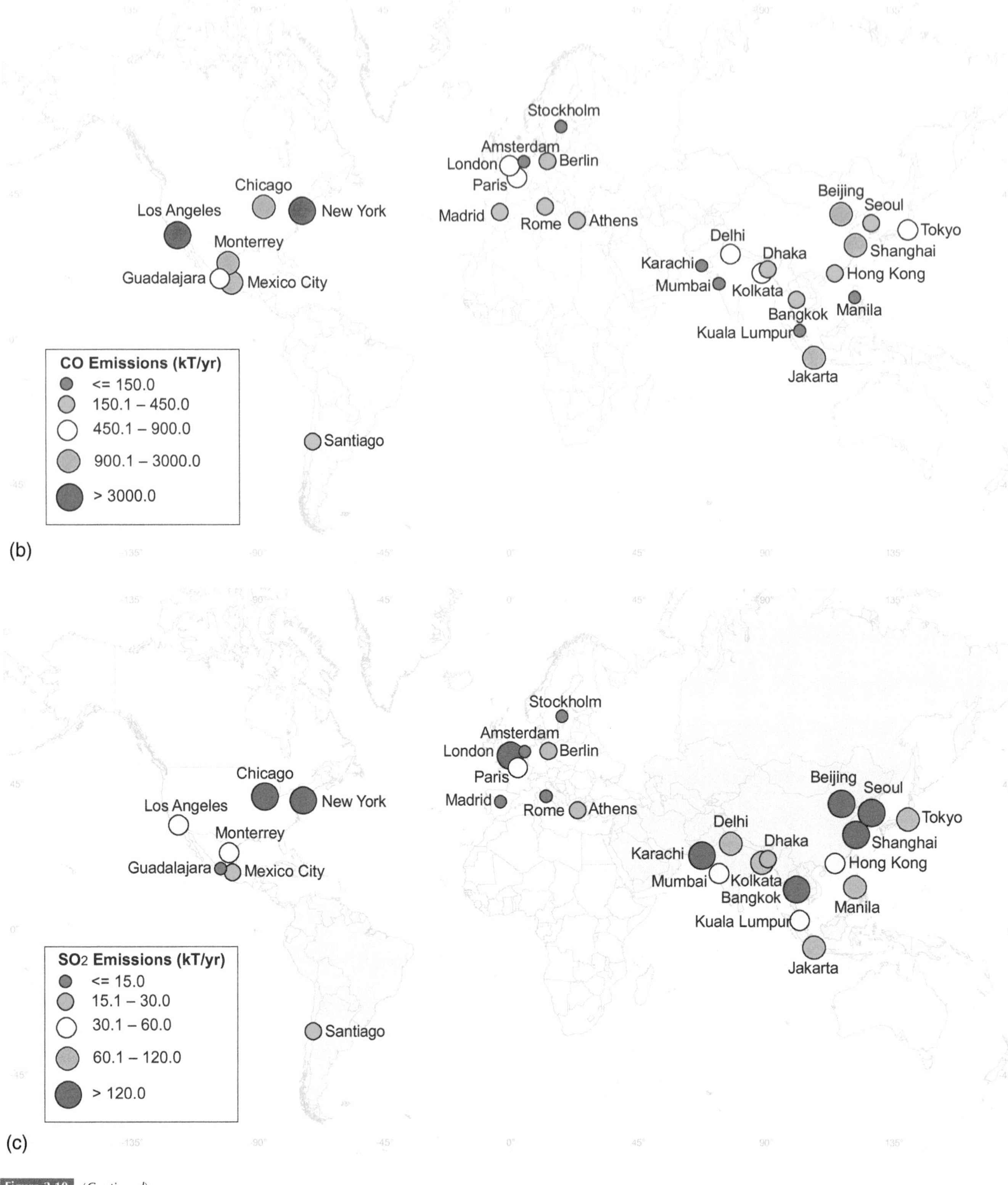


**Figure 2.10** *(Continued).*

followed by Asian cities. In Latin America, Mexico City is a major contributor to traffic related pollutants (CO and $NO_x$), whereas in Europe, Greater London is one of the largest emitter of air pollutants due to its size in terms of population and high levels of road traffic. Nearly 60 per cent of the total air emissions of $NO_x$ in London is due to road traffic. Asian and far eastern cities are significant emitters of $SO_2$, which results from industrial sources as well as road traffic. Gurjar et al. (2004), however, have investigated the air pollutant emissions, from Delhi for the period 1990–2000. The study showed that the majority of $SO_2$ and total suspended particles (TSP) originated from power generators, while road traffic was mainly responsible for CO and $NO_x$.

Figure 2.10 showed the total pollutant emissions in a city, but for air quality research and assessment

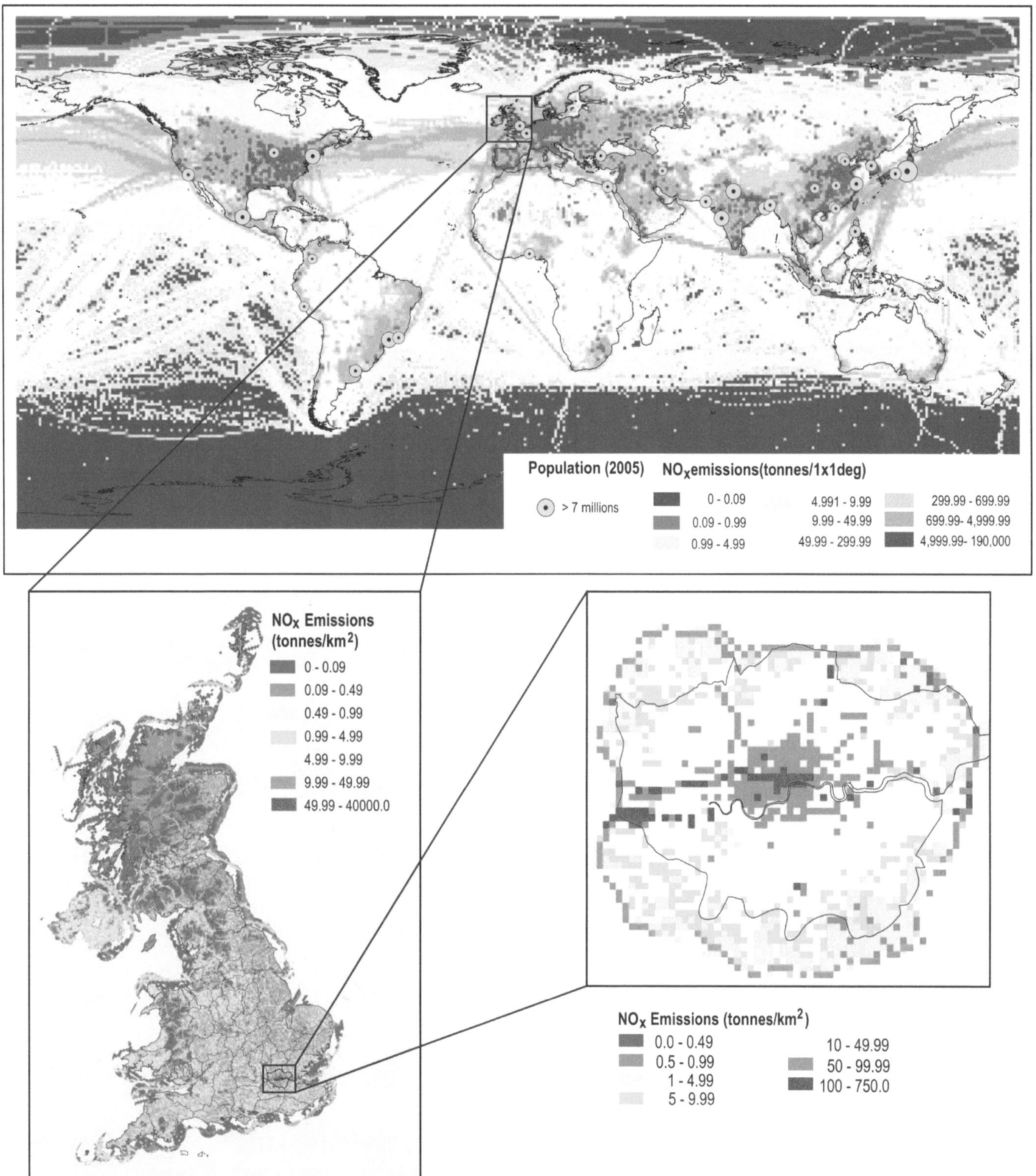


Figure 2.11 Emission inventories of $NO_x$ at global, national and city scales. Global map is of emissions for year 2005 from EDGAR (2011) at a spatial resolution of 1° × 1°. Cities with population greater than 7 million (2005) are also indicated on the global map (data from UN 2009). UK emissions are for 2008 (NAEI 2011) and for London the emissions are for 2008 (LAEI 2010). The UK and London emissions are shown at a spatial resolution of 1 × 1 km.

studies it is more useful to have detailed emission inventories. Although inventories are usually annual aggregates of emissions, they often provide spatially resolved data across the city domain. They are a vital prerequisite to undertaking air quality impact studies and allow policy makers and regulators to review and assess the trends in air pollution emissions and the resulting ambient concentrations. These trends can then be used to test the effectiveness of pollution reduction measures. Some recent examples of the use of emission inventories as part of urban air quality impact research can be found in Sánchez-Ccoyllo et al. (2006), Sokhi (2006) and Bell et al. (2006).

Global emissions inventories of air pollutants are produced typically at spatial resolutions of 1°×1° which is too coarse for urban studies. An example of such an annual emission inventory for $NO_x$ is shown in Figure 2.11 along with the location of major cities with population greater than 7 million (EDGAR 2011). For studies at a city scale, inventories of much higher resolution are required. The figure also shows two good examples of high resolution annual emission inventories for $NO_x$ for the UK (NAEI 2011) and London (LAEI 2010), both at a spatial resolution of 1×1km. A much higher level of detail can be observed from the 1×1km inventories with urban areas, and even major roads, being highlighted in the UK map. Similarly, the London emissions map clearly shows the variations in emissions which result from the spatial distribution of $NO_x$ sources. Higher emissions are observed in the centre of the cities where traffic density is highest. For London, the emissions from Heathrow, one of the largest airports in the world, are clearly identifiable (far left of the London map).

## 2.4 Air Quality in Cities

Figures 2.12a, b and c compare the annual concentrations for particulate matter, $NO_2$ and $SO_2$. Data has been extracted for the period 1990–2002 to provide sufficient spatial coverage, but in some cases, values are only shown for the latest year for which the data was available. As stated earlier, datasets are limited for many cities, and often measurements are not available for the same year. Data has been used from urban stations to reflect the overall pollution levels of the city. With regard to particles, datasets on total suspended particles (TSP) are generally more readily available than for $PM_{10}$, and hence have also been used for this figure to enable a wider geographical comparison. This is particularly the case for Asian cities. Given these limitations in the datasets, the figures provide a qualitative global overview rather than an accurate quantitative comparison of city air pollution across the world.

$PM_{10}$ concentrations are shown in Figure 2.12a for North American cities except for Montreal, where the TSP value is given (OECD 2002; Baldasano et al. 2003). Data have been presented for 1999 (Montreal, Chicago, Guadalajara, Los Angeles, New York, Toronto and Vancouver) and 2000 (Mexico City). In the case of European cities, $PM_{10}$ concentrations are presented from urban background stations, except for Reykjavik, which is an urban traffic site. $PM_{10}$ values were not available for Moscow and St Petersburg and TSP data are shown instead. To provide sufficient spatial coverage across Europe, data have been extracted from different sources for the years 1995 (Moscow), 1998 (St Petersburg), 1999 (Bratislava, Reykjavik), 2000 (Madrid, Oslo, Paris), 2001 (Athens) and 2002 (Berlin, London, Prague, Rome) (Baldasano et al. 2003; AIRBASE 2005; OECD 2002). The following TSP data has been used for the cities in the other world regions: 1990 (Jakarta), 1994 (Delhi, Kolkata, Mumbai), 1995 (Beijing, Seoul, Shanghai, Shenyang, Tokyo, Rio de Janeiro, Nairobi, Johannesburg), 1998 (Bangkok, Sydney), 1999 (Cairo) and 2000 (São Paulo) (WRI 1998; OECD 2002; Baldasano et al. 2003; Molina and Molina 2004).

Most cities shown in the map have experienced exceedances of $PM_{10}$ levels over the WHO annual mean guideline of 20 $\mu g/m^3$ apart from the cities in Canada (Vancouver and Toronto) and northern Europe such as Oslo. For cities in developed countries such as in the United States of America and in most European countries, the levels of $PM_{10}$ remain within or near the EU annual mean standard of 40 $\mu g/m^3$. In some EU cities, exceedances often occur due to an increasing use of diesel vehicles and increasing traffic volumes (EEA 2007).

The highest concentrations of particulate matter (as TSP) were observed in less developed countries, particularly Asian cities such as Delhi, Kolkata, Beijing and Shenyang with levels

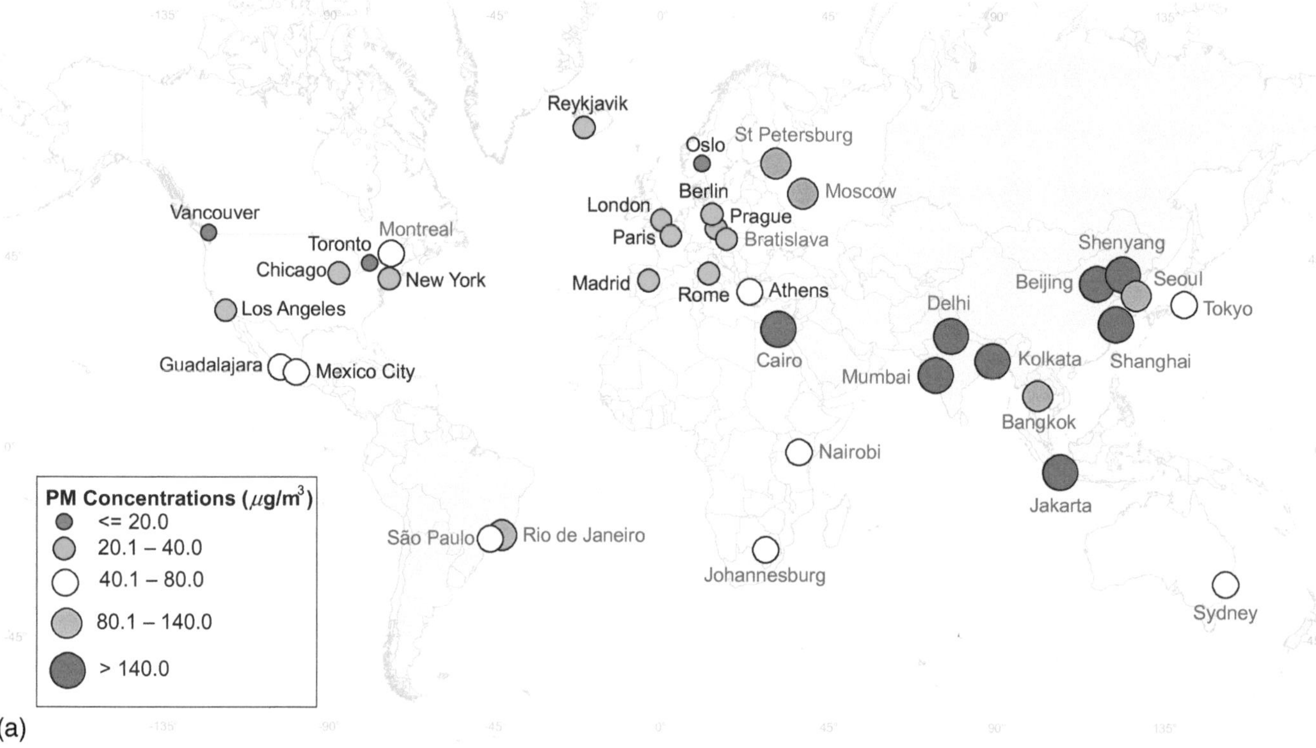


**Figure 2.12** Annual concentrations of (a) particulate matter as $PM_{10}$ or TSP (city names shown in red), (b) nitrogen dioxide ($NO_2$) and (c) sulphur dioxide ($SO_2$) for selected cities (data from WRI 1998; OECD 2002; Baldasano et al. 2003; AIRBASE 2005).

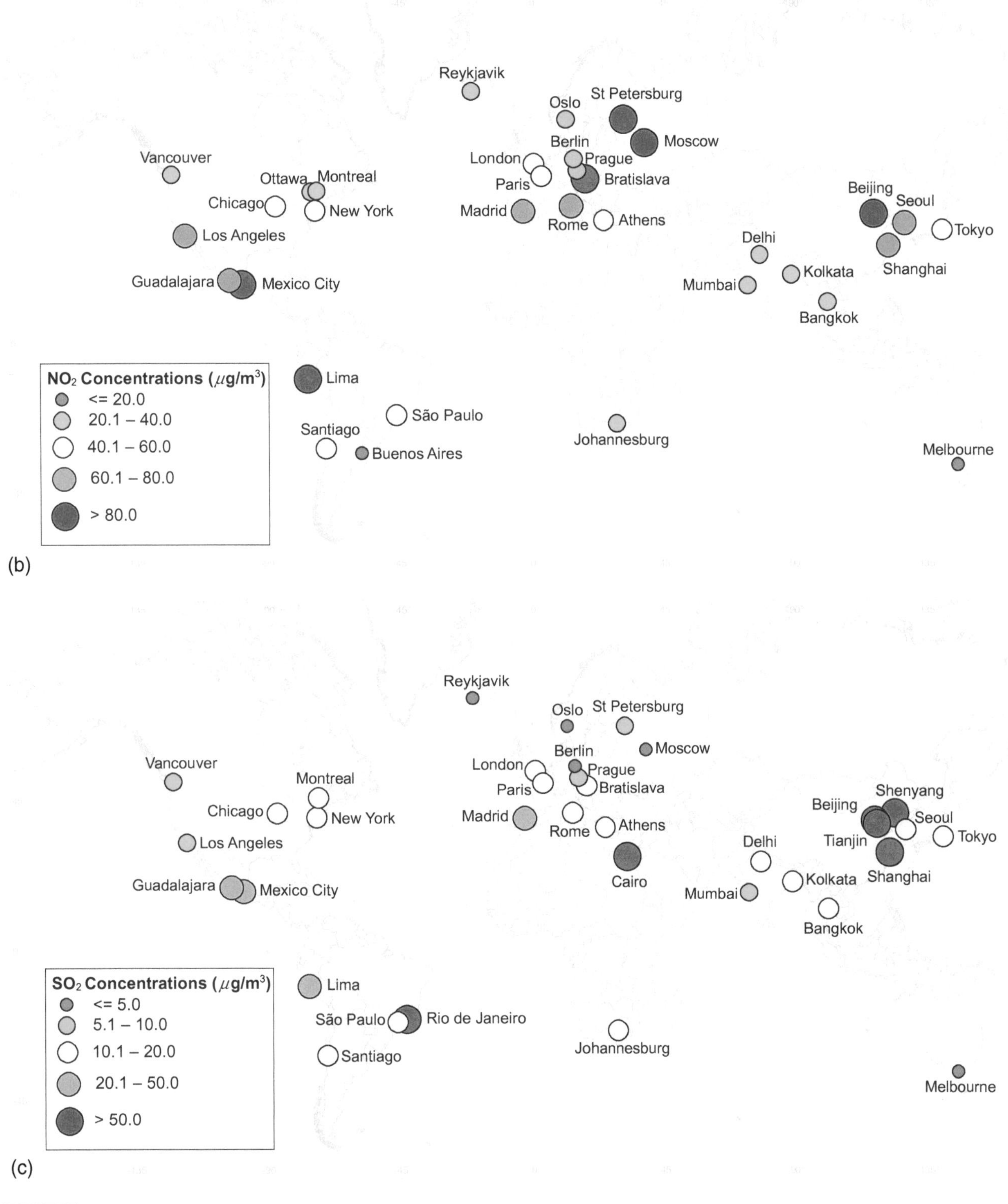


**Figure 2.12** *(Continued).*

reaching 300–400 $\mu g/m^3$. For Indian cities such as Delhi, the impact of abatement measures is small and power plants were observed to be the main contributor of TSP during 1990–2000 (Gurjar et al. 2004). For cities like Beijing, particulate matter remains higher than the national standard despite the introduction of several measures, and this is attributed to an increase in traffic in the past decade and the continuing use of coal burning as the main energy source. The pollution reducing measures that have been introduced include the use of low-sulphur coal, the partial replacement of coal with natural gas or liquefied petroleum gas, the use of unleaded gasoline, and the transfer of highly polluting industries outside of the city.

$NO_2$ is mainly formed when nitric oxide (NO) reacts with oxidants in the atmosphere ($O_3$ being the most important). Ozone is formed through photochemical reactions involving solar radiation, $NO_x$ (sum of NO and $NO_2$) and volatile organic compounds (VOCs). Under certain meteorological conditions (such as stable low wind speed conditions), a dense haze (photochemical smog) can form above a city. Under the action of sunlight, $NO_2$ can photodissociate and contribute to the formation of $O_3$. Figure 2.12b shows that high levels of $NO_2$ are a problem for most of the main cities of the world due to rising road traffic. Most cities exceed the EU and WHO annual health protection standard value of 40 $\mu g/m^3$ for $NO_2$ particularly for cities in developed countries with high levels of traffic vehicle usage. Within Europe, the highest concentrations are observed for southern and eastern European countries. EU abatement of traffic related $NO_x$ emissions play an important part in $NO_2$ reduction in EU cities, but full attention to traffic related emissions is still required in order to maintain the $NO_2$ levels within the international or national standards (EEA 2007). South American cities exhibit higher $NO_2$ levels compared to North American cities. Within Asia, some of the highest levels of $NO_2$ are observed in Chinese, Korean and Japanese cities.

For $SO_2$ the main polluted cities are Rio de Janeiro, Cairo, Shenyang, Beijing, Shanghai and Tianjin, with concentrations ranging from 75 to more than 125 $\mu g/m^3$. Many of the cities in developed countries show much lower $SO_2$ levels as strict industrial pollution controls have been implemented for several years.

## 2.5 Changes in Air Quality in Urban Areas

In most developed countries there has been a substantial reduction in the overall emissions resulting from industry including the power generation sector. In particular, levels of $SO_2$ have markedly improved over the past two decades. In less developed countries there have also been gradual reductions in industrial emissions leading to improvements in air quality. A study of 20 Asian cities by Schwela et al. (2006) has shown a decrease in all the key pollutant between 1994–2004 although $SO_2$ levels in Indian cities such as Delhi and Kolkata are still rising (Guttikunda et al. 2003). A downward trend of $SO_2$ emissions in most cities such as Beijing, Tokyo, Bangkok and São Paulo has been observed, particularly after 1990. For cities in developed countries such as London, Paris and New York, the levels of $SO_2$ have decreased markedly (e.g. Baldasano et al. 2003).

Unlike $SO_2$, traffic related pollutant (e.g. $PM_{10}$ and $NO_2$) concentrations are still above international air quality limit values in most cities. Globally, pollutants such as $NO_2$ and $PM_{10}$ still exhibit high levels in urban areas due mainly to the continual increase of road traffic. However, the annual $PM_{10}$ concentrations of cities in developed countries such as London are much lower than the $PM_{10}$ levels of cities in less developed countries, for example Santiago and Delhi. For cities in less developed countries, the level of $PM_{10}$ can often be many times higher than the WHO standard. For example, the annual mean $PM_{10}$ level in Delhi has been observed as high as 200 $\mu g/m^3$ which has been attributed to road traffic increases (doubling every seven years) (APMA 2002).

For $NO_2$, the downward trend in mega-cities such as São Paulo, Mexico City, London, Paris, Los Angeles and New York has been observed particularly before 2000 (Baldasano et al. 2003; AIRBASE 2005). An exception to this downtrend is seen in the case of some Indian cities such as Delhi where the level of $NO_x$ emissions has increased approximately 50 per cent over the years 1990–2000 (Gurjar et al. 2004). Although new technologies have been adopted, such as catalytic converters or the introduction of more stringent emission controls, an increase of $NO_x$ has been attributed in some part to local emissions from cooking gas. Beyond 2000, the levels of $PM_{10}$ and $NO_2$ have generally remained stable for most mega-cities but are still higher than the WHO recommended guidelines (OECD 2002).

Overall, most cities in developed regions such as North America and western Europe have been more successful in reducing levels of air pollution than cities in less developed countries. This success is mainly due to more effective implementation of environmental legislation on energy consumption, industry and transportation, all of which are subject to local, regional and international pollution management frameworks (APMA 2002). Figure 2.13 shows the trends of air quality for European cities, as an example. The trends in annual means of hourly $SO_2$, $NO_2$, $PM_{10}$ and of maximum daily 8-h mean for CO and $O_3$ concentrations have been derived from AIRBASE (EEA 2011). The datasets were extracted from urban background stations between 1990 and 2009. Dash and solid red lines represent WHO and EU guidelines, respectively. Since 1990 the implementation of air quality control policies have led to a general downward trend of most primary pollutants. $O_3$ trends rise slightly and this could be due to the lower levels of $NO_x$. The maximum daily 8-h mean concentrations are, however, below the EU limit value as shown in Figure 2.13. The levels of almost all pollutants remain stable after 2000 although $PM_{10}$ concentrations tend to rise in some cities. Exceedances of $PM_{10}$ over the EU limit value in many cities is observed (e.g. Kukkonen et al. 2005). The causes of such high levels can be due to multiple reasons including stagnant meteorological conditions, contributions from incoming polluted air masses as well as increases in local emissions.

Air pollution levels in several cities in developing countries such as Beijing, Bangkok, Mumbai and Metro Manila are still high but remains stable while a downward trend of primary air pollution (e.g. CO, $NO_x$, $SO_2$) in some cities such as Tokyo has been observed APMA (2002). Nevertheless, photochemical smog, ozone and transboundary air pollution from neighbouring countries has become more of a concern. The emissions from Asian cities are predicted to rise and this will continue to have an impact on hemispheric background ozone level as well as global climate (Gurjar and Lelieveld 2005). Effective environmental control strategies are urgently required since there is no sign of a slow down in the growth of these cities. There is also an urgent need for good quality data to improve our understanding of the problems of air pollution in large cities in developing nations. Furthermore, investment in traffic management strategies has been relatively modest and the use of older vehicles with low grade fuel is continuing which is all adding to the air pollution burden of the cities (World Bank 2003).

## 2.6 Indoor Air Pollution

Although the number of studies on indoor air pollution is increasing in the literature, the data reported on pollution levels in urban buildings is still quite disparate. Studies differ in the type of indoor environment investigated, methods used, pollutants analysed as well as duration and time of sampling. This makes a common comparison or a graphical representation of indoor levels for cities difficult. However, a short review is presented in the section below to indicate the extent of indoor air quality in different city environments. Indoor levels of particulate matter and nitrogen dioxide cited in recent literature are listed in Table 2.3.

People spend most of their time (approximately 80 per cent) indoors and consequently the levels of indoor pollution can be critical in determining the total exposure of a person to air pollutants (see for example, Wallace 1996, Anderson et al. 1999 and Pluschke 2004). This can be higher for some groups, such as the elderly or the ill. Cultural and social habits can also influence how much time we spend indoors. The type of cooking or heating fuel, for example, will directly determine how much pollution is produced in a home. Warwick and Doig (2004) cite that around 200 million people rely on biomass fuels, and when combined with inefficient stoves for cooking this means they can be exposed to very high levels of pollutants. High levels of indoor air pollution, especially in poorly ventilated areas, can lead to adverse health impacts, especially for women and children (Bruce et al. 2000). In addition to emissions sources, the chemical reactions within indoor areas can affect the quality of air. Weschler (2004) has reviewed studies on the chemistry of indoor air and has highlighted how short-lived and highly reactive products can be produced through oxidation reactions which can then lead to further impacts on the health of the occupants.

In the case of direct emissions, Smith (2002) has concluded that indoor air pollution from cooking and heating can lead to substantial ill-health in developing countries where the majority of households rely on solid fuels in the form of coal or biomass. A study by Dasgupta et al. (2006) of homes in Bangladesh has shown the high variation that exists between indoor $PM_{10}$ concentrations and the type of cooking fuel used. Mean concentrations were measured to be 313 $\mu g/m^3$ inside kitchens which were part of the living space when using solid fuel for cooking and 134 $\mu g/m^3$

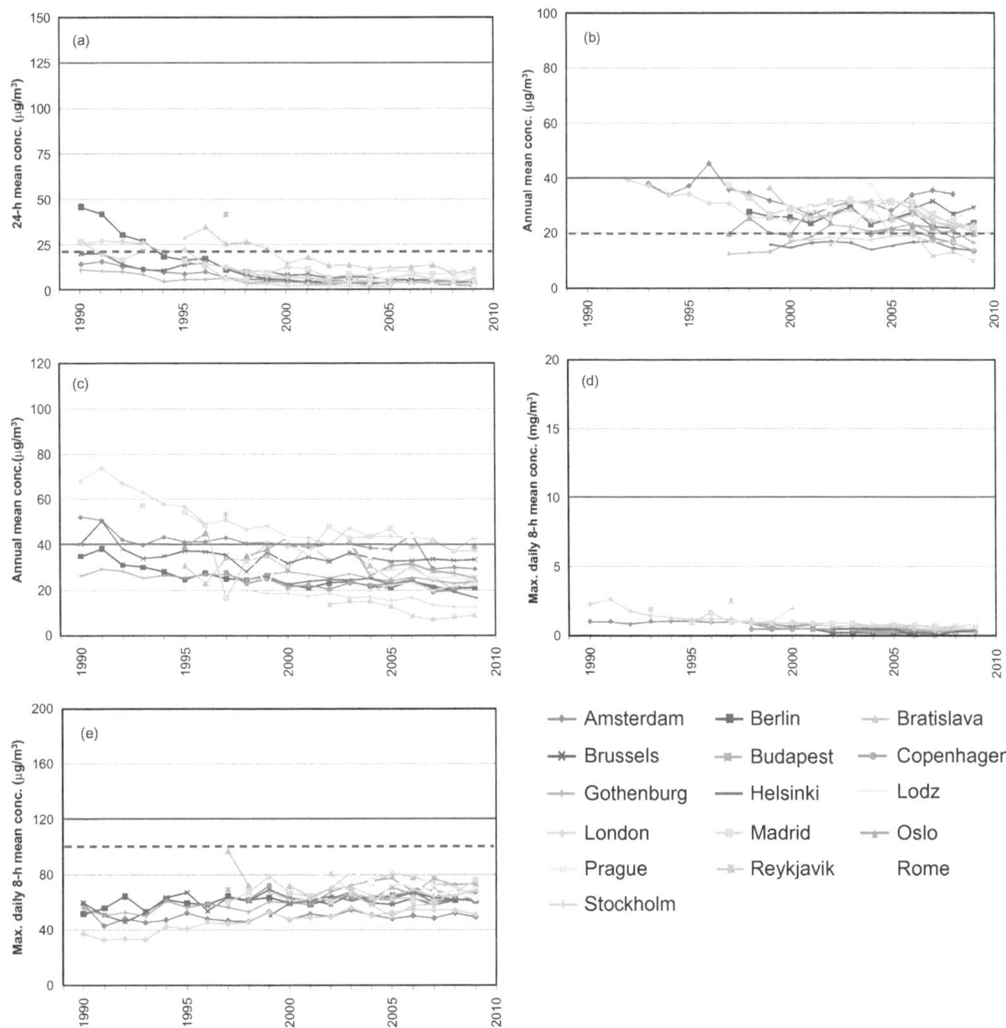


Figure 2.13 Trends of air pollutants for selected European cities (1990–2009): (a) annual mean of $SO_2$ concentrations, (b) annual mean of $PM_{10}$ concentrations, (c) annual mean of $NO_2$ concentrations, (d) maximum daily 8-h mean of CO concentrations and (e) maximum daily 8-h mean of $O_3$ concentrations (data extracted from EEA 2011). Red lines indicate WHO and EU guidelines for pollutants.

when using gas. The mean ambient urban concentration was cited as 89 $\mu g/m^3$. The authors also cite a World Bank study where significantly higher concentrations were measured for similar indoor environments in a study of homes in India (666 $\mu g/m^3$ compared to ambient level of 91 $\mu g/m^3$). Summer time measurements of $PM_{10}$ and $PM_{2.5}$ levels in hospitals (Guangzhou, China) has been reported by Wang et al. (2006). Overall, the indoor concentrations ranged from 41 to 215 $\mu g/m^3$ with a mean value of 99 $\mu g/m^3$ for $PM_{2.5}$ and from 62 to 250 $\mu g/m^3$ with a mean of 128 $\mu g/m^3$ for $PM_{10}$. The corresponding outdoor level means were 98 $\mu g/m^3$ for $PM_{2.5}$ and 144 $\mu g/m^3$ for $PM_{10}$. Similarly high indoor concentrations have been observed in other cities. For example, Vallejo et al. (2004) reports $PM_{2.5}$ levels of 68 $\mu g/m^3$ (median) compared to an outdoor value of 90 $\mu g/m^3$ in Mexico City.

In contrast to the above levels, indoor concentrations observed in some European cities are significantly lower (see Table 2.3). It can be seen that typically levels in European indoor environments range from 10–30 $\mu g/m^3$ for $PM_{2.5}$ from the EXPOLIS study reported by Lai et al. (2006). The ratio of $PM_{2.5}/PM_{10}$ is typically around 0.6–0.7 and hence the majority of $PM_{10}$ levels consist of particles in the fine fraction.

Indoor air quality is not only determined by indoor sources but also by the quality of outdoor air (Wallace 2000). A study by Lazaridis et al. (2006) conducted in Oslo not only showed that the indoor $PM_{10}$ levels were correlated to specific indoor sources (e.g. cooking and smoking) but also that they depended on the nearby outdoor concentrations. Another study in Delhi showed the strong influence of outdoor pollution (mainly of vehicular origin) on indoor suspended particle concentrations (Srivastava and Jain 2007). Lawrence et al. (2004) have shown the relationship between indoor and outdoor concentrations of nitric oxide (NO) and nitrogen dioxide ($NO_2$). They report that although the living room levels (urban) were lower than the outdoor values for nitrogen oxides, positive correlation between indoor and outdoor levels was observed.

## 2.7 Control of Air Pollution in Cities

In order to control air pollution and reduce its impact, guidelines and standards have been set across the world. WHO has stipulated guidelines, but these are not legally binding on any nation. Guidelines for several pollutants have been updated recently (WHO 2005) and again in October 2006 (WHO 2006). The EU limit values, however, are binding on the member states (see Directives 96/62/EC, 99/30/EC, 2000/69/EC and 2002/3/EC). Tables 2.4 a, b shows the current list of EU limit values and WHO guidelines for several key pollutants. The USA has a standard for $PM_{2.5}$ and a guideline has been introduced by WHO (2005, 2006). On the European level, a limit value for $PM_{2.5}$ has also been proposed. As Table 2.4 shows, standards can differ in terms of averaging times and can also be different from one region to another.

Table 2.3 Indoor levels of key air pollutants reported for selected cities.

| Cities | $PM_{10}$ (μg/m³) | $PM_{2.5}$ (μg/m³) | $NO_2$ (μg/m³) | Sampling Period | Subjects | Type of Micro-Environment | Result Cited | Reference |
|---|---|---|---|---|---|---|---|---|
| Agra | | | 487 | October 2002–February 2003 | 15 houses: 5 rural, 5 urban and 5 roadside | Living room | Average value | Lawrence et al. 2004 |
| Amsterdam | | 14 | | 24-h average measured biweekly from 2 November 1998 to 18 June 1999. | 37 non-smoking elderly (aged 50–84 years) | | Median | Janssen et al. 2005 |
| Athens | | 28 | | Two consecutive sampling days | | Home of adult participants | Geometric means | Lai et al. 2006 |
| Baltimore | 57 | 45 | 60 | Samplings over a 72-h period | Asthmatic child | In the sleeping room of the asthmatic child | | Breysse et al. 2005 |
| Basel | | 19 | 25 | Two consecutive sampling days | | Home of adult participants | Geometric means | Lai et al. 2006 |
| Beijing | 110 | | | Mean 12-h (December 2002 to January 2003) | 4 residential homes | 2 smoking and 2 non-smoking houses (living room) | | Houyin et al. 2005 |
| Erfurt | | | 15 | 1 week sampling period between June 1995–November 1996 | 204 dwellings | Living room and bed room | Average value | Cyrys et al. 2000 |
| Greater Lille | | | 36 | Two 24-h sampling periods (from Thursday 12:00 to Friday 12:00 on working days and from Saturday 12:00 to Sunday 12:00 during weekends), 2 campaigns (winter 2001 and summer 2001) | Winter 2001: 13 participants<br>Summer 2001: 31 Participants | At home and various other indoor places including workplace, shops and restaurants | Average value | Piechocki-Minguy et al. 2006 |
| Guangzhou | 128 | 99 | | 2 August–10 September 2004 | 4 hospitals: rural, urban, children and specialist hospitals | Treatment room, in-patient department, out-patient department, emergency treatment department and doctor office | Average values | Wang et al. 2006 |
| Hamburg | | | 18 | 1 week sampling period between June 1995–November 1996 | 201 dwellings | Living room and bed room | Average value | Cyrys et al. 2000 |
| Helsinki | | 9 | 15 | Two consecutive sampling days | | Home of adult participants | Geometric means | Lai et al. 2006 |
| London | 18 | | | Hourly average, April–October 1998 | 4 rooms located in two buildings on Marylebone Road in Central London | 4 unoccupied offices on different floors with different ventilation types | Average value from all sites | Ní Riain et al. 2003 |
| Mexico City | | 68 | | 13-h period starting at 09:00. Working day in rainy season (April–August 2002) | 40 Non-smoker volunteers (age 21–40 years) | At home, at work, at school or indoor public places, e.g., theatres. | Median | Vallejo et al. 2004 |
| Milan | | 32 | | Two consecutive sampling days | | Home of adult participants | Geometric means | Lai et al. 2006 |
| Munich | 88 | 18 | | Sampling was done about 5 hours of one school day in each classroom between December 2004 to March 2005 and May to July 2005 | 64 primary and secondary schools in the city of Munich and in a neighbouring rural district | 58 classrooms were measured for both periods (December 2004 to March 2005 and May to July 2005) | Average value from laser aerosol spectrometer monitoring method | Fromme et al. 2007 |
| Oslo | 2–14 | | | Hourly average, June 2002, August–September 2002 and January 2003 | 2 houses: 1 residential area (in suburbs) and 1 apartment, 1st floor in city centre close to busy road | Furnished places and no smokers, ground floors with well mixed and homogeneous air circulation in the absence of indoor sources (house in suburbs), bed room (house in city centre) | Minimum value (with no indoor activities, weekends August 2002) to maximum value (with indoor activities, working days January 2003), overall average of $PM_{10}$ is 8.56 μg/m³) | Lazaridis et al. 2006 |
| Oxford | | 12 | 23 | Two consecutive sampling days | | Home of adult participants | Geometric means | Lai et al. 2006 |
| Prague | | 28 | 37 | Two consecutive sampling days | | Home of adult participants | Geometric means | Lai et al. 2006 |
| Quebec City | | | 8 | Averaged over 7 days between January and April 2005 | 96 dwellings | At home | Geometric means | Gilbert et al. 2006 |
| Santiago | 104 | 69 | 68 | 24-h period | 20 children | Non-smoking households, main activity room of the house excluding the kitchen | | Rojas-Bracho et al. 2002 |

(Continued).

**Table 2.3** *(Continued)*.

| Cities | $PM_{10}$ ($\mu g/m^3$) | $PM_{2.5}$ ($\mu g/m^3$) | $NO_2$ ($\mu g/m^3$) | Sampling Period | Subjects | Type of Micro-Environment | Result Cited | Reference |
|---|---|---|---|---|---|---|---|---|
| Seoul | | | 105 | Samplings during working hours (average 10h) | 32 Shoe stalls of participants aged between 40–69 from 32 districts | Shoe stall located within 15m distance from the roadways, 21 participants were smokers with 6 of these reported smoking in the workspace | Geometric Means ± Geometric Standard Deviation | Bae et al. 2004 |
| Tokyo | 95.5–272.6 | | | November 20 (11:30–13:43h) and 24 (13:18–15:21h), 1997. | | Smokey rooms | Range given | Sakai et al. 2002 |
| Toronto | 30 | 21 | | $PM_{10}$: Average of 2 summer months in 1995<br>$PM_{2.5}$: Average of September 1995 to August 1996 | | | | Pellizzari et al. 1999 |
| Utrecht | | | 99 | 48-h period repeated 4 times for each participant, spread over 9 months | 4 schools for children between 10–12 years 2 schools were within 100m of a major freeway and 2 schools were at urban background close to one of the busy road schools. | At schools | Median value | Van Roosbroeck et al. 2007 |

**Table 2.4a** Air quality limit and guideline values. (Source: EU Directives 96/62/EC, 99/30/EC, 2000/69/EC; WHO 2005, WHO 2006.)

| Pollutant | WHO Air Quality Guideline | | EU limit Values | | Date by which limit is to be met |
|---|---|---|---|---|---|
| | Guideline (Time-Weighted Average) | Averaging Time | Limit Value | Averaging Time | |
| Particulate matter ($PM_{10}$) | 50 $\mu g/m^3$ | 24 hours | 50 $\mu g/m^3$ not to be exceeded 35 times in a calendar year | 24 hours | 1 January 2005 |
| | 20 $\mu g/m^3$ | Annual | 40 $\mu g/m^3$ | Annual | 1 January 2005 |
| Particulate matter ($PM_{2.5}$) | 25 $\mu g/m^3$ | 24 hours | | | |
| | 10 $\mu g/m^3$ | Annual | | | |
| $SO_2$ | 500 $\mu g/m^3$ | 10 minutes | – | – | – |
| | – | – | 350 $\mu g/m^3$ not to be exceeded more than 24 times in a calendar year | 1 hour | 1 January 2005 |
| | 20 $\mu g/m^3$ | 24 hours | 12 $\mu g/m^3$ not to be exceeded more than 3 times a calendar year | 24 hours | 1 January 2005 |
| | 50 $\mu g/m^3$ (WHO 2005) | Annual | – | – | – |
| $NO_2$ | 200 $\mu g/m^3$ | 1 hour | 200 $\mu g/m^3$ not to be exceeded more than 18 times in a calendar year | 1 hour | 1 January 2010 |
| | 40 $\mu g/m^3$ | Annual | 40 $\mu g/m^3$ | Calendar year | 1 January 2010 |
| | 120 $\mu g/m^3$ | 8-h | – | – | – |
| Carbon monoxide | 100 $\mu g/m^3$ | 15 minutes | 10 $\mu g/m^3$ | Max. daily 8–hour mean | 1 January 2005 |
| | 60 $\mu g/m^3$ | 30 minutes | | | |
| | 30 $\mu g/m^3$ | 1 hour | | | |
| | 10 $\mu g/m^3$ | 8-h | | | |
| Ozone | 100 $\mu g/m^3$ | 8-h | 120 $\mu g/m^3$ not to be exceeded on more than 25 days per calendar year averaged over 3 years | Max. daily 8–hour mean | 2010 |
| Lead | 0.5 $\mu g/m^3$ | Annual | 0.5 $\mu g/m^3$ | Calendar year | 1 January 2005 |
| Benzene | – | – | 5 $\mu g/m^3$ | Calendar year | 1 January 2010 |

**Table 2.4b** Limit values for the protection of ecosystems.

| Pollutant | WHO Air Quality Guideline | | EU Limit Values | | Date by which limit is to be met |
|---|---|---|---|---|---|
| | Guideline (Time-Weighted Average) | Averaging Time | Limit Value | Averaging Time | |
| $SO_2$ | 10–30 $\mu g/m^3$ depending on type of vegetation | Annual and winter mean | 20 $\mu g/m^3$ | Calendar year and winter (1 October to 31 March) | 19 July 2001 |
| $NO_2$ | 30 $\mu g/m^3$ | 1 year | 30 $\mu g/m^3$ | Calendar year | 19 July 2001 |
| Ozone | – | – | 18,000 $\mu g/m^3$–h averaged over 5 years | AOT40, calculated from 1 hour values from May to July | 2010 |

In order to meet the guideline or limit values a range of pollution reduction measures have to be considered. On a strategic level, there is considerable scope to share experience between the major regions of the world (Haq et al. 2002). However, any effective air quality management strategy will require a combination of measures on local, national and regional, if not global levels which are cooperative in nature. More importantly, it requires a strong desire on the part of governments and individuals to strike the appropriate balance between growth, development and sustainability.

A strategic framework for managing air pollution in cities (see for example, SEI 2004; Schwela et al. 2006) needs to be underpinned with coherent and verifiable assessment procedures which can be implemented locally. These should include:

- Development of reliable and detailed emission inventories, coupled with a sound understanding and knowledge of source distributions.
- Establishment of a monitoring network aimed at identifying hotspots as well as temporal trends in pollution levels, allowing the effectiveness of control measures to be evaluated.
- Assessment of factors that affect personal exposure to air pollution (from all sources, indoor and outdoor), particularly for critical groups.
- A programme to introduce pollution control technologies and management strategies.
- Use of current, scientifically sound modelling methods to evaluate the pollution reduction measures and to improve the understanding of factors that influence the air quality of a region.

On a more detailed level, there are several steps that could be taken in order to reduce air pollution problems. The effectiveness of such measures will inevitably depend on different forces operating at local and national levels, but the following list provides some examples that can lead to improved air quality:

- Fuel quality:
  - Replace leaded petrol.
  - Introduce low-sulphur diesel.
  - Investigate the role and effectiveness of new fuels.
- Road transport:
  - Make public transport more accessible and affordable.
  - Improve maintenance of the roads.
  - Ensure regular inspection and maintenance of vehicles.
  - Identify and reduce gross polluters.
  - Restrict traffic in congested areas.
- Technology:
  - Encourage uptake of improved technology vehicles, such as catalytic converters.
  - Improve fuel and engine efficiency of vehicles.
- Public:
  - Educate, raise awareness and provide training.
  - Introduce public air quality information systems.
- Other:
  - Reduce burning of biomass and improve agricultural burning methods.
  - Improve cooking stoves, reduce indoor sources and increase ventilation.

# LONG-RANGE TRANSPORT OF ATMOSPHERIC POLLUTANTS AND TRANSBOUNDARY POLLUTION

*S Trivikrama Rao, Christian Hogrefe, Tracey Holloway and George Kallos*



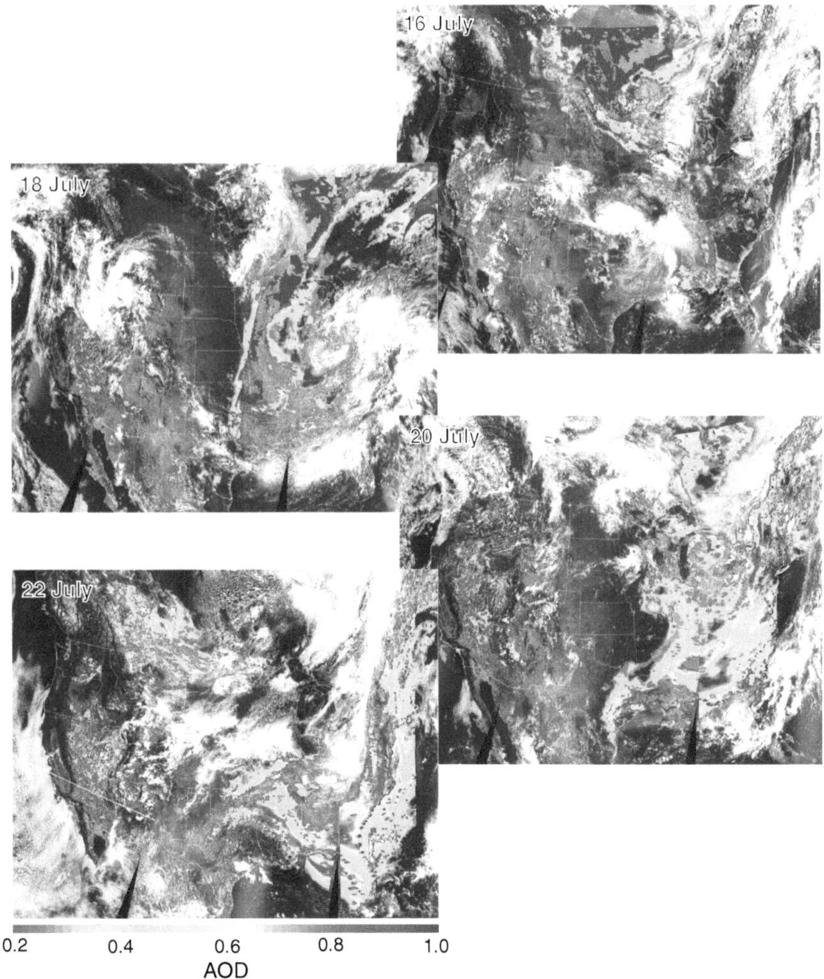


During the 2004 summer, the largest Alaskan wild fire event on record occurred in late June-July and consumed 2.72 million hectares of boreal forest. The Figure shows the aerosol optical depth (AOD) data from the MODIS instrument aboard the Terra satellite for a series of days in July 2004. The MODIS AOD is plotted over the MODIS Terra true color image for each day. These series of days show high aerosol loading associated with long-range transport of the Alaskan wild fire plume as it crosses over the northern border of the United States on July 16. This aerosol plume was advected south-eastward behind the cold front (evident in the clouds captured in the MODIS true color) over the following days, eventually affecting surface $PM_{2.5}$ levels along the Eastern United States.

Early air pollution control efforts were prompted by urban episodes due to local emissions, such as the 1952 London smog associated with sulphur from burning coal (see Chapter 1). Although local areas typically experience the highest levels of health- and ecosystem-damaging air pollution, many species remain in the atmosphere for days, months, or even years. The longer a pollutant stays in the atmosphere, the farther from its original source it travels. For example, it takes about five days for a pollutant to cross the Pacific Ocean, but over a year for pollution to cross from the Northern Hemisphere mid-latitudes to the Southern Hemisphere. An example of such a pollutant plume as seen by satellite is shown in Figure 3.1.

## 3.1 Regional Air Pollution Transport

Although early scientific and regulatory efforts focused on local emissions and local effects, since the late 1970s, the geographic scale on which pollutant transport is studied and regulated has expanded. In Europe, the UNECE Convention on Long-Range Transboundary Air Pollution (LTRAP, www.unece.org/env/lrtap) came into force in 1983, driven by a concern about acid rain in Europe, and it has since expanded to address nitrogen deposition, ozone, heavy metals, particulate matter, and persistent organic pollutants (POPs). Scientific support for the Convention is provided through the Cooperative Programme for Monitoring and Evaluation of the Long-Range Transmission of Air Pollutants in Europe (EMEP), which assesses air pollution impacts on Europe. Figure 3.2 shows the annual mean regional concentrations of particulate matter (PM) in the size fractions $PM_{10}$ and $PM_{2.5}$ over Europe based on Unified EMEP model calculations and EMEP monitoring station observations for the year 2008 (Yttri et al. 2010). The model typically predicts annual mean concentrations of PM from all sources (local and long range) in the range from 5 to 20 $\mu g/m^3$ for $PM_{10}$ and 5 to 15 $\mu g/m^3$ for $PM_{2.5}$ over most of Europe. The levels of air pollutants vary spatially and temporally year to year according to several factors including changes to emissions and the variability in meteorological conditions. For example, the particulate matter levels shown in Figure 3.2 were generally lower than those modelled for the year 2007 (Yttri et al. 2010). Inter-annual variability in meteorology (e.g. warm temperatures and stagnant conditions) have also led to elevated air pollution concentrations across many parts of Europe for other years such as 2003 (Yttri and Tørseth 2005).

In the eastern USA, regional-scale transport aspects of ozone, particulate matter and their precursors have been addressed through the US Environmental Protection Agency's (EPA) 2005 Clean Air Interstate Rule (www.epa.gov/cair), following earlier analyses by the Ozone Transport Assessment Group (OTAG, www.epa.gov/ttn/naaqs/ozone/rto/otag/finalrpt/), the Ozone Transport Commission (www.otcair.org/about.asp), and other studies. Figure 3.3a shows non-attainment (exceedance) areas for $PM_{10}$, ozone ($O_3$), carbon monoxide (CO) and sulphur dioxide ($SO_2$) which are all criteria pollutants. The figures represent the situation as of March 2007. An overall picture can be produced of the extent of US areas that do not meet the National Ambient Air Quality Standards (NAAQS) and is shown in Figure 3.3b. Nitrogen dioxide ($NO_2$) and lead (Pb) are also criteria pollutants in the US but there were no significant exceedances for $NO_2$ and only a few areas showed non-attainment for Pb. Although the EPA regulates six 'criteria' air pollutants, concentrations of ozone and PM consistently remain above health-based standards in many US counties. Both ozone and PM can be formed through a complex set of physical, chemical and meteorological interactions in the atmosphere over regional scales and thus their effective management requires national policies.

Like Europe and the US, regional air pollution has been a concern for Asia. To date, most efforts to estimate source-receptor relationships have focused on acid deposition (e.g. Carmichael et al. 2002; Holloway et al. 2002), illustrated in Figure 3.4. However, in recent years attention has turned to the health-relevant species of ozone and PM. Emissions of sulphur dioxide contribute both to acid deposition and PM and emissions of nitrogen oxides contribute to acid deposition, PM and ozone. Thus all three issues are all closely linked. Following the success of the EMEP monitoring network in Europe, the Acid Deposition Monitoring Network in East Asia (EANET) began monitoring activities in 2001, and regional modelling research is advancing through national and research efforts, as well as through the Model Inter-Comparison Study for Asia (MICS-Asia).

It is important to understand the specific mechanisms associated with long-range air pollution transport. The significance of air pollution caused by long-range transport is illustrated by examining problems affecting two regions of the world: ozone in the eastern US and sulphur in the Mediterranean.

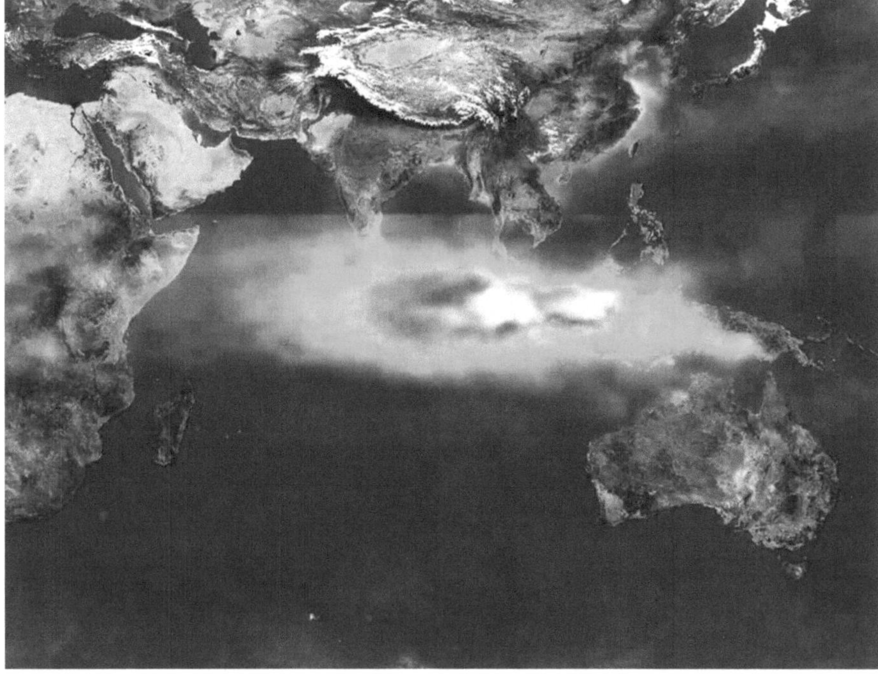

**Figure 3.1** A pollutant plume as seen by satellite. Researchers have discovered that pollutants move in different ways through the atmosphere. A series of unusual events several years ago created a blanket of pollution over the Indian Ocean. In the second half of 1997, smoke from Indonesian fires remained stagnant over Southeast Asia while smog, which is tropospheric, spread more rapidly across the Indian Ocean toward India. Researchers tracked the pollution using data from NASA's Earth Probe Total Ozone Mapping Spectrometer (TOMS) satellite instrument. The Figure shows the pollution over Indonesia and the Indian Ocean on October 22, 1997. White represents the aerosols (smoke) that remained in the vicinity of the fires. Green, yellow, and red pixels represent increasing amounts of tropospheric ozone (smog) being carried to the west by high-altitude winds. (Source: NASA 1997.)

### Transport of Ozone in the Eastern United States

Ozone is a secondary pollutant formed in the atmosphere from emitted hydrocarbons and nitrogen oxides reacting in the presence of sunlight. The extent of $O_3$ transport is difficult to assess with direct measurements since it may be directly affected by transport or through a combination of imported precursors with local emissions. To tackle this issue, some researchers have introduced the notion of an 'airshed' for $O_3$ (Civerolo et al. 2003), following the analogy of watersheds at the surface (Dennis 1997). The analogy must not be taken too literally, however. Whereas transport through a watershed is limited to rivers and other bodies of water and the surrounding land surfaces, pollutant transport through the atmosphere can occur over much longer distances and is strongly influenced by the meteorological conditions (Eder et al. 1994; Vukovich 1995; OTAG (www.otcair.org/about.asp); Rao et al. 2003). For example, a

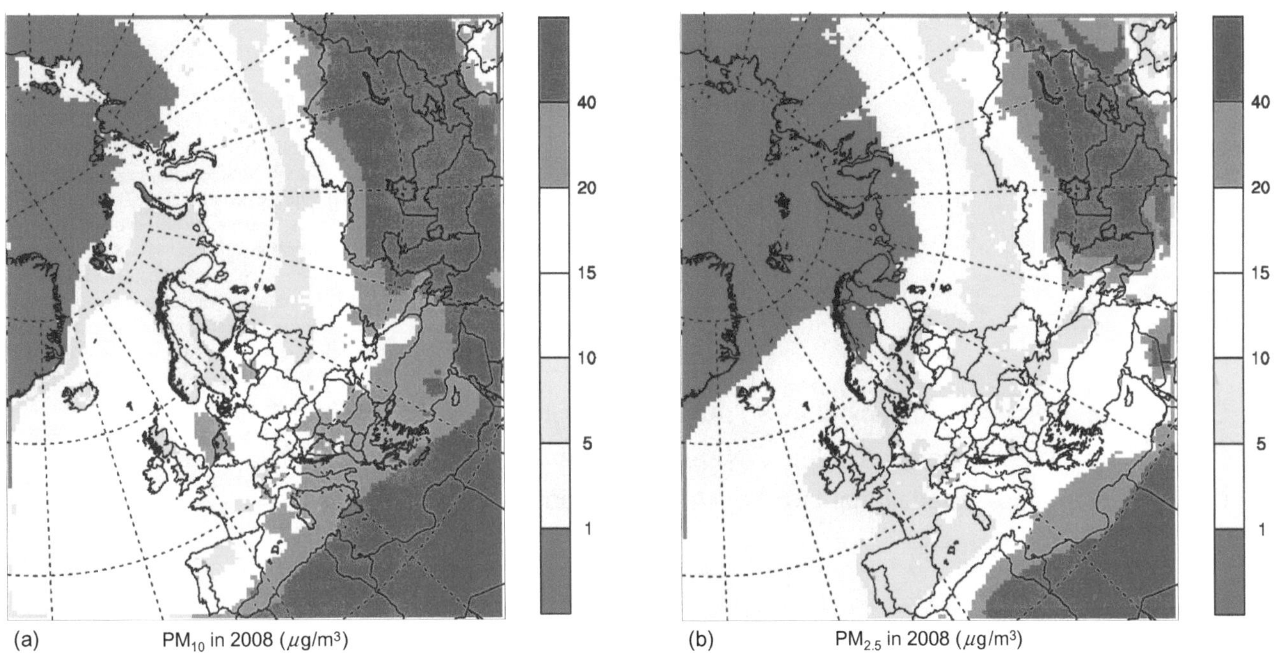


Figure 3.2 Regional air pollution concentrations across Europe. Annual mean concentrations of (a) $PM_{10}$ and (b) $PM_{2.5}$ over Europe for 2008 based on the EMEP model calculations and observations from EMEP monitoring stations. (Source: Yttri et al. 2010.)

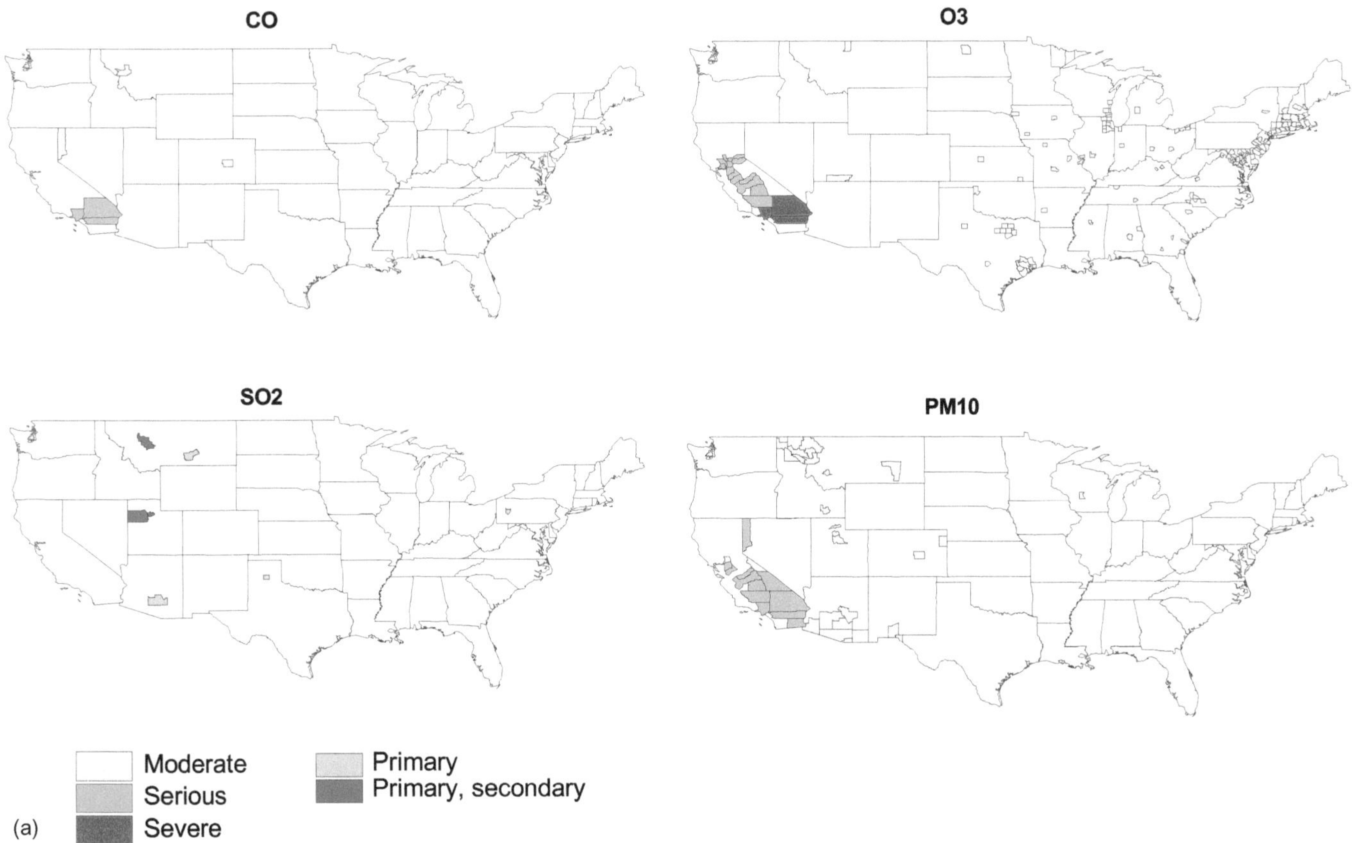


Figure 3.3 (a) Areas of non-attainment in the USA of the US EPA National Ambient Air Quality Standards (NAAQS) for CO, $O_3$, $PM_{10}$ and $SO_2$ criteria pollutants as of March 2007. (b) Combined non-attainments across USA showing areas where one, two and three pollutant concentrations exceeded the NAAQS threshold. Note that marginal classified areas have not been shown. In the case of $SO_2$, non-attainment of the primary and secondary standards are shown. *Primary standards* set limits to protect public health, including the health of 'sensitive' populations such as asthmatics, children and the elderly. *Secondary standards* set limits to protect public welfare, including protection against decreased visibility, damage to animals, crops, vegetation and buildings. There were no $NO_2$ non-attainments and Pb was excluded from the figures. (Source: www.epa.gov/air/oaqps/greenbook.)

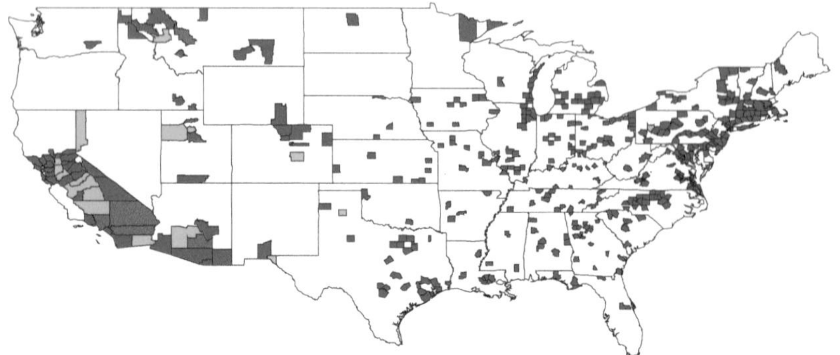


**Figure 3.3** (*continued*).

common synoptic-scale feature associated with $O_3$ episodes over the eastern USA is the presence of a high pressure system aloft (500 mb), usually accompanied by subsidence, clear skies, strong shortwave radiation, high temperatures and stagnant air masses near the ridge line of the sea-level high pressure region (Gaza 1998; Zhang and Rao 1999; Schichtel and Husar 2001). Westerly and south-westerly, nocturnal, low-level jets during these episodic events facilitate the transport of pollutants over long distances (Mao and Talbot 2004). These synoptic conditions augment local photochemical production and contribute to elevated levels of pollutant concentrations, which blanket much of the north-eastern USA for several days (Zhang et al. 1998). Figures 3.5a, b and c illustrate the role of synoptic-scale meteorological features in regional-scale transport of $O_3$ (Fishman and Balok 1999).

### Transport of Sulphur Compounds in the Mediterranean

In the Mediterranean region, air pollutants may be transported from Europe to North Africa and other areas of the Middle East due to differential heating between the land of North Africa and southern Europe and the Mediterranean waters. The transport paths and scales of air pollution transport in the Mediterranean region has been the subject of various projects during the last two decades (e.g. SECAP, T-TRAPEM). For example, Luria et al. (1996) found that sulphate amounts monitored in Israel could not be explained by emissions from local sources only. The temporal scales of transport, about 90 hours, from Europe to the Middle East are comparable to the chemical transformation scales of emitted $SO_2$ to sulphate particles, which is the primary constituent of acid deposition and an important source of secondary PM. Despite similar climatological characteristics, the western and eastern Mediterranean vary significantly in the typical dispersion and photochemical processes affecting oxidant formation and transport (Kallos et al. 1997a, b, 1998). Urban plumes from various locations in southern Europe can be transported over the Mediterranean, maintaining most of their characteristics.

During the warm period of the year, the Intertropical Convergence Zone (ITCZ) is shifted north (over Egypt, Libya and Algeria). Due to the trade wind system across the Aegean and strong sea breezes, polluted air masses from Europe can be transported southward and enter the ITCZ within a few days (Kallos et al. 1998). Once entrained within the ITCZ, sulphate particles may affect rainfall patterns and, hence, water availability.

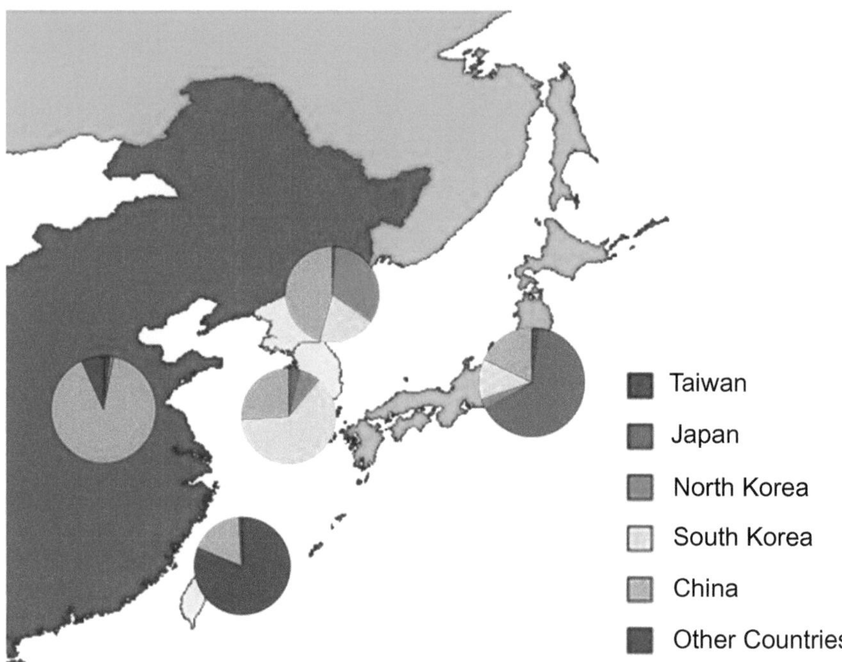


**Figure 3.4** Long-range pollution contributions in East Asia. Annual average contribution (one year) of $NO_x$ emissions from selected countries in East Asia to neighbouring nations, based on the ATMOS Lagrangian model. (Source: Holloway et al. 2002.)

## 3.2 Hemispheric Air Pollution Transport

Because domestic emission controls in many countries have reduced the contribution of local sources to air quality problems, the relative impact of long-range transport is growing in many areas. Air pollution tends to move between continents in two ways: (1) via episodic advection, where distinct polluted air masses may be traced from source to receptor; and (2) by increasing the global background level of pollutants, which, in turn, increases surface concentrations far from the emission source regions (Figure 3.6). The emission strength, transport duration, degree of photochemical processing and wet and dry deposition during transit will ultimately determine the species concentrations that reach surface air over a receiving continent. Some pollutants stay in the atmosphere over a year, long enough to mix between the Northern and Southern Hemispheres. Mixing, however, is much more rapid from west-to-east in the Northern Hemisphere mid-latitudes, where westerly winds create a 'conveyer belt', transporting species among North America, Europe and Asia (www.htap.org).

## 3.3 Methods for Analysing Long-Range Transport of Air Pollution

Earlier sections presented a description of typical pollutant transport patterns and introduced some concepts useful in understanding regional and global pollution. Here we discuss some techniques to assess the long-range transport problem.

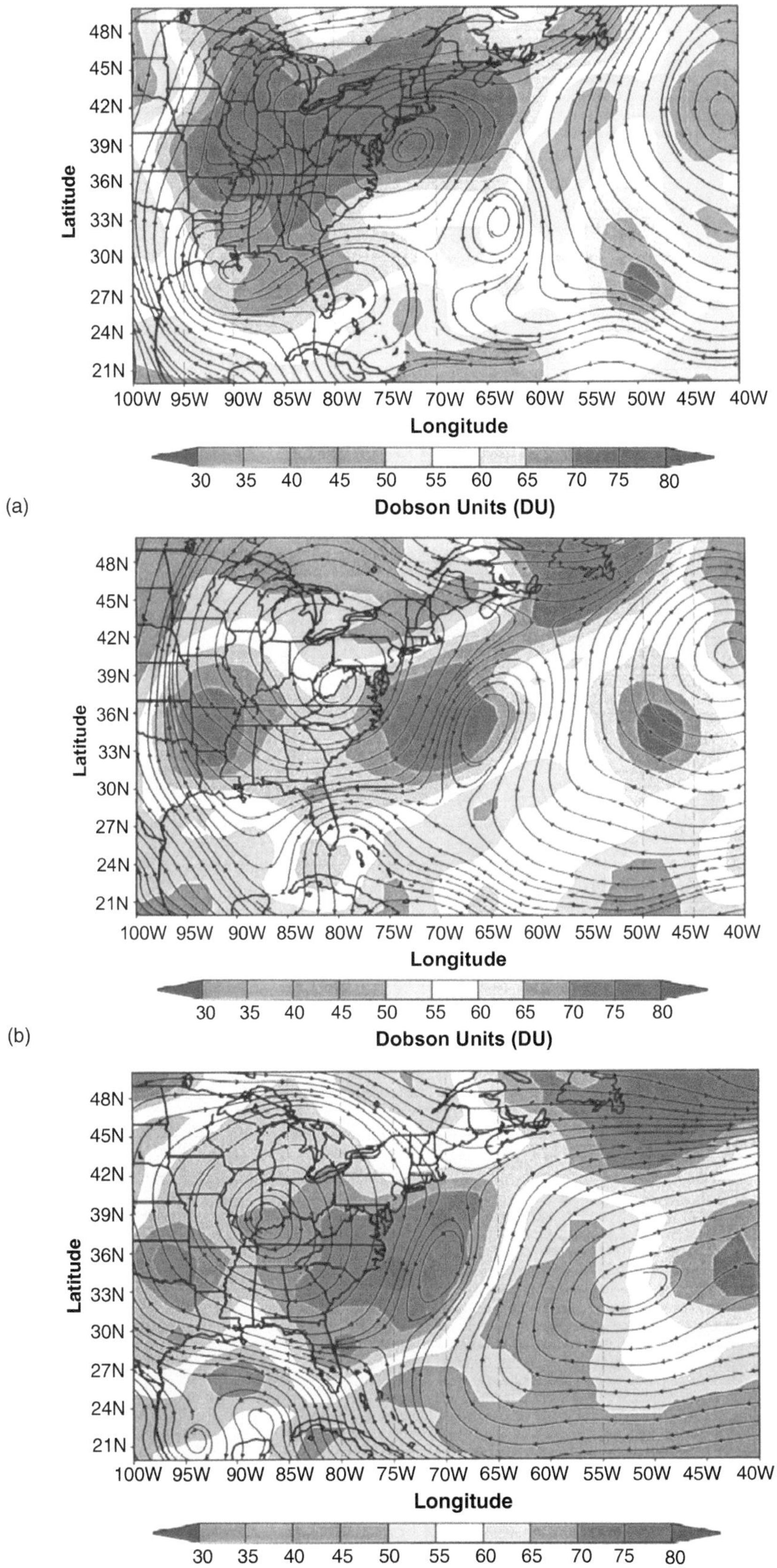

(a)

(b)

(c)

Figure 3.5 Synoptic-scale meteorological features in regional-scale transport of $O_3$. Tropospheric ozone residuals and 850 mbar wind streamlines for (a) July 4, 1988, (b), July 6, 1988, and (c) July 8, 1988. (Source: Fishman and Balok 1999.)

## Satellite Observations

Satellite measurements have greatly advanced scientists' ability to study episodic transport on a global scale. Asian dust events occur most often in springtime, and Figure 3.7 illustrates a large dust storm that occurred in early April 2001. The sequence of images (Figures 3.7a–e) shows the Aerosol Index measured by Earth Probe TOMS (Total Ozone Mapping Spectrometer) during this event (NASA 2001a). The dust cloud originated between 6 and 9 April 2001, when strong winds from Siberia kicked up millions of tons of dust from the Gobi and Takla Makan deserts in Mongolia and China, respectively. Air currents then carried the dust eastward. The leading edge of the cloud reached the US west coast on 12 April, and two days later it had crossed the east coast shoreline and begun heading out into the Atlantic Ocean. Dust clouds blowing east from Asia are a common occurrence in the springtime, and satellite images of these clouds can be used to study the atmospheric flow patterns that can also govern the transport of invisible, anthropogenic emissions. It has been shown through air quality measurements at Cheeka Peak in Washington State and aeroplane-based measurements that pollution from Asian sources can affect the air quality in the western USA, although the level of transport of pollutants varies widely (NASA 2001b; Husar et al. 2001; Vaughan et al. 2001; McKendry et al. 2001).

In addition to elucidating the atmospheric flow patterns that govern global pollutant transport, satellite images help to characterise anthropogenic and biogenic emissions. An example of satellite data useful for both objectives is NASA's Terra spacecraft, which directly measures atmospheric CO concentrations. Figures 3.8a and b present images of carbon monoxide concentrations in the lower atmosphere, ranging from about 50 parts per billion to 390 parts per billion. Carbon monoxide is a gaseous by-product from the burning of fossil fuels, in industry and automobiles, as well as burning of forests and grasslands. Notice that in the 30 April 2000 image levels of carbon monoxide are much higher in the Northern Hemisphere, where human population and human

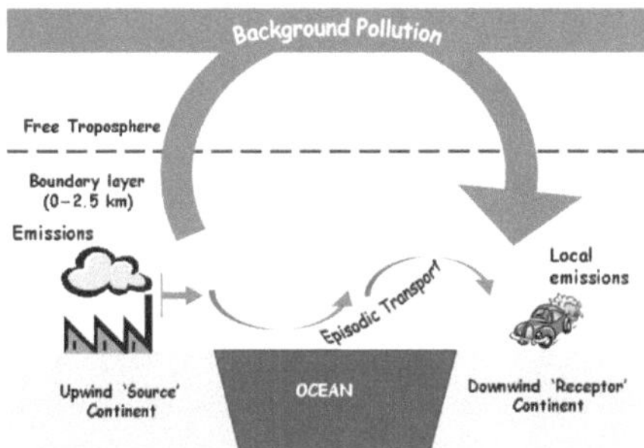


Figure 3.6 A simple schematic of intercontinental air pollution transport. Emissions from the upwind 'source' continent are advected to the downwind 'receptor' continent through episodic transport events and/or by enhancing the global background pollution concentration. Emissions may be mixed vertically into the free troposphere for rapid long-range transport or transported within the boundary layer. The degree of photochemical processing and deposition that occurs during transport controls the air pollutant concentrations that are ultimately detected on the receptor continent. (Source: Holloway et al. 2003.)

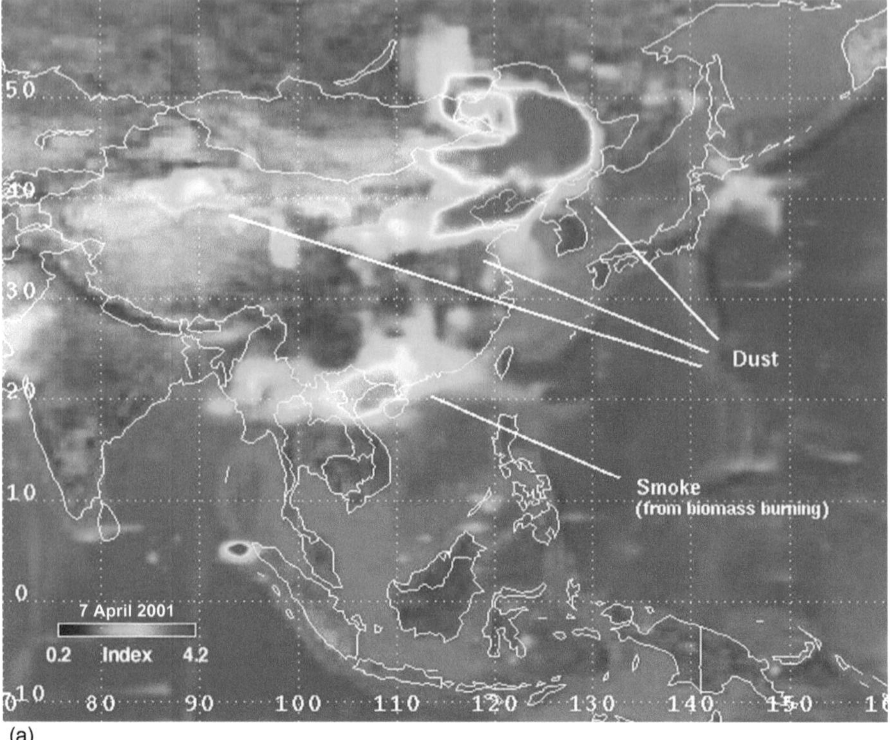

(a)

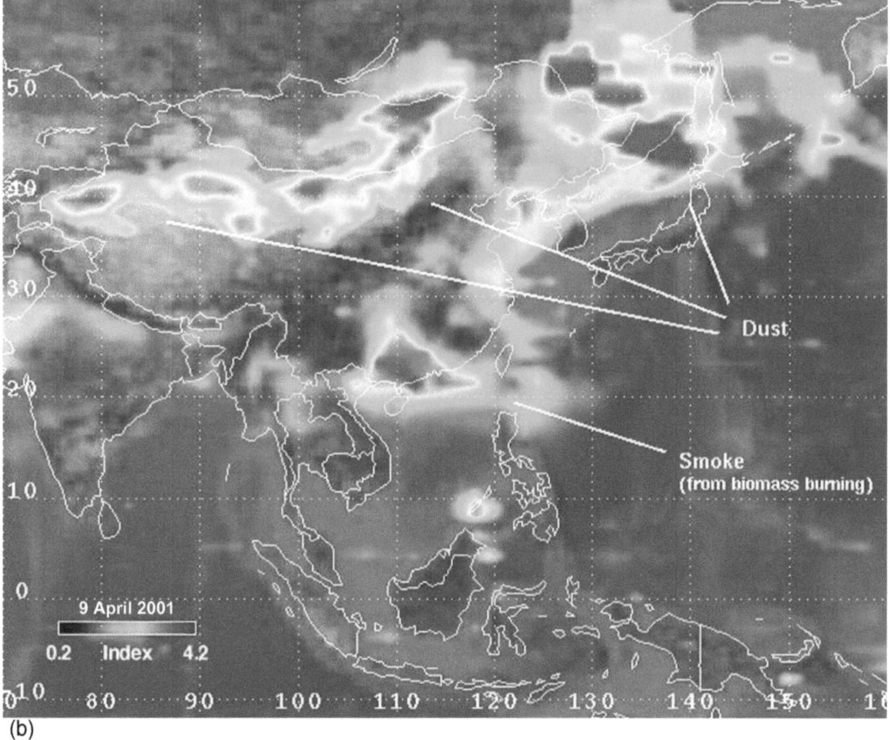

(b)

Figure 3.7 (a)–(e): Aerosol Index measured by Earth Probe TOMS (Total Ozone Mapping Spectrometer) during the Asian dust storm of April 2001. (Source: NASA 2001a.)

industry is much greater than in the Southern Hemisphere. However, in the 30 October 2000 image, notice the immense plumes of gas emitted from forest and grassland fires burning in South America and southern Africa (NASA 2000).

## Statistical Analysis of Measurements

The spatial extent of a pollutant airshed – the domain over which significant regional transport occurs – may be estimated through statistical analysis of observed values at different times and measurement locations. For example, correlating time series of observed daily maximum 1- or 8-hour $O_3$ concentrations at different stations, repeating the analysis for all possible station pairs within a domain of interest and plotting the decay of correlation between stations as a function of distance between the stations, one can obtain a measure of the coherence in pollutant levels among different air monitoring stations embedded within the same synoptic-scale weather pattern. Over the north-eastern USA, this type of analysis indicates that the characteristic scale for $O_3$ transport is about 600 km along the direction of the prevailing wind (Figure 3.9). Further, one can perform a time-lagged correlation analysis in order to assess the characteristic one- to two-day transport distances associated with the synoptic-scale $O_3$ component. (Brankov et al. 1998). Figure 3.10 shows an example of such an analysis using an ozone monitor in Pittsburgh, PA as the reference station against which other $O_3$ monitors were correlated at lags of zero and one day.

Statistical analyses cannot establish causal relationships, but this approach offers a powerful tool to estimate the spatial scales of pollutants. Results from the case study in the United States presented here suggest that $O_3$ levels in a region from Virginia to Maine can potentially be affected by emissions in the Pittsburgh area within one day, whereas Pittsburgh may be affected by emissions in a region from Michigan to the western Ohio Valley to the Carolinas (Civerolo et al. 2003).

## Trajectory Analysis

While a statistical approach provides important insights into understanding observations, it does not explicitly take physical transport processes into account. To assess the effects of the synoptic-scale atmospheric transport patterns on observations at a specific site, the pathways on which air masses have travelled may be analysed to examine which emission sources may have contributed to measured levels.

For example, when analysing $O_3$ measurements taken from the CN tower in Toronto, Canada, Brankov et al. (1998) employed the trajectory-clustering methodology. This approach entails calculating a large number of back-trajectories from the observational site over a long period of time. The Hybrid Single Particle Lagrangian Integrated Trajectories model (Draxler 1992) was used to calculate 24-hour back-trajectories for every summer day (June, July and August) over a period of seven years, from 1989 to 1995.

Applying a trajectory-clustering technique, trajectories close to each other and with similar directions are grouped together, producing a more manageable number of representative groups to reflect the behaviour of a large number of trajectories. Statistical procedures can then be used to test for statistically significant differences in the chemical composition of the clusters (Brankov et al. 1999). The back-trajectory clustering-methodology applied on CN tower back-trajectories resulted in eight clusters of trajectories whose average trajectories are shown in

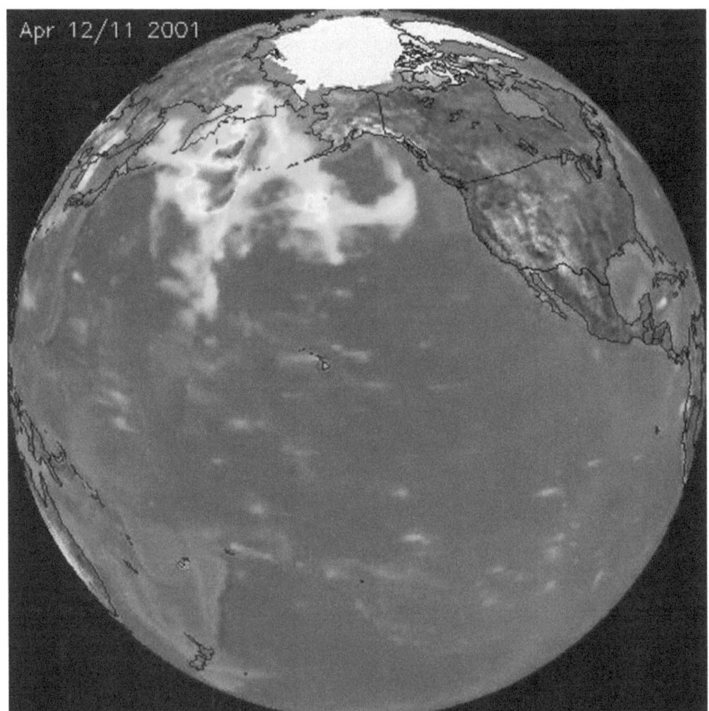


(c)

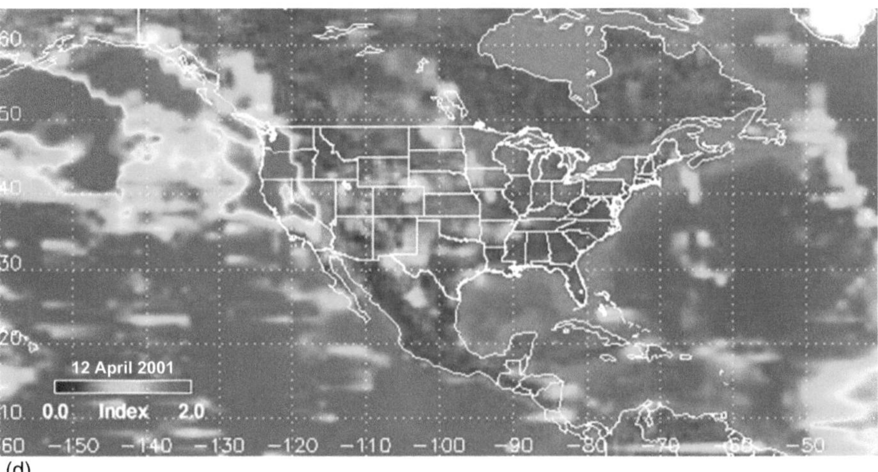


(d)

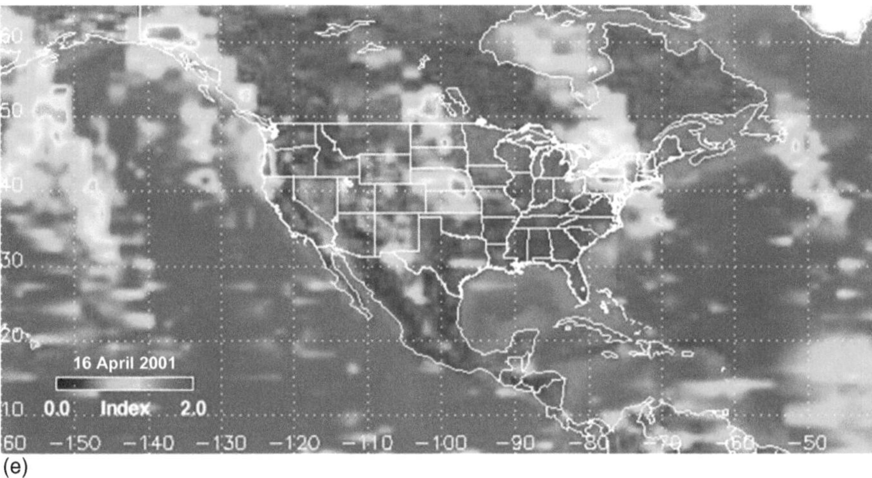


(e)

**Figure 3.7** *(Continued).*

Figure 3.11a. Of all summer trajectories arriving to the CN tower, 54 percent are associated with air masses almost exclusively travelling over Canada, and 46 per cent of the airflow regimes bring air from the USA. Figure 3.11b shows box-whisker plots of '$O_3$ clusters' obtained by segregating short-term $O_3$ concentration data according to the clusters in Figure 3.11a. Each box-whisker displays five percentiles (tenth, twenty-fifth, fiftieth, seventy-fifth and ninetieth), as well as the minimum and maximum concentrations of $O_3$ assigned to one particular cluster. Thus, this methodology can be used to identify distinct atmospheric transport patterns associated with high levels of $O_3$ concentrations, illustrating the effects of transboundary pollution exchange and potential source regions for this pollutant.

Another example of trajectory analysis to assess regional $O_3$ transport is shown in Figures 3.12a–d (Schichtel and Husar 1996). These illustrations show the merging of a simulation of the atmospheric flow (particles) and measured ozone data from over 600 monitoring stations. In this example, a summertime air mass over the industrial Midwest raised afternoon $O_3$ concentrations from approximately 70 ppb throughout the region to more than 100 ppb in parts of the Ohio River Valley (Figure 3.12b). As the $O_3$-laden air mass was transported east-north-east, afternoon $O_3$ concentrations in parts of western Pennsylvania increased over 40 ppb from the previous day's levels, producing levels higher than 100 ppb (Figure 3.12c). Such an illustration provides strong evidence of the role of atmospheric transport in determining ozone concentration in the north-east of USA (Schichtel and Husar 1996).

On the global scale, tracer models have also been used to study the pathways and timescales of intercontinental transport patterns (Stohl et al. 2002). Chemically inert particles are released in source regions of interest and their fate is then tracked by the model as they undergo horizontal and vertical transport and mixing, as determined by the meteorological fields used as input.

## Dynamic Air Quality Models

The mechanisms responsible for air pollution transport may be examined independently of any particular set of observations. Air quality models describe atmospheric chemistry and transport mathematically and then solve the relevant equations with high-speed computers. These models have two basic structures: Lagrangian (e.g. Draxler 1992) and Eulerian (e.g. Byun and Schere 2006). Conceptually, Lagrangian models solve the equations for each moving air mass, whereas Eulerian models solve the equations on a fixed grid. Both types of models allow researchers to build a 'virtual atmosphere', useful for testing our understanding of atmospheric processes and analysing 'what-if' scenarios valuable for environmental policy analysis.

Building on the case study of $O_3$ in the north-eastern USA, a three-dimensional Eulerian model, the Urban Airshed Model-Variable grid version

(a)

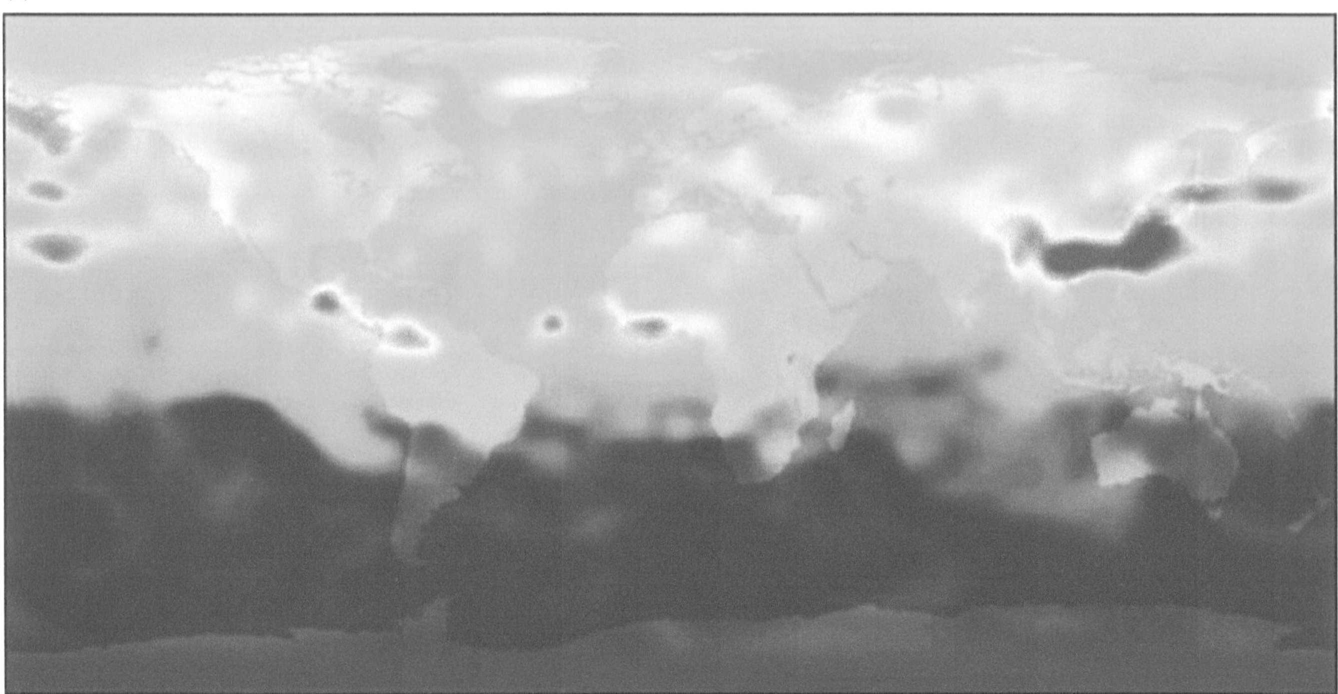

30 April 2000

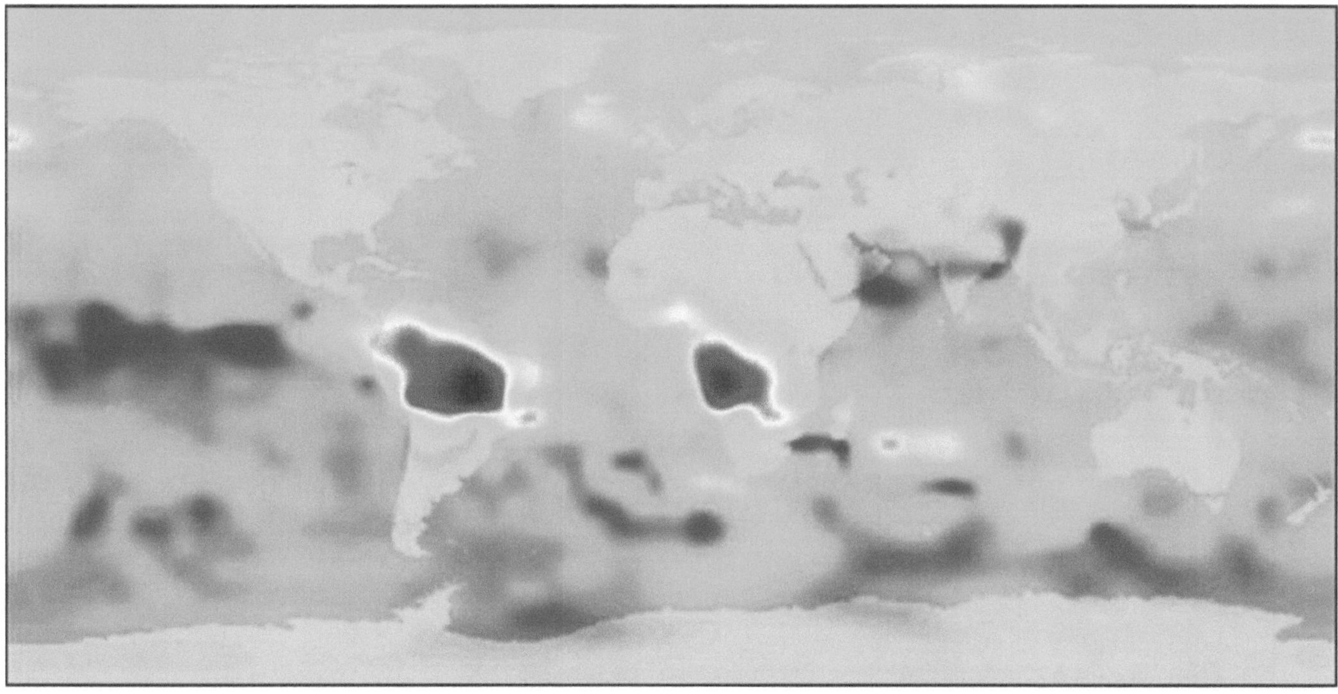

30 October 2000

(b)

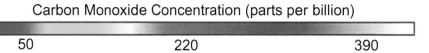

**Figure 3.8** (a) and (b) Carbon monoxide concentrations in the lower atmosphere measured by NASA's Terra spacecraft. (Source: NASA 2000.)

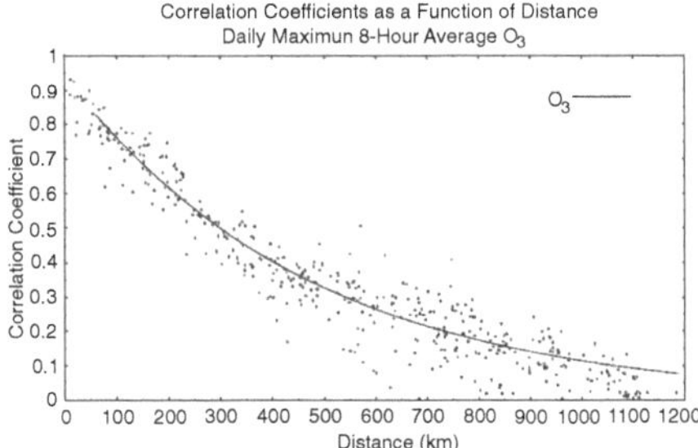


**Figure 3.9** Analysis of the characteristic scale for $O_3$ transport over the north-eastern USA. Correlation coefficients between summertime synoptic forcings in $O_3$ between Philadelphia and all other sites along prevailing flow direction are shown as a function of distance from Philadelphia. Both the data points and a best-fit line are shown. (Source: Civerolo et al. 2003.)

(UAM-V) (SAI 1995), has been used. Employing the 1995 meteorological data, and emissions from man-made and natural sources, the model simulated summer $O_3$ over much of the eastern USA and southern Canada. Since the model offers a 'virtual atmosphere', a researcher can turn off selected emission sources to examine how individual reductions affect total regional $O_3$. In this illustrative case, researchers examined how total $O_3$ would be affected by reductions in anthropogenic emissions in New York State versus those in the Canadian province of Ontario (Brankov et al. 2003).

Reducing Ontario emissions led to improvements of 15 per cent or greater in the near-field, and 6 per cent or greater throughout most of New York State. The dramatic NOx reductions near Toronto actually led to increased $O_3$ in the urban core area. The situation was similar in the New York emissions reduction case, where $O_3$ improvements within New York ranged from about 3 per cent to 15 per cent. Even along southern Ontario, $O_3$ decreased by up to about 6 per cent. It should be emphasised that these percentage reductions are seasonal averages; the percentage reduction at a grid cell on any one day may be quite large. In addition, the sign of the change may vary from day to day, depending on prevailing winds. In a similar study, Rao et al. (1998) showed that the decay of the ozone reductions stemming from the elimination of emissions in one region has a spatial scale dependence that is consistent with that of the decay of correlations in ozone observations shown in Figure 3.9.

Atmospheric chemistry models are especially useful for examining global transport patterns where few measurement data are available and where large-scale transport phenomena require detailed analysis. For example, results from modelling studies indicate that $O_3$ produced from Asian emissions can enhance $O_3$ concentrations in surface air over the western USA by 3–10 ppb; that $O_3$ produced from North American emissions can enhance European ozone concentrations by 2–15 ppb; and that European emissions raise East Asian $O_3$ concentrations by 3 ppb on average in spring (Holloway et al. 2003 and references therein).

Although a few studies have diagnosed $O_3$ enhancements from intercontinental transport via analyses of air mass origin, transport of $O_3$ primarily occurs through increases in background concentrations, making it difficult to observe events directly on a receptor continent.

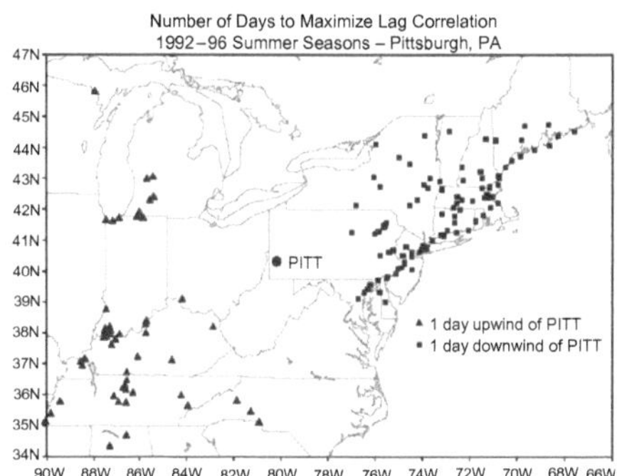


**Figure 3.10** Analysis of the characteristic one- to two-day transport of the synoptic-scale $O_3$ component. The Figure shows the number of days needed to maximize the summertime synoptic-scale $O_3$ correlations between Pittsburgh (large dot) and various locations throughout the eastern US. Only the sites which Pittsburgh lags by 1 day (triangles) or leads by 1 day (squares) are shown, and only the statistically significant (95%) correlation coefficients were considered. (Source: Civerolo et al. 2003.)

Determining the sources of air pollution is an important precursor to any large-scale international management effort, and a number of global air quality models have been used to estimate such source contributions. Jacob et al. (1999) used a global atmospheric chemistry model to forecast how future economic growth in Asia – through increased emissions of $O_3$ precursors – could affect $O_3$ concentrations over the USA. The group concluded that a tripling of anthropogenic emissions from Asia could increase monthly mean surface $O_3$ over the USA by 1–6 ppb (minimum in

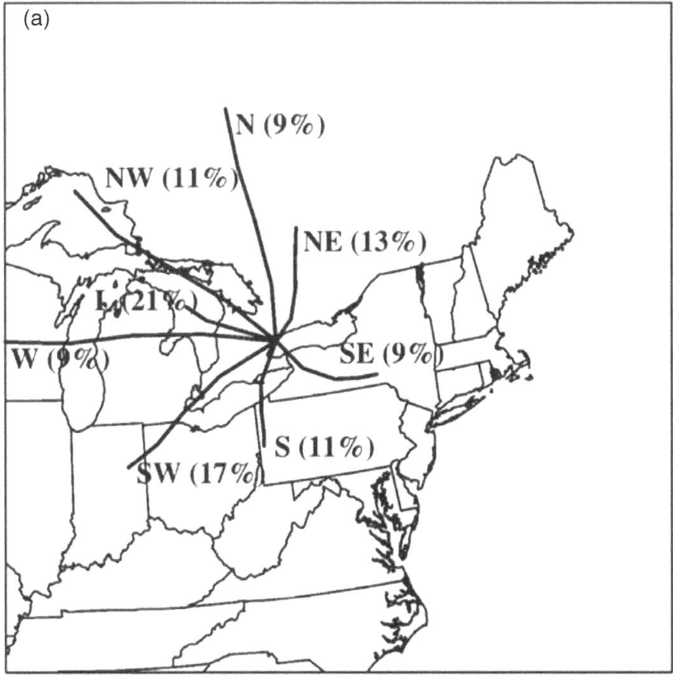


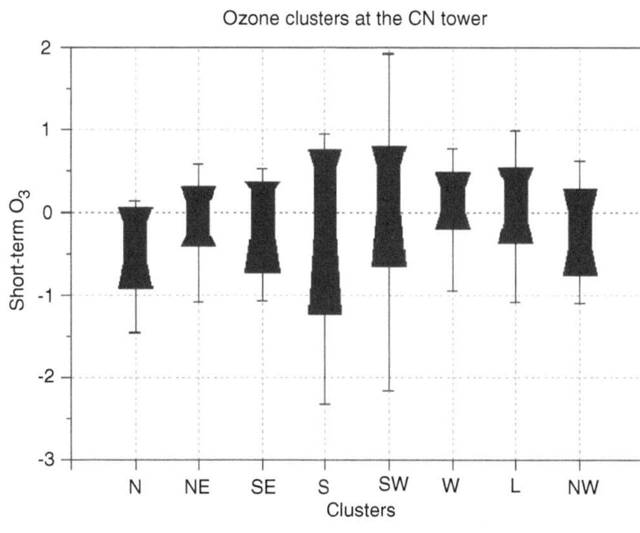


**Figure 3.11** Back-trajectory clustering at the Canadian National (CN) tower. (a) Group of eight clusters of average back-trajectories for the CN tower receptor site. The clusters are labelled according to the origin of the airmass: north-west (NW), north (N), north-east (NE), south-east (SE), south (S), south-west (SW), west (W), and local circulation patterns (L). The percentage of all trajectories belonging to each cluster is also shown in the figure. (b) Box-whisker plots of the strength of the synoptic forcing for each of the clusters shown in (a). (Source: Brankov et al. 2003.)

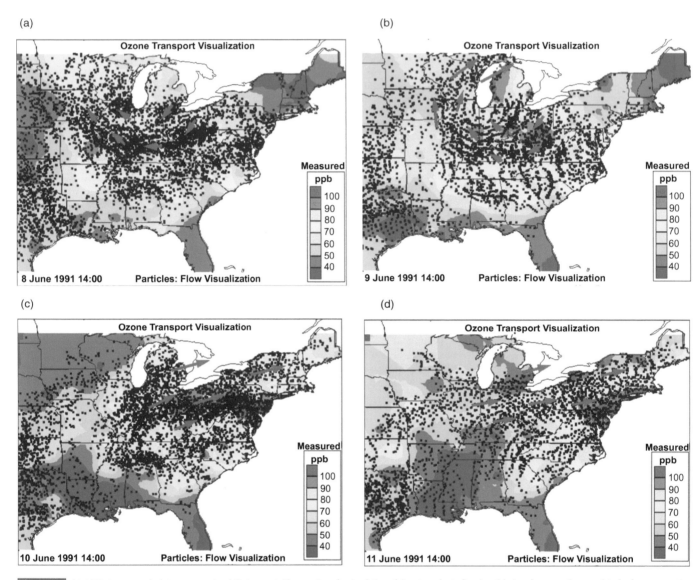

**Figure 3.12** (a)–(d) Trajectory analysis to assess regional $O_3$ transport. The merging of a simulation of the atmospheric flow (particles) and measured ozone data for four consecutive days. The ozone has been spatially interpolated from over 600 monitoring sites. The arrows represent the direction and speed of transport. (Source: Schichtel and Husar 1996.)

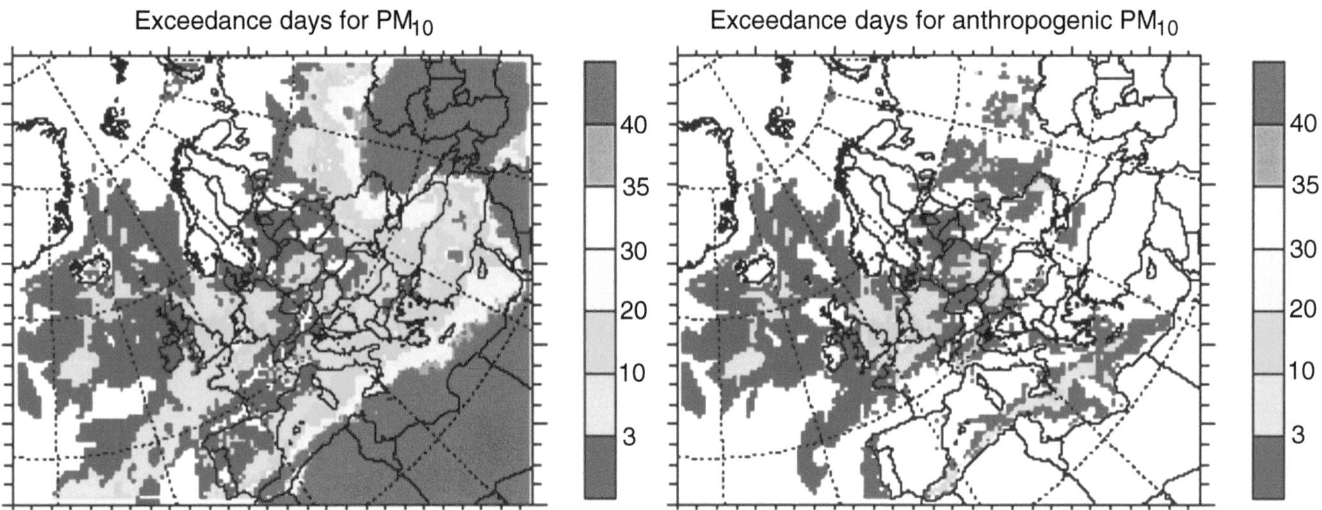


**Figure 3.13** Calculated number of days with regional $PM_{10}$ concentrations exceeding the EU limit value of 50 $\mu g/m^3$ in 2004 for $PM_{10}$ including both anthropogenic and natural particles (left panel) and for anthropogenic $PM_{10}$ concentrations (right panel). (Source: Yttri and Aas 2006.)

the east, maximum in the west during spring). While the magnitude of this increase appears small, it would more than offset the benefits of 25 per cent reductions in domestic anthropogenic emissions in the western USA (Jacob et al. 1999). Another study by the same group concluded that with the subset (20 per cent) of violations in the 8-hour average, some part of these ozone exceedances are due to anthropogenic emissions from North America (Li et al. 2002).

As in the case of other regions of the world, particulate matter is one of the main pollutants of concern in Europe. Figure 3.13 shows the predictions calculated with the Unified EMEP model of number of days exceeding the EU daily limit value for $PM_{10}$ of 50 $\mu g/m^3$ over Europe for the year 2004 (Yttri and Aas 2006). The modelled maps show the number of exceedance days calculated for $PM_{10}$ including both anthropogenic and natural particles and for anthropogenic $PM_{10}$ only. The $PM_{10}$ concentrations exceeded 50 $\mu g/m^3$ for more than 35 days in regions of Belgium, the Milan area and the Moscow area. The modelling study also showed that, compared to 2003, the number of calculated $PM_{10}$ exceedances were lower for 2004 partly due to the reduced emissions of oxides of sulphur and nitrogen and $PM_{10}$, and because of the differences in the meteorological conditions experienced in 2004 compared to 2003 (Yttri and Tørseth 2005).

The above examples demonstrate that a range of methods is available to researchers for investigating the spatial scales associated with air pollutant transport. While each methodology has its own limitations, a combination of observational and modelling approaches consistently shows that $O_3$ and aerosol pollution is a regional, multi-state and even international issue, and not a problem existing only at local or urban scales.

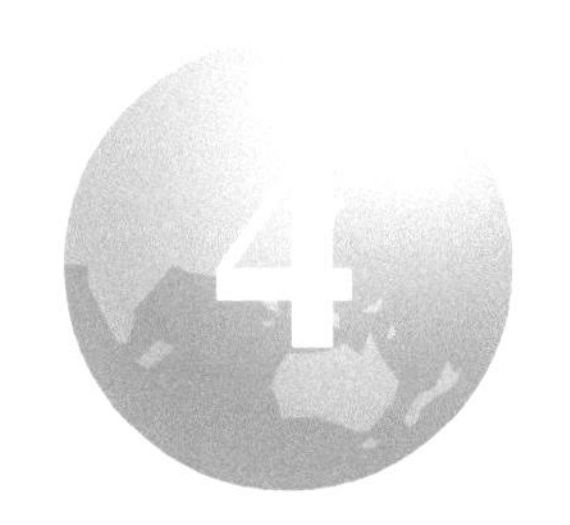

# GLOBAL AIR POLLUTION AND CLIMATE CHANGE

*Ding Yihui and Ranjeet S Sokhi*



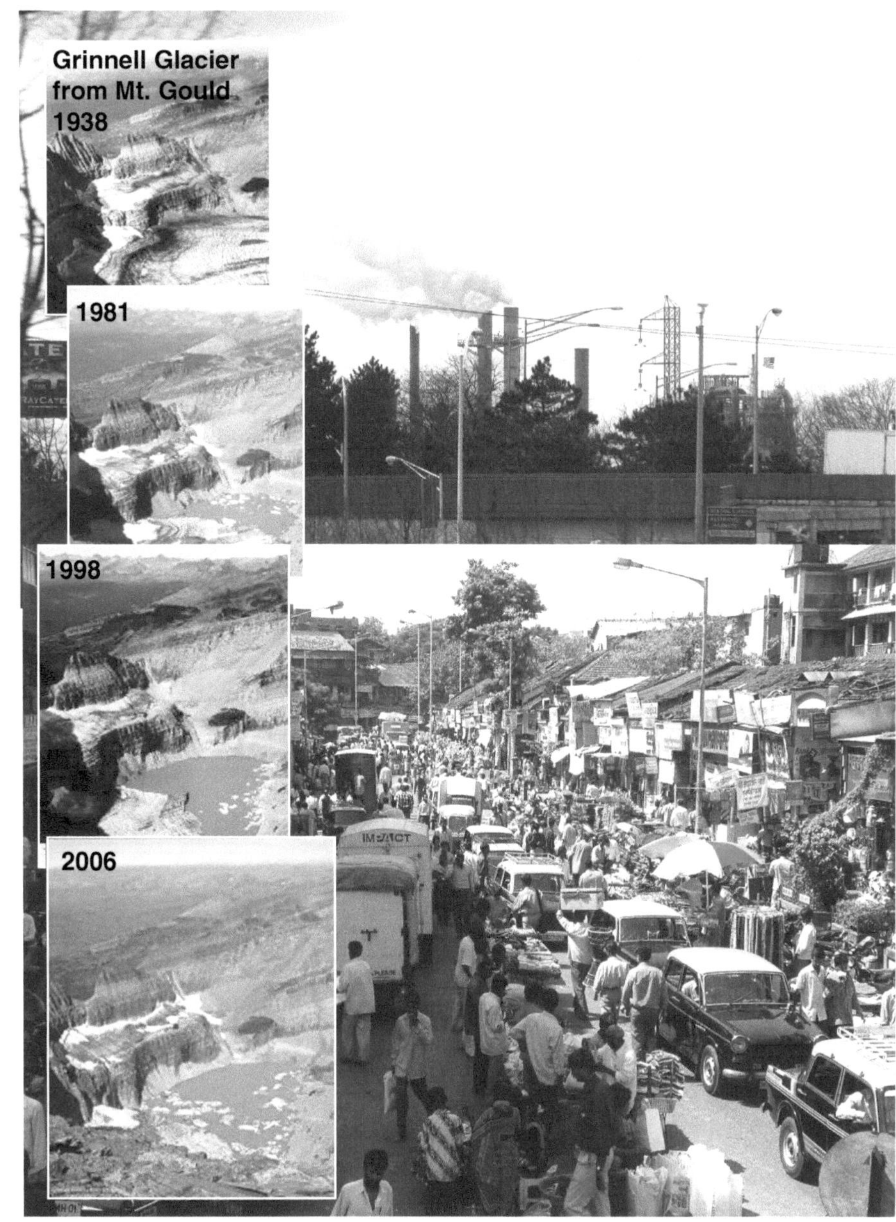


Melting Grinnell Glacier from Mount Gould (1938–2006) responding to climate change and images of industrial plant and road traffic (suburb region of Mumbai) which are major sources of greenhouse gases. (Source: UNEP 2007; Rakesh Kumar 2007 personal communication.)

The increasing body of observations has shown that we are faced with a warming world and other changes in the climate system brought about by increasing anthropogenic (human induced) pollution of our atmosphere. The global average surface temperature has increased over the twentieth century by about 0.6 °C (IPCC 2001a). The Fourth Assessment Report (AR4) of the International Panel on Climate Change (IPCC) has re-estimated the global average surface temperature in light of new research and derived a slightly higher warming magnitude due to inclusion of several particularly warm years in this century (2003, 2004, 2005 and 2006) (IPCC 2007; Meehl et al. 2007). The changes in climate, with global warming as their major characteristic feature, are known to occur as a result of internal variability within the climate system and external factors (both natural and anthropogenic). Based on the conclusions derived by IPCC in their Third Assessment Report or TAR in short form (IPCC 2001b) and now in AR4 (IPCC 2007), there is new and stronger evidence that most of the warming observed over the last 50 years is attributable to human activities. As a result of these anthropogenic activities, concentrations of atmospheric greenhouse gases and their radiative forcing have continued to increase since 1750 (the pre-industrial era), thus leading to global warming and significant environmental consequences. Now the emissions of greenhouse gases and aerosols are of major environmental concern throughout the world as an issue of global air pollution.

There is a large amount of literature available on the topics of climate change and global air pollution. As the importance of these topics is increasingly recognized, this body of published works is also continually growing. Inevitably, this chapter can only discuss and show results from selected pieces of published work and it does not attempt to provide a comprehensive synthesis of this area which can be found in works such as that published by IPCC (e.g. IPCC 2001b, IPCC 2007). Consequently, this chapter features some of the key results reported recently by IPCC as well as those in the wider literature, for example, on changes in atmospheric composition.

## 4.1 Basic Concepts of Climate Change

Climate directly influences our environment, ecosystems and our way of life. Over the past decade it has become clearer that we are also having a direct influence on the delicate balance of our climate and causing it to change. The consequences of this 'climate change' in terms of environmental, economic and societal impacts are now becoming more apparent. In order to understand the causes of climate change and how it may be reduced, a large number of components and interactions have to be considered. These are schematically represented in Figure 4.1 (CCSP 2003).

The following section provides a brief explanation of key terms and concepts that are generally associated with global air pollution and climate change.

### Greenhouse Gases and Aerosols

Greenhouse gases (GHGs) are those gaseous constituents of the atmosphere, both natural and anthropogenic, that absorb and emit radiation at specific wavelengths within the spectrum of infrared radiation emitted by the earth's surface, the atmosphere and clouds. This radiative property of these gases causes the greenhouse effect. Water vapour ($H_2O$), carbon dioxide ($CO_2$), nitrous oxide ($N_2O$), methane ($CH_4$) and ozone ($O_3$) are the primary greenhouse gases in the earth's atmosphere. Moreover, there are a number of entirely

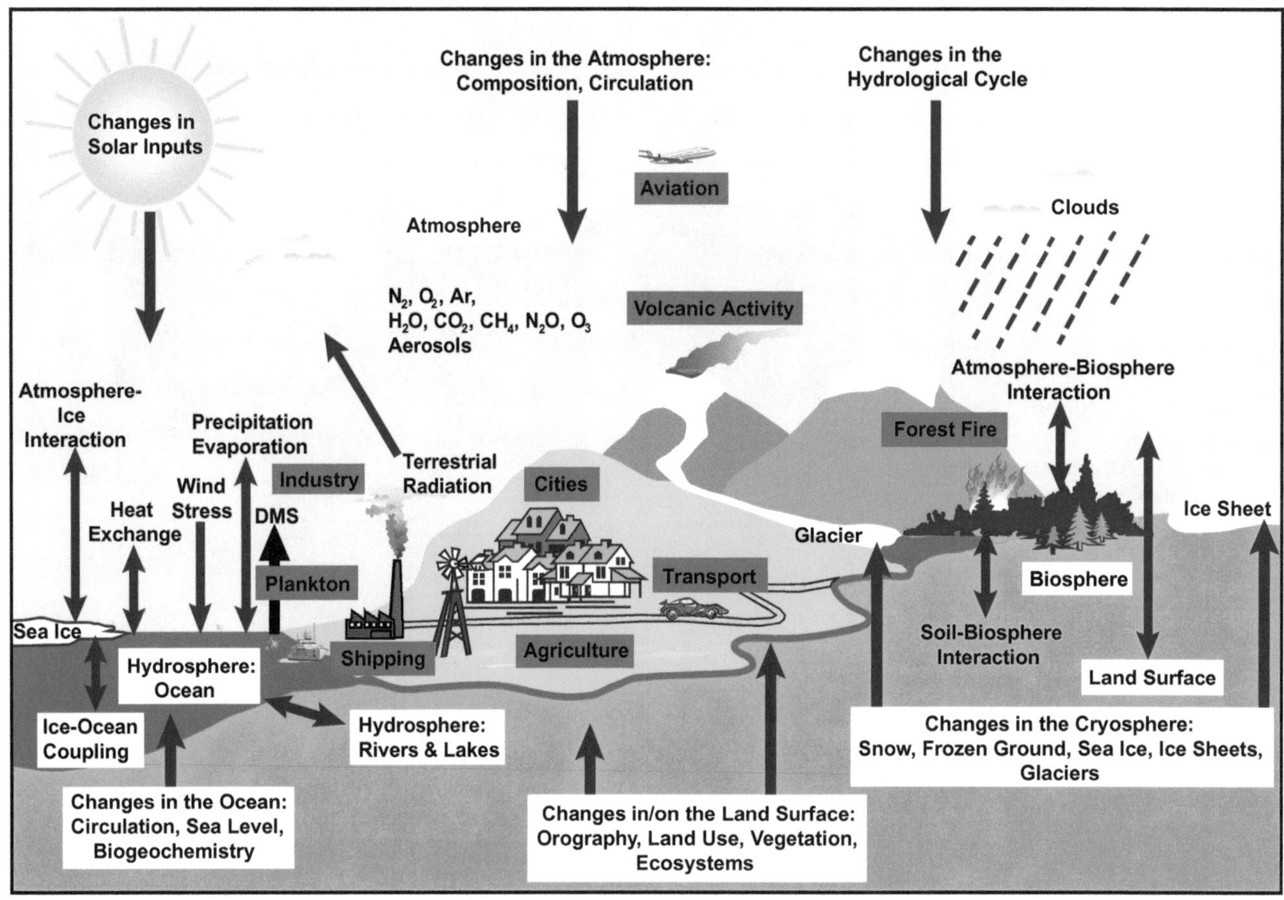


Figure 4.1 Major components of the climate system and their interactions. The arrows indicate the interactions between the subsystems (white boxes) and the red outlined boxes indicate the main emission sources (Source: IPCC 2007; see also CCSP 2003). Chemical species shown in the figure are nitrogen ($N_2$), oxygen ($O_2$), argon (Ar), water ($H_2O$), carbon dioxide ($CO_2$), methane ($CH_4$), nitrous oxide ($N_2O$), ozone ($O_3$) and dimethylsulphide (DMS).

human-made greenhouse gases in the atmosphere, such as the halocarbons and other substances containing chlorine and bromine, which are dealt with under the Montreal Protocol. Beside $CO_2$, $N_2O$ and $CH_4$, the Kyoto Protocol deals with the greenhouse gases sulphur hexafluoride ($SF_6$), hydrofluorocarbons (HFCs) and perfluorocarbons (PFCs). $CO_2$ has the most significant effect on the global climate change, accounting for 63 per cent of the global warming effect, with atmospheric lifetime of 50 to 200 years and being well mixed on the global scale. The gas $CH_4$ takes the second place for its warming effect.

Greenhouse gases effectively absorb infrared radiation which is emitted by the earth's surface, the atmosphere itself and by clouds. Atmospheric radiation is emitted to all sides, including downward to the earth's surface. Thus greenhouse gases trap heat within the surface-troposphere system. This is called the natural greenhouse effect. If the atmospheric concentration of greenhouse gases increases due to human activities, this can lead to an increased infrared opacity of the atmosphere, and therefore to an effective radiation transfer into space from a higher altitude at a lower temperature. This causes a radiative forcing, an imbalance that can only be compensated for by an increase in the temperature of the surface-troposphere system. This is the enhanced or anthropogenic greenhouse effect (Baede et al. 2001).

The major sources of anthropogenic aerosols are fossil fuel and biomass burning. These sources are linked to degradation of air quality and acid deposition. Atmospheric aerosols such as sulphate aerosols can reflect solar radiation, which leads to a cooling tendency in the climate system, while black carbon (soot) aerosols tend to warm the climate system because they can absorb solar radiation. In most cases, tropospheric aerosols produce a cooling effect, with a much shorter lifetime (days to weeks) than most greenhouse gases. As a result, their climate effects are short-lived and regional in scale.

## Radiative Forcing

The term 'radiative forcing' has been employed in the IPCC Assessment to denote an externally imposed perturbation in the radiative energy budget of the earth's climate system. Such a perturbation can be brought about by secular changes in concentrations of the radiatively active species (e.g. $CO_2$ and aerosols), changes in the solar irradiance incident on the planet or other changes that affect the radiative energy absorbed by the surface (e.g. changes in surface reflection properties). This imbalance in the radiation budget has the potential to lead to a change in climate parameters and thus result in a new equilibrium state of the climate system (IPCC 2001a).

As pointed out above, increases in the concentrations of greenhouse gases will reduce the efficiency with which the earth's surface radiates to space. More of the outgoing terrestrial radiation from the surface is absorbed by the atmosphere and re-emitted at higher altitudes. This results in a positive radiative forcing that tends to warm the lower atmosphere and surface, because less heat escapes to space. The amount of radiative forcing depends on the size of the increase in concentration of each greenhouse gas, the radiative properties of the gases involved, and the concentrations of other greenhouse gases already present in the atmosphere. In contrast, atmospheric aerosols mostly produce negative radiative forcing. Figure 4.2 gives the global mean radiative forcing of the climate system for the year 2005, relative to 1750. The Figure shows the global-average radiative forcing (RF) estimates in 2005 for anthropogenically emitted gases, $CO_2$, $CH_4$, $N_2O$, as well as $O_3$, aerosols and other important agents. An indication is given below of the confidence that exists in the level of scientific understanding. Long-lived species can be distributed on global scales whereas aerosols tend to exhibit variations over local to regional (or continental) scales. Ozone is naturally present in the stratosphere (see Chapter 5) and has a global impact but is also produced in the troposphere as a result of precursor species (nitrogen oxides, carbon monoxide and volatile organic compounds) being transported and transformed over regional scales. There is now a high level of scientific understanding of how the long-lived species such as $CO_2$ and $CH_4$ influence our climate, but this level decreases in the case of ozone and aerosols. The level of understanding of how cloud-aerosol interactions, linear contrails and solar irradiance impact on climate change is particularly low (Solomon et al. 2007).

The radiative forcing due to the increase of the well-mixed long-lived greenhouse gases ($CO_2$, $CH_4$, halocarbons and $N_2O$) from 1750 to 2005 is estimated to be 2.64 $Wm^{-2}$. In the case of $CO_2$, the radiative forcing has increased by + 13% between 1998 to 2005 (Forster et al. 2007). Direct radiative forcing of aerosols is estimated to be − 0.5 $Wm^{-2}$ and results in a cooling effect whereas for black carbon it results in a warming effect. The IPCC has stated that there is a 'very high confidence' that the overall net radiative forcing caused by human activity since 1750 is positive 1.6 $Wm^{-2}$ (range of + 0.6 to + 2.4 $Wm^{-2}$).

## Climate Change and Variability

Climate change in IPCC usage refers to any change in climate over time, whether due to natural variability or as a result of human activity. This usage differs from that in the United Nations Framework Convention on Climate Change (UNFCCC), where climate change refers to a change of climate that is attributed directly or indirectly to human activity that alters the composition of the global atmosphere and that is in addition to natural climate variability observed over comparable time periods. Any human-induced changes in climate will be embedded in a background of natural climatic variations that occur on a whole range of time- and space-scales. Climate variability can occur as a result of natural changes in the forcing of the climate system, for example, variations in the strength of the incoming solar radiation and changes in the concentrations of aerosols arising from volcanic

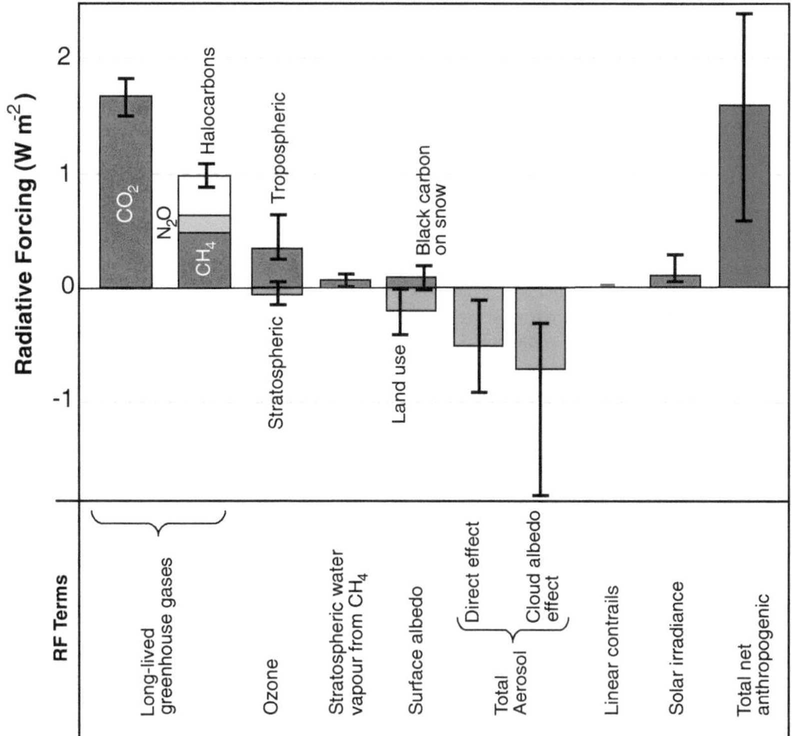


**Figure 4.2** Comparison of main global mean radiative forcing (RF) estimates and ranges in 2005 relative to 1750 for anthropogenically derived carbon dioxide ($CO_2$), methane ($CH_4$), nitrous oxide ($N_2O$), black carbon (BC), ozone ($O_3$), aerosols, land use and linear contrails. It also shows RF associated with natural solar irradiance along with the net anthropogenic radiative forcing. Contributions to natural forcing associated with aerosols from volcanoes are not included in this figure due to their episodic nature. Range for linear contrails does not include other possible effects of aviation on cloudiness (adapted from Solomon et al. 2007).

eruptions. Natural climate variations can also occur in the absence of a change in external forcing, as a result of complex interactions between the atmosphere and the ocean. The El Niño-Southern Oscillation (ENSO) phenomenon is an example of such natural 'internal' variability on inter-annual timescales. To distinguish anthropogenic climate changes from natural variations, it is necessary to identify the anthropogenic 'signal' against the background 'noise' of natural climate variability.

## 4.2 Global Emission Sources and Sinks

### Global Trends of Greenhouse Gas Emissions

The global percentage share of the main GHG emissions is shown in Figure 4.3a (Olivier et al. 2005, 2006). $CO_2$ makes up 77% of the total anthropogenic emissions followed by $CH_4$ (14%), $N_2O$ (8%) and then fluorinated gases (1%). Fossil fuels are the dominant source of GHG such as $CO_2$ (Olivier et al. 2006). As discussed by Solomon et al. (2007), since the pre-industrial era, anthropogenic $CO_2$ has dominated the radiative forcing relative to all other agents. When total GHG emissions are compared for each source sector (Figure 4.3b) the largest contributing sectors are the energy supply (26%), industry (19%), forestry (17%) and total transport (13%). In the case of global emissions of $CO_2$ since 1970, the largest growth has come from the electricity sector (see Figure 4.4). There is also a significant increase in $CO_2$ emissions from the transport sector. Other sectors show a slow growth or are relatively stable in their contribution to global $CO_2$ emissions.

The emission trends of GHGs as $CO_2$-equivalent from 1970 to 2004 are compared in Figure 4.5. Since 1970 $CO_2$ emissions from fossil energy have risen by about 90%, $CH_4$ emissions by about 40%, $N_2O$ emissions by about 50% and fluorinated gases by nearly four times. Overall, GHGs when weighted by the global warming potential, have risen by about 75% from 1970 to 2004 (Olivier et al. 2006).

### Global $CO_2$ Budget, Sources and Sinks

The atmospheric concentration of $CO_2$ has increased by 31 per cent since 1750, due to fossil fuel burning and land-use change, especially deforestation. The present $CO_2$ concentration has not been exceeded during the past 420,000 years and is likely not to have been exceeded during the last 20 million years. The current rate of increase is unprecedented during at least the past 20,000 years. The measurements made by two baseline stations at Mauna Loa, Hawaii and Antarctica have shown a rapid increase in the atmospheric $CO_2$ concentration from 305 ppm in 1957 to 368.5 ppm in 2000. Figure 4.6 shows the monthly $CO_2$ concentrations since 1960 as measured at the Mauna Loa station. Seasonal variations in the concentrations are superimposed on the general increasing trend of the levels. During 2001, mean

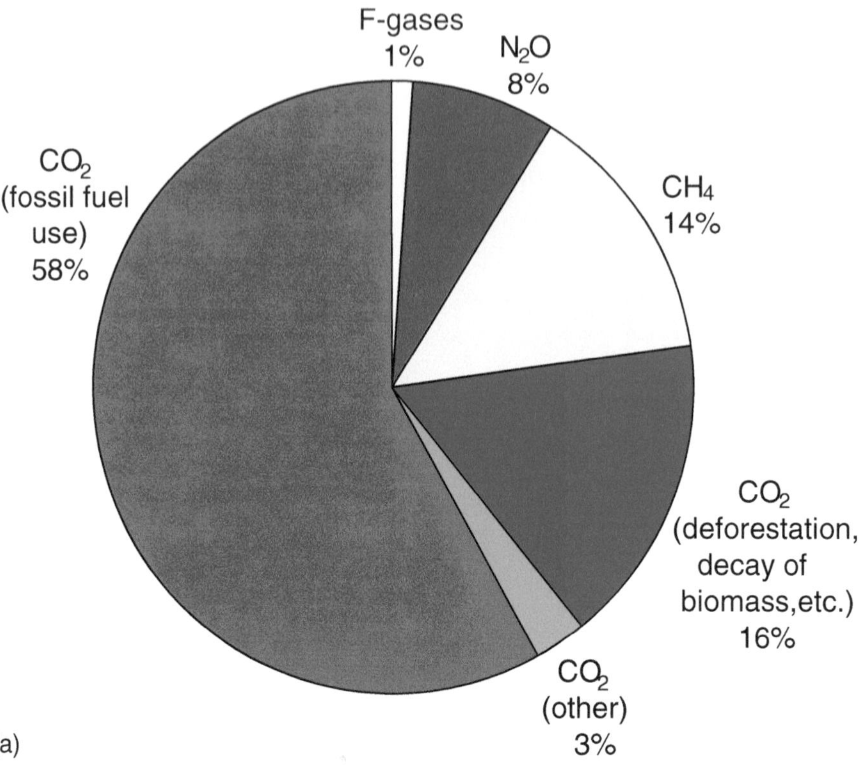


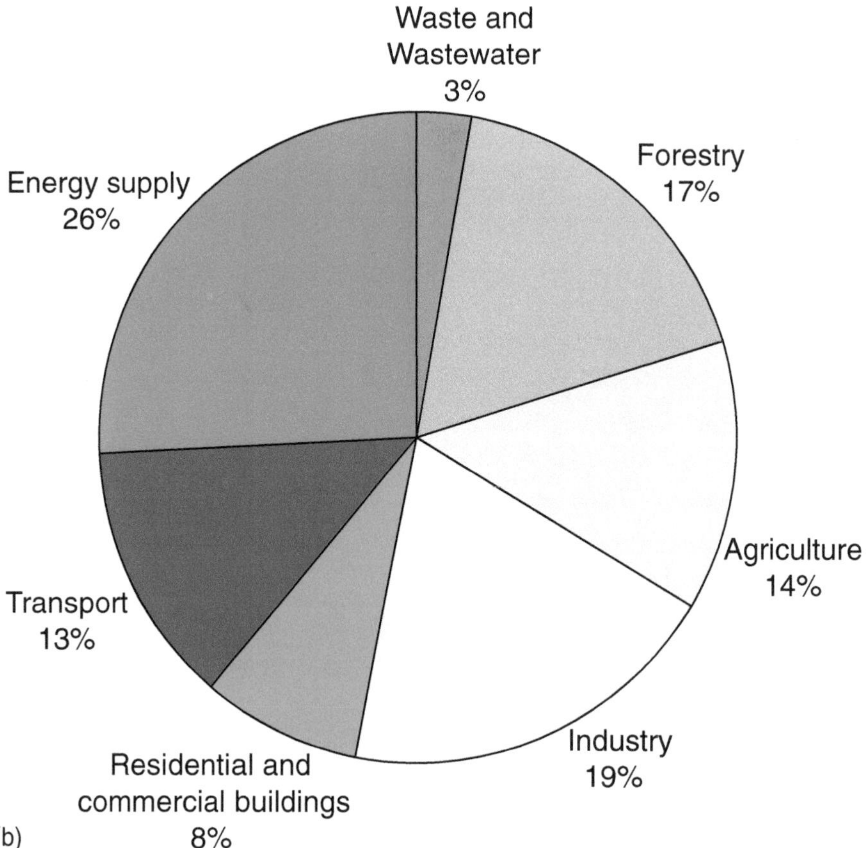


Figure 4.3 (a) Global anthropogenic greenhouse gas (GHG) emissions and (b) GHG emissions by sector for 2004. (Source: Olivier 2007 personal communication; Olivier et al. 2005, 2006.)

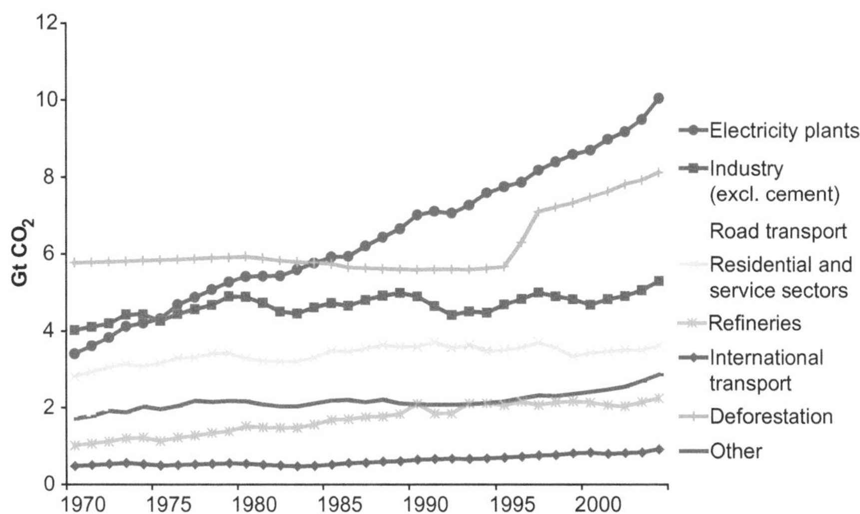


**Figure 4.4** Contributions from different source sectors to direct global anthropogenic $CO_2$ emissions between 1970 and 2004. (Source: Olivier 2007 personal communication; Olivier et al. 2005, 2006.) 'Other' includes domestic surface transport, non-energetic use of fuels, cement production and venting/flaring of gas from oil production, and 'international transport' including aviation and marine transport.

atmospheric concentrations of $CO_2$ reached 370 ppm, an increase of 1.5 ppm relative to the previous year (Meteorological Service of Canada 2003). More recent records show that the 2005 mean concentration was 379 ppm (Keeling and Whorf 2005) and reached 390 ppm in 2010 (see Figure 4.6). The distribution of atmospheric $CO_2$ measurement stations established on the different sites so far is uneven and severely under-represents the continents. This under-representation is due partly to the problem of finding continental stations where measurement will not be overwhelmed by local sources and sinks. A global monitoring programme, known as FLUXNET (http://www.eosdis.ornl.gov/FLUXNET), has been developed to provide an improved database to support related carbon research (Baldocchi et al. 2001).

The Waliguanshan baseline station, located in Tsinghai-Xijiang plateau (Tibetan plateau), with an elevation of about 3500 m above sea level, also measures a similar variation of the atmospheric $CO_2$ concentration. Palaeo-atmospheric data from ice cores and firn for several sites in Antarctica and Greenland, supplemented with the data from direct atmospheric samples over several decades, reveals the concentration changes occurring in earlier millennia for $CO_2$ including an unprecedented growth over the industrial era since 1750 (e.g. Etheridge et al. 1996; Monnin et al. 2001; Monnin et al. 2004; Siegenthaler et al. 2005; Tans and Convay 2005; Keeling and Whorf 2005; MacFarling Meure et al. 2006). Many of these studies have been synthesized in the latest IPCC report on Palaeoclimate (Jansen et al. 2007).

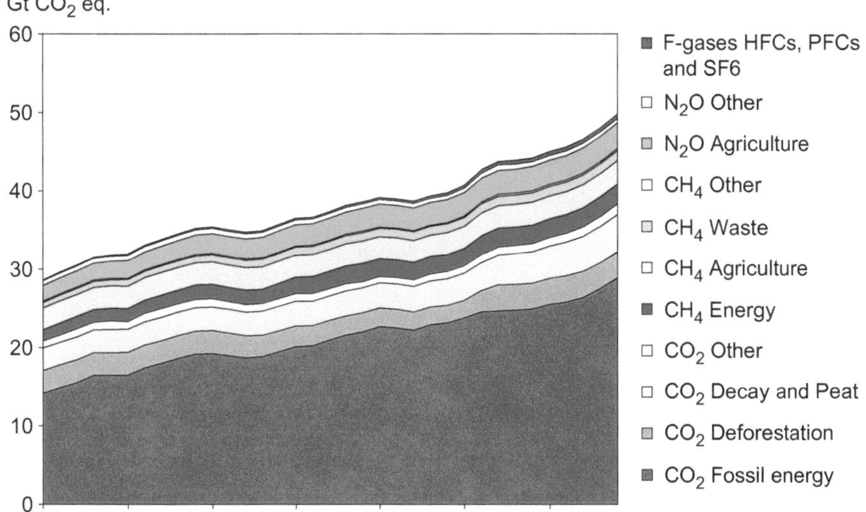


**Figure 4.5** Contributions from different source sectors to direct global anthropogenic greenhouse gas (GHG) emissions between 1970 and 2004 in units of Gt $CO_2$-equivalent. (Source: Olivier 2007 personal communication; Olivier et al. 2005, 2006.)

The issue of $CO_2$ sources and sinks is most important in the global $CO_2$ budget. However, the estimates made by various investigators are not fully consistent. One key problem is the lack of reliable and quantitative data for uptake of $CO_2$ by the terrestrial ecosystem. The major sources of $CO_2$ are burning of fossil fuels, land-use change and industrial production. Figure 4.7 shows the global anthropogenic emissions for $CO_2$ for the year 2000. The figure shows the spatial distribution of $CO_2$ emissions across the continents along with the emissions from the major shipping lanes.

Table 4.1 shows a comparison of global $CO_2$ budgets from the IPCC estimate (TAR) with previous IPCC estimates (SAR) (IPCC 2001a). The table is also updated with data from the IPCC AR4 report (Denman et al. 2007). SRLULUCF represents the estimates from IPCC Special Report on Land Use, Land Use Change and Forestry (IPCC 2000) and SRRF represents the estimates from the Special Report on Radiative Forcing (Schimel et al. 1995). It can be seen that for $CO_2$ sources of fossil fuel and industry, the total emission was on average $5.4 \pm 0.3$ PgC/yr (1 ton carbon=3.7 tons $CO_2$), during the 1980s while during the 1990s the emission rate increased to $6.3 \pm 0.4$ PgC/yr. The emission due to land-use change (e.g. deforestation) was 1.7 PgC/yr (0.6 to 2.5), with a large uncertainty (on average, 1.0 PgC/yr). The estimate of ocean-atmosphere flux was $-1.9 \pm 0.6$ PgC/yr; therefore the ocean is an important sink region, with much less uncertainty than the sources of land-use change. The land-atmosphere flux is estimated as residuals of the sum of the above terms (i.e. the difference between the total emission sources and the ocean uptake). Thus, the resulting estimate of land-atmosphere flux includes all the measurements and computational errors. Its flux was $-0.2 \pm 0.7$ PgC/yr during the 1980s. During the 1990s this flux increased significantly. If one partitions the land-atmosphere flux into the part of land-use change (source term) and the part of the uptake term by terrestrial ecosystem (sink term), $CO_2$ emission due to land-use change would be 1.7 (0.6 to 2.5), and the terrestrial sink would be $-1.9$ ($-3.8$ to 0.3). However, this part of the $CO_2$ sink is the most uncertain. On the other hand, $3.2 \pm 1$ (for 1980s) or $3.3 \pm 0.1$ (for 1990s) PgC/yr have been estimated to have entered the atmosphere to increase the atmospheric $CO_2$ content. As indicated the uncertainty is relatively small, only with 0.1 PgC/yr.

## Global Budget, Sources and Sinks of Methane, Nitrous Oxide and Fluorinated Gases

As with significant increase in $CO_2$ levels, the atmospheric concentrations of methane ($CH_4$) and nitrous oxide ($N_2O$) have also increased. Methane has increased by more than 1060 ppb (150 per cent) since 1750. There is now sufficient confidence to conclude that the present $CH_4$ concentrations have not been exceeded during the past 420,000 years. The atmospheric concentration of nitrous oxide ($N_2O$) has increased by 16 per cent since 1750 to 319 ppb in 2005 (Denman et al. 2007) and also continues to increase. The present $N_2O$ concentration

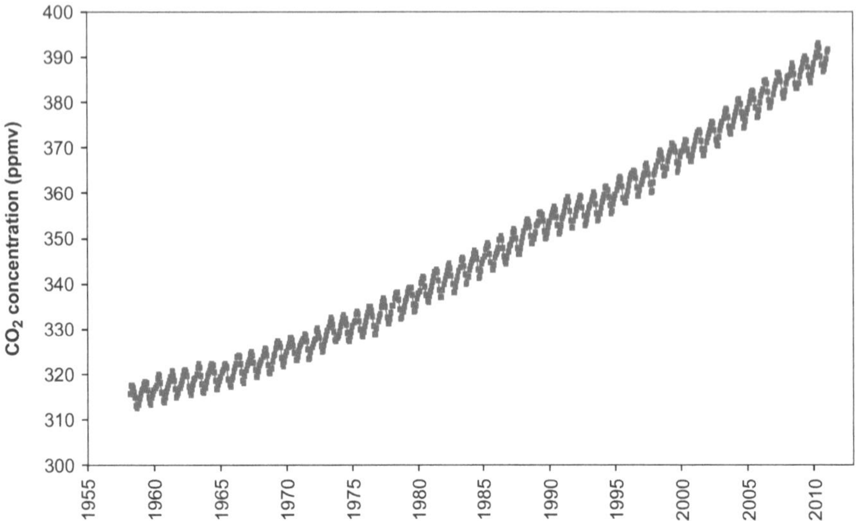


**Figure 4.6** Atmospheric $CO_2$ concentration measured by baseline stations in Hawaii, Mauna Loa. (Source: http://cdiac.ornl.gov/trends/co2/ and http://www.esrl.noaa.gov/gmd/ccgg/trends/.)

has not been exceeded during at least the past thousand years. Slightly more than half of current $CH_4$ emissions are anthropogenic (e.g. use of fossil fuels, cattle, rice agriculture and landfills). About a third of current $N_2O$ emissions are anthropogenic (e.g. agricultural soil, cattle feed lots and chemical industry).

The global distribution of anthropogenic $CH_4$ emissions is shown in Figure 4.8 for the year 2000. Over the recent years new data has become available to enable $CH_4$ source emissions to be recalculated and now estimates of total global pre-industrial emissions of $CH_4$ are in the range of 200 to 250 Tg/yr (see Denman et al. 2007 and references therein). Emissions from natural $CH_4$ sources (of which wetlands is the largest) account for the dominant fraction of between 190 and 220 Tg/yr with the anthropogenic sources (rice agriculture, livestock, biomass burning and waste) contributing to the remaining amount. Anthropogenic emissions, however, dominate the present-day $CH_4$ budgets and consist of more than 60% of the total global budget. IPCC (AR4) has given the total estimate of $CH_4$ emissions as 582 Tg ($CH_4$)/yr (Denman et al. 2007) for the period 2000–2005.

The removal of methane is accomplished mainly through chemical destruction processes. It can react with tropospheric hydroxyl radical (OH) through the reaction $OH+CH_4 \rightarrow CH_3+H_2O$ and then disappears.

With regard to $N_2O$ emissions, agriculture is the single largest anthropogenic source with changes caused by land-use practices (Smith and Conen 2004; Del Grosso et al. 2005; Neill et al. 2005). The total estimated emissions are 17.7 Tg(N)/yr with anthropogenic contributions being 6.7 Tg(N)/yr (Denman et al. 2007). Other estimates are consistent with this value (e.g. Hirsch et al. 2006). Natural sources of $N_2O$ come from the ocean, atmosphere and soils, and the anthropogenic sources result from agricultural soils, biomass burning, stationary and mobile combustion, industrial production (e.g. nylon and nitric acid production) and cattle feed lots. The main sink is stratospheric loss through photochemical dissociation. The global map of anthropogenic $N_2O$ emissions for 2000 is shown in Figure 4.9.

Since 1995 the atmospheric concentration of many of those halocarbon gases that are both ozone-depleting and greenhouse gases (e.g. $CFCl_3$ and $CF_2Cl_2$) has been either increasing more slowly or decreasing, both in response to reduced emissions under the regulations of the Montreal Protocol and its Amendments. The total amount of $O_3$, in the troposphere however, is estimated to have increased by 136 per cent since 1950, due primarily to anthropogenic emissions of several $O_3$-forming gases.

Fluorinated gases, such as, hydrofluorocarbons (HFCs), perfluorocarbons (PFCs) and sulphur hexafluoride (SF6) have relatively low emissions, as shown previously in Figure 4.5, but their lifetime and radiative forcing in the atmosphere is very large. So their role as the greenhouse gases cannot be underestimated. In addition, the emission increase of these F-gases is quite rapid, in particular for HFCs. In 1995, HFCs emission increased most rapidly. $SF_6$ emissions from the electricity sector are

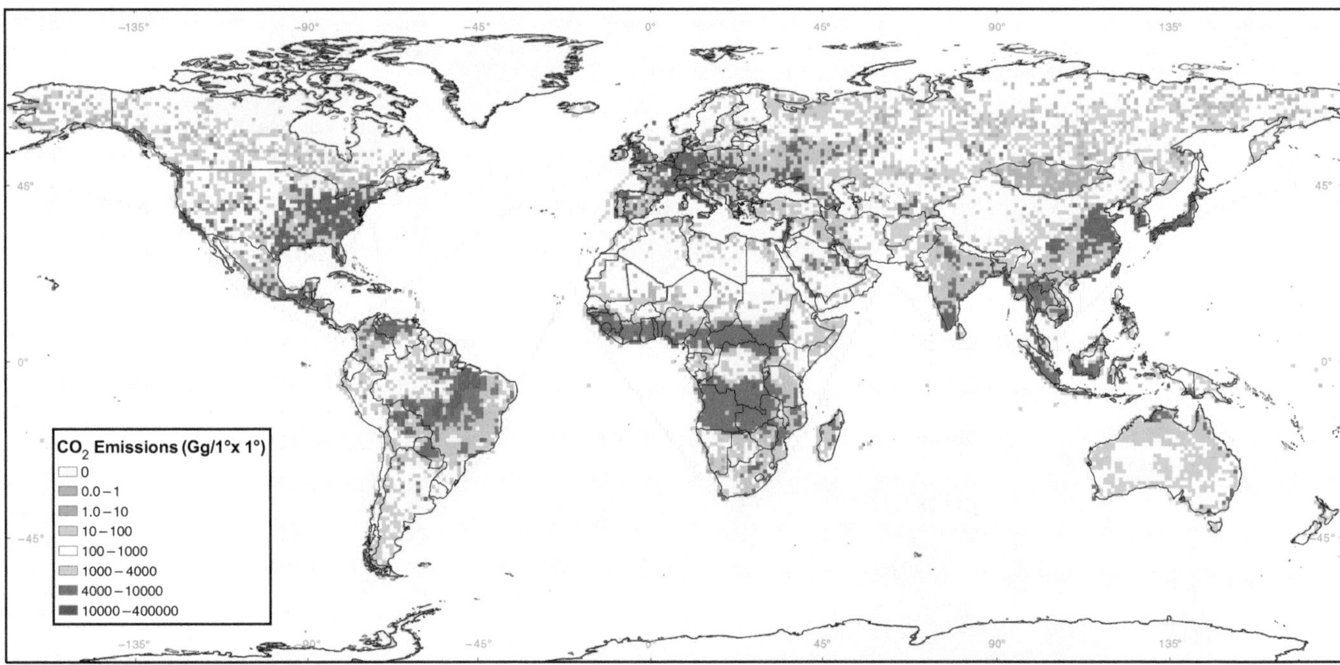


**Figure 4.7** Global map of anthropogenic $CO_2$ emissions for the year 2000. Emissions from the major shipping lanes are also shown. (Source: EDGAR 3.2 Fast Track 2000 database.) The Emissions Database for Global Atmospheric Research (EDGAR; http://edgar.jrc.it) is a joint effort of the European Commission Joint Research Centre, and the Netherlands Environmental Assessment Agency (MNP).

**Table 4.1** Comparison of the global $CO_2$ budgets from IPCC estimates (units are PgC/yr) (IPCC 1996; IPCC 2001a; Denman et al. 2007). The data has been extracted from the IPCC Second Assessment Report (SAR), Third Assessment Report (TAR), Fourth Assessment Report (AR4), Special Report on Land Use, Land Use Change and Forestry (SRLULUCF) and the Special Report on Radiative Forcing (SRRF).

| | 1980s | | | | 1990s | 1989 to 1998 | 2000 to 2005 |
|---|---|---|---|---|---|---|---|
| | TAR | SRLULUCF | SAR | SRRF | TAR | SRLULUCF | AR4 |
| Atmospheric increase | 3.3 ± 0.1 | 3.3 ± 0.1 | 3.3 ± 0.1 | 3.2 ± 0.1 | 3.2 ± 0.1 | 3.3 ± 0.1 | 4.1 ± 0.1 |
| Emissions (fossil fuel, cement) | 5.4 ± 0.3 | 5.5 ± 0.3 | 5.5 ± 0.3 | 5.5 ± 0.3 | 6.4 ± 0.4 | 6.3 ± 0.4 | 7.2 ± 0.3 |
| Ocean-atmosphere flux | −1.9 ± 0.6 | −2.0 ± 0.5 | −2.0 ± 0.5 | −2.0 ± 0.5 | −1.7 ± 0.5 | −2.3 ± 0.5 | −2.2 ± 0.5 |
| Land-atmosphere flux | −0.2 ± 0.7 | −0.2 ± 0.6 | −0.2 ± 0.6 | −0.3 ± 0.6 | −1.4 ± 0.7 | −0.7 ± 0.6 | −0.9 ± 0.6 |
| Partitioned as follows | | | | | | | |
| Land-use change | 1.7 (0.6 to 2.5) | 1.7 ± 0.8 | 1.6 ± 1.0 | 1.6 ± 1.0 | Insufficient data | 1.6 ± 0.8 | n.a. |
| Residual terrestrial sink | −1.9 (−3.8 to 0.3) | −1.9 ± 1.3 | −1.8 ± 1.6 | −1.9 ± 1.6 | | −2.3 ± 1.3 | n.a. |

becoming the largest sources of F-gases of all sources, overtaking aluminium production's number one position around 1990. HFC-23 emissions as a by-product during the manufacture of HCFC-22 (chlorodifluoromethane, a hydrochlorofluorocarbon compound which is a common refrigerant), follow the $SF_6$ trends resulting from electricity application but have tended to grow less strongly since 1990. $SF_6$ emissions from magnesium production and 'other $SF_6$ use' are also significant.

## Role of Ozone

Tropospheric $O_3$ is a direct greenhouse gas. The past increases in tropospheric $O_3$ is estimated to provide the third largest growth in direct radiative forcing since the pre-industrial era. In addition, through its chemical impact on the hydroxyl radical (OH), it modifies the lifetimes of other greenhouse gases, such as $CH_4$. Its budget, however, is much more difficult to derive than that of a long-lived gas (IPCC 2001a). The sources and sinks of tropospheric ozone are even more difficult to quantify than the burden. Influx of stratospheric air is a source. Near the ground level, photochemical production of ozone is tied to the abundance of primary pollutants, oxides of nitrogen ($NO_x$) and volatile organic compounds (VOCs). The dominant photochemical sinks for tropospheric $O_3$ are the catalytic destruction cycle involving the $H_2O + O_3$ reaction and photolytic destruction. Ozone also plays an important role near the surface where it reacts with nitric oxide (NO) which is emitted from road vehicles and other combustion processes. The ozone oxidises NO into nitrogen dioxide ($NO_2$) which itself can be photodissociated into NO and O. The oxygen atom (O) then combines with an oxygen molecule ($O_2$) to form ozone ($O_3$). The other large sink is surface loss mainly to vegetation.

## Sources of Atmospheric Aerosols

Aerosols are liquid or solid particles suspended in the air. They may be emitted from various sources or formed in the atmosphere from gaseous precursors. In general, they have two types of sources: natural sources and anthropogenic sources. The significant importance of atmospheric aerosols for climate change was realized in the late twentieth century, with direct radiative forcing and indirect radiative forcing (IPCC 2001a). Direct radiative forcing by aerosols is produced by scattering and absorption of solar and infrared radiation in the atmosphere, depending mainly on their optical characteristics. On the other hand, aerosols can alter the formation and precipitation efficiency of liquid water, ice and mixed-phase cloud formation processes by increasing droplet number concentration and ice particle concentration, thus causing changes in cloud properties, which is called indirect radiative forcing.

Table 4.2 lists the global sources for major aerosol types (IPCC 1995; Seinfeld and Pandis 1998; Satheesh and Moorthy 2005). Soil dust, volcanic emissions and sea salt (sea spray from the oceans) are natural, while anthropogenic emissions consist of industrial emissions (e.g. dust) and black carbon emissions including those from biomass burning. The aerosol precursors are indicated in the table as secondary sources. They include sulphates from biogenic gases, sulphates from $SO_2$, organic matter from biogenic VOCs and nitrates from $NO_x$.

Soil dust is a major component of aerosol loading and optical thickness, especially in subtropical and tropical regions (e.g. Zender et al. 2004; Hara et al. 2006; Goudie and Middleton 2006). Estimates of its global source strength range from 1000 to 5000 Mt/yr (IPCC 2001a; Tanaka 2007) and shows very high spatial and temporal variability. Dust source

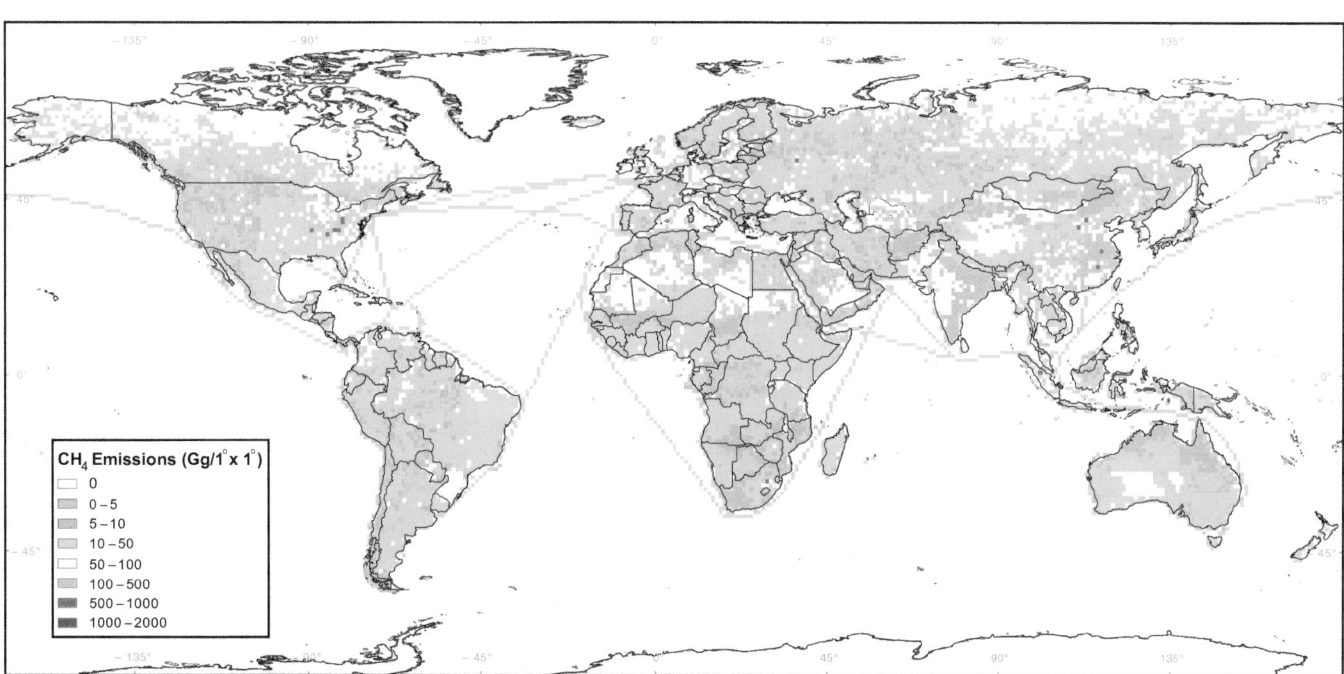


**Figure 4.8** Global map of anthropogenic $CH_4$ emissions for the year 2000. Emissions from the major shipping lanes are also shown. (Source: EDGAR 3.2 Fast Track 2000 emissions database.) The Emissions Database for Global Atmospheric Research (EDGAR; http://edgar.jrc.it) is a joint effort of the European Commission Joint Research Centre, and the Netherlands Environmental Assessment Agency (MNP).

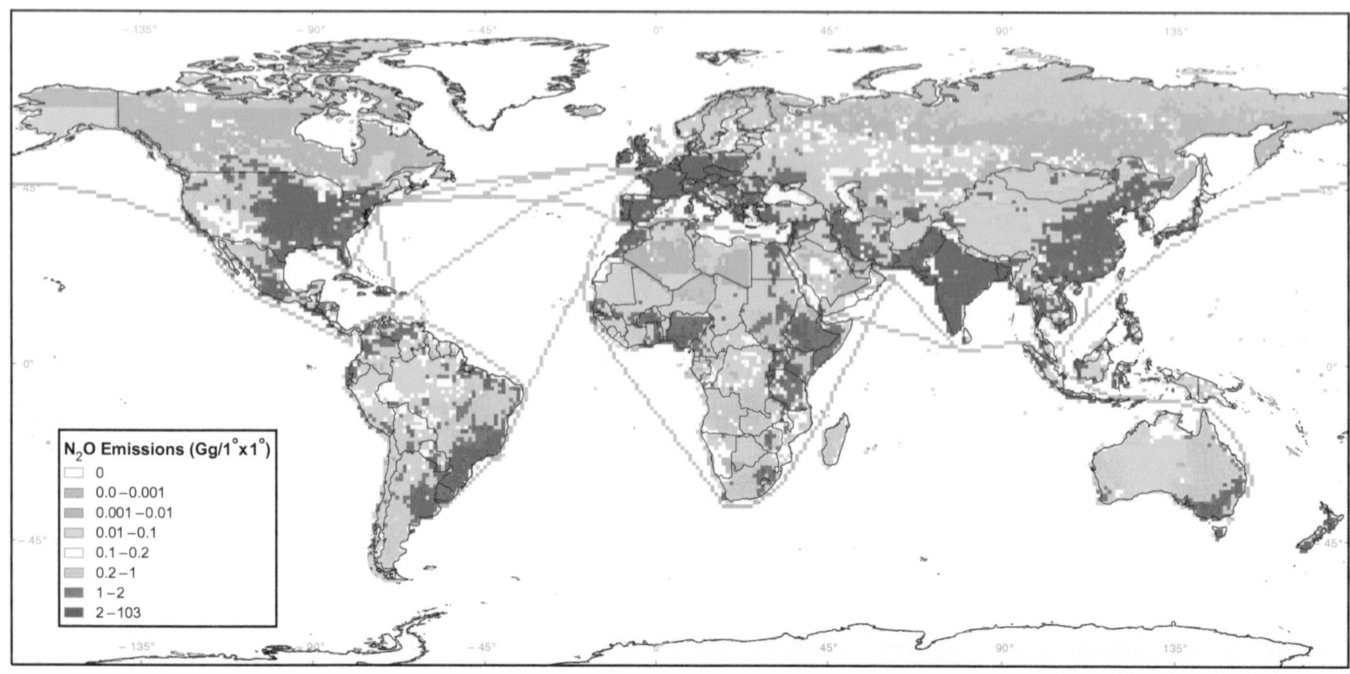


Figure 4.9 Global map of anthropogenic $N_2O$ emissions for the year 2000. Emissions from the major shipping lanes are also shown. (Source: EDGAR 3.2 Fast Track 2000 emissions database.) The Emissions Database for Global Atmospheric Research (EDGAR; http://edgar.jrc.it) is a joint effort of the European Commission Joint Research Centre, and the Netherlands Environmental Assessment Agency (MNP).

regions are mainly deserts, dry lake beds and semi-arid desert fringes, but also drier regions where vegetation has been reduced or soil surfaces have been disturbed by human activities.

Salt aerosols are generated by various physical processes, especially the bursting of entrained air bubbles during whitecap formation. For the present-day climate, the total sea salt flux from ocean to atmosphere is estimated to be 3300 Tg/yr (IPCC 2001a).

Volcanic aerosol emission consists of solid particles (ash) and gases (mainly $H_2O$ vapour, $SO_2$, and $CO_2$) in very variable concentrations. $SO_2$ is a precursor to aerosol formation by gas to particle conversion. Most of the known active volcanoes are in the Northern hemisphere.

Natural organic aerosols, represented by natural emission of particulate organic carbon (POC), are produced by marine and continental sources. The ocean is a large source of POC owing to the injection of naturally derived marine surfactants from bubble-bursting processes, while terrestrial sources of primary POC include natural products emitted from vegetation. POC may also be emitted by combustion processes.

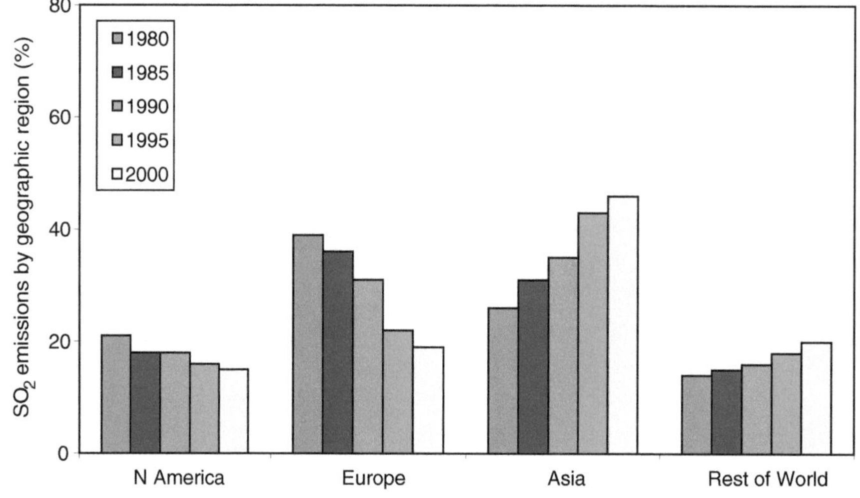


Figure 4.10 Percentage global anthropogenic sulphur emissions for four regions of the world. (Source: Smith et al. 2001.)

Table 4.2 Global sources for major aerosol types (e.g. IPCC 1995; Seinfeld and Pandis 1998; Satheesh and Moorthy 2005).

| | | |
|---|---|---|
| Natural | Primary | Soil dust (mineral aerosol), sea salt, volcanic dust, primary organic aerosols |
| | Secondary | Sulphates from biogenic gases, sulphates from volcanic $SO_2$, organic matter from biogenic VOCs, nitrate from $NO_x$ |
| Anthropogenic | Primary | Industrial dust (except soot), black carbon (includes biomass burning) |
| | Secondary | Sulphate from $SO_2$, nitrates from $NO_x$, organic matter from biogenic VOCs |

Among the anthropogenic aerosol sources is the industrial dust which originates from the incombustible material present as inorganic impurities in fuel (mainly coal and oil) during the combustion process and incomplete fuel combustion.

Carbonaceous aerosols consist predominantly of organic substances and various forms of black carbon. Their main sources are biomass and fossil fuel burning, and atmospheric oxidation of biogenic and anthropogenic volatile organic compounds (VOCs). Often a distinction is made in size classes, that is, smaller than 10 $\mu m$ aerodynamic diameter ($PM_{10}$) and smaller than 2.5 $\mu m$ ($PM_{2.5}$). These types of aerosols may significantly affect visibility and human health.

Anthropogenic and natural sulphate aerosols are produced by chemical reactions in the atmosphere from gaseous precursors (with the exception of sea salt sulphate and gypsum dust particles). The global distribution of sulphate aerosols results from anthropogenic $SO_2$ and from natural sources, primarily dimethylsulphide (DMS). It should be pointed out that anthropogenic sulphur emissions for Europe and North America are declining as shown in Figure 4.10, but those for Asia have been increasing significantly over the past decade, resulting in relatively constant global emissions during recent decades (Smith et al. 2001). Recent work of Stern (2006) generally support this trend although the study indicates a decline in global sulphur emissions over the last decade. According to Smith et al. (2005), the emissions of sulphur will continue to decrease in Europe and North America but will peak in Asia, Africa and South America in around 2020 before decreasing.

Owing to the short lifetime of aerosol particles in the troposphere and the non-uniform distribution of sources, their geographical distribution is also highly non-uniform and the resulting radiative forcing is short-lived and localized.

Most aerosols are also important pollutants in the atmosphere, as particulate matter (PM) suspended in the air. One of the most significant impacts of these pollutants is on human health. Since the turn of the new century, focus has been attached to the atmospheric brown cloud (ABC) phenomenon, which was first detected during the Indian Ocean Experiment (Ramanathan et al. 2001). Pollution from the region is forming a thick layer of aerosols high up in the atmosphere and this covers a large area of Asia, including the Indian Ocean, South East Asia and China (Kuylenstierna and Hicks 2002). Figure 4.11 (http://earthobservatory.nasa.gov/) shows a MODIS image of a layer of thick brownish haze over north of India. The haze consists of aerosol particles all along the southern edge of the Himalayan Mountains, and streaming southward over Bangladesh and the Bay of Bengal. In contrast the air over the Tibetan Plateau to the north of the Himalayas is very clear. Studies have shown that the aerosols associated with the haze not only represent a health hazard but can also have a significant impact on the region's hydrological cycle and climate (e.g. Meywerk and Ramanathan 2002).

## 4.3 Climate Consequences and Environmental Effects of Greenhouse Gases and Aerosols

### Greenhouse Effect and Global Warming

The energy source that drives the climate system is radiation emitted from the sun. Figure 4.12 shows schematically the radiation balance of the earth (IPCC 2007; Kiehl and Trenbreth 1997). Each square metre at the top of the earth's atmosphere receives about 342 watts (W) of incoming solar radiation, averaged for the whole globe for an entire year. About 31 per cent (107 $Wm^{-2}$) of this amount of solar energy is reflected back to space by clouds, the atmosphere and the earth's surface. The remaining 235 $Wm^{-2}$ is partly absorbed by the atmosphere (67 $Wm^{-2}$), but most (168 $Wm^{-2}$) warms the land and ocean surface. For a stable climate, a balance is required between incoming solar radiation and the outgoing radiation emitted by the climate system. Therefore, the climate system itself must radiate on average 235 $Wm^{-2}$ back into space, but all this heat is radiated as infrared back to outer space due to low temperature. Most of it is absorbed by molecules of greenhouse gases (including water vapour) and clouds in the lower atmosphere. These re-radiate the energy in all directions, some back towards the surface and some upwards, where other molecules higher up can absorb the energy again. This process of absorption and re-emission is repeated until, finally, the energy does escape from the atmosphere to space. However, because much of the energy has been recycled downwards, the surface temperature becomes much warmer than if the greenhouse gases were absent from the atmosphere. As mentioned previously, this natural process is known as the greenhouse effect. Without greenhouse gases, the earth's average temperature would be −19 °C instead of + 14 °C, or 33 °C colder.

However, various human activities have been increasing atmospheric concentrations of key greenhouse gases such as $CO_2$, $CH_4$ and $N_2O$ on a global scale. The increased atmospheric concentrations of greenhouse gases are absorbing more infrared energy and causing a progressive warming of the earth's lower atmosphere. The portion of the warming caused by human activities is often called the 'anthropogenic' or 'enhanced' greenhouse effect in contrast to the natural greenhouse effect.

### Observed Global Climate Change and Its Impact

The observed variations of the temperature indicators and the hydrological and storm-related indicators at the global scale provide a measure

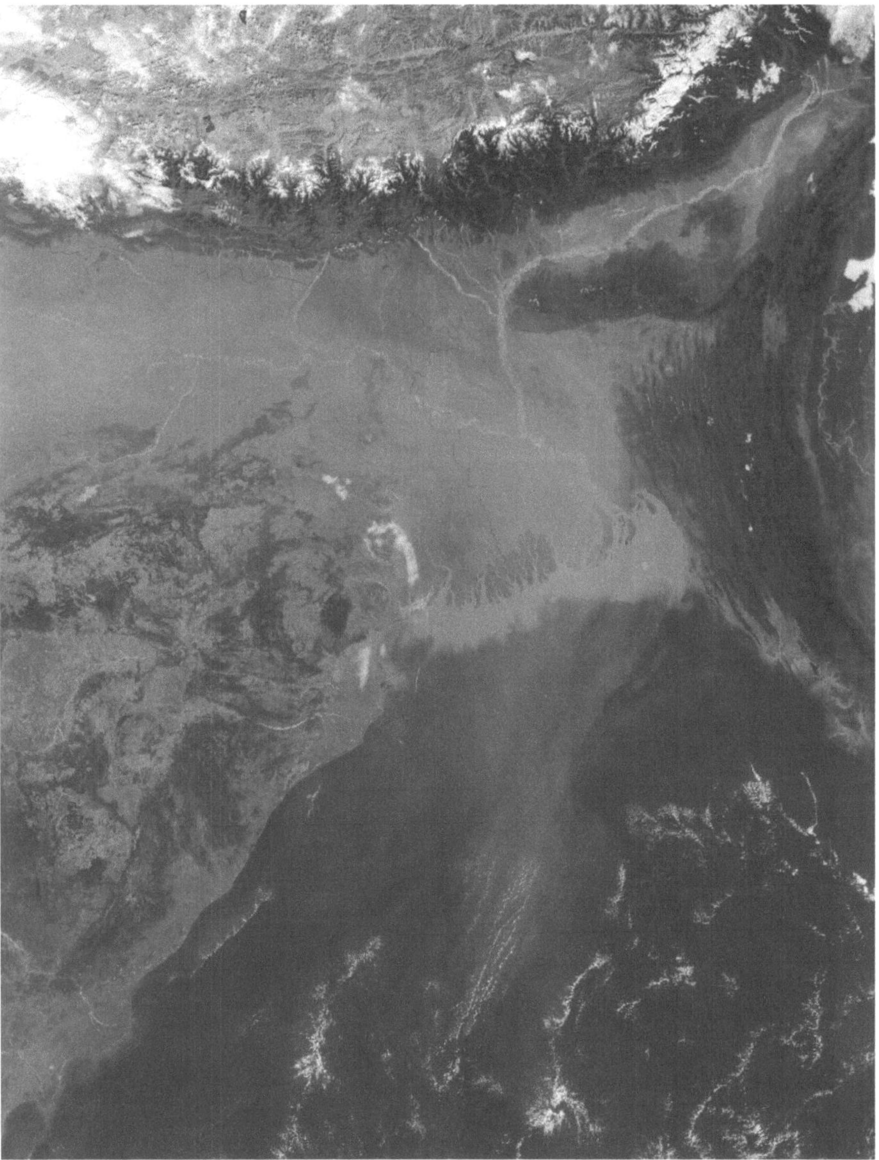

Figure 4.11 Satellite image of pollution haze over North India. This true-color image of 14 January 2002, was generated by the Moderate-resolution Imaging Spectroradiometer (MODIS), flying aboard NASA's *Terra* satellite. Image courtesy of Jacques Descloitres, *MODIS Land* Rapid Response Team at NASA GSFC (http://earthobservatory.nasa.gov/).

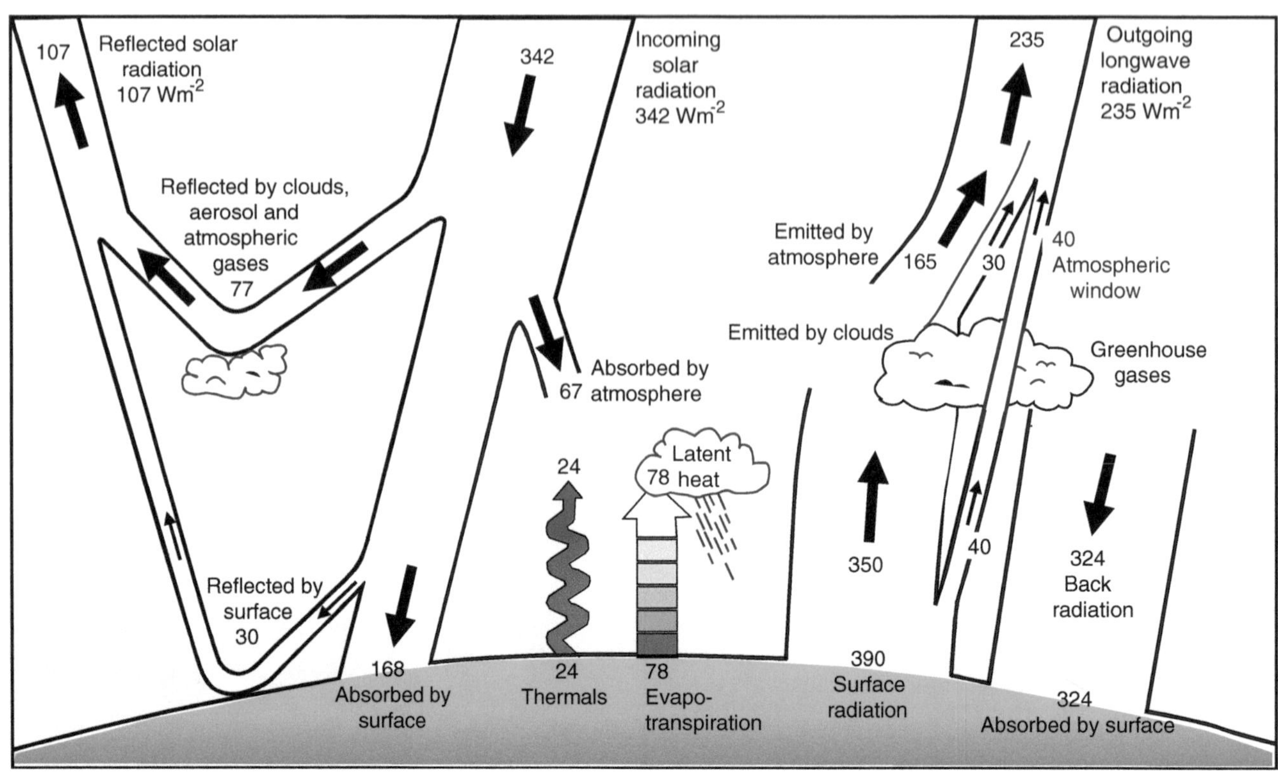


Figure 4.12 Estimate of the Earth's annual and global mean energy balance showing the main incoming and outgoing radiation pathways. (Source: IPCC 2007; Le Treut et al. 2007; Kiehl and Trenberth 1997.)

of confidence about each change (IPCC 2001a). The global average surface temperature (the average of near-surface air temperature over land and sea surface temperature) has increased since 1861. Over the twentieth century the increase has been 0.6 ± 0.2 °C. There is an emerging tendency for the global land-surface air temperature to warm faster than the global ocean-surface temperature. The instrumental record shows a great deal of variability: for example, most of the warming that occurred during the twentieth century can be attributed to two periods, 1910–45 (0.35C) and with stronger warming for 1976–present (0.55C) as shown in Figure 4.13.

Analysis of proxy data for the Northern hemisphere indicates that this increase in temperature in the twentieth century is likely to have been the largest of any century during the past thousand years. Globally, it is likely that the 1990s was the warmest decade and 1998 the warmest year according to the instrumental records, since 1861, although more recent years (e.g. 2003 and 2005) have also been recorded to be especially warm years (Trenberth et al. 2007).

Since the late 1950s (the period of adequate observations from weather balloons), the overall global temperature increase in the lowest 8 km of the atmosphere and in surface temperature have been similar, at 0.1 °C per decade (Trenberth et al. 2007). Satellite measurements, starting from 1979, show a similar increase. In the lower stratosphere, there has been a 0.5–2.5 °C temperature decrease since 1979. The seasonal variations in global warming are shown in Figure 4.14 for 1979 to 2005.

There is considerable spatial variation in the decadal rate of increase of temperature over continental areas of Asia, north-western North America, South East Brazil and over some mid-latitude ocean regions of the Southern hemisphere showing the strongest warming (Trenbreth et al. 2007). Seasonal differences are observed such as warming being the strongest over western North America, northern Europe and China during winter periods and over Europe and northern and eastern Asia

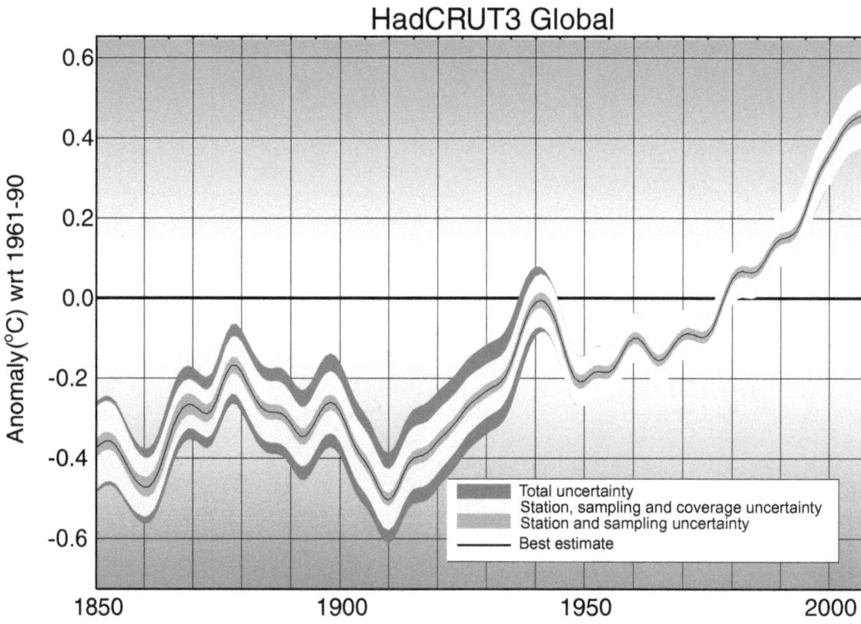


Figure 4.13 Variations in the Earth's surface temperature for the past 150 years. Annual series smoothed with a 21-point binomial filter. The dataset is HadCRUT3 which is the third release of the historical surface temperature analysis by the Hadley Centre of the UK Met Office and the Climate Research Unit (CRU), University of East Anglia, UK. (Source: Brohan and Kennedy 2007 personal communication, Crown Copyright 2007; Brohan et al. 2006.)

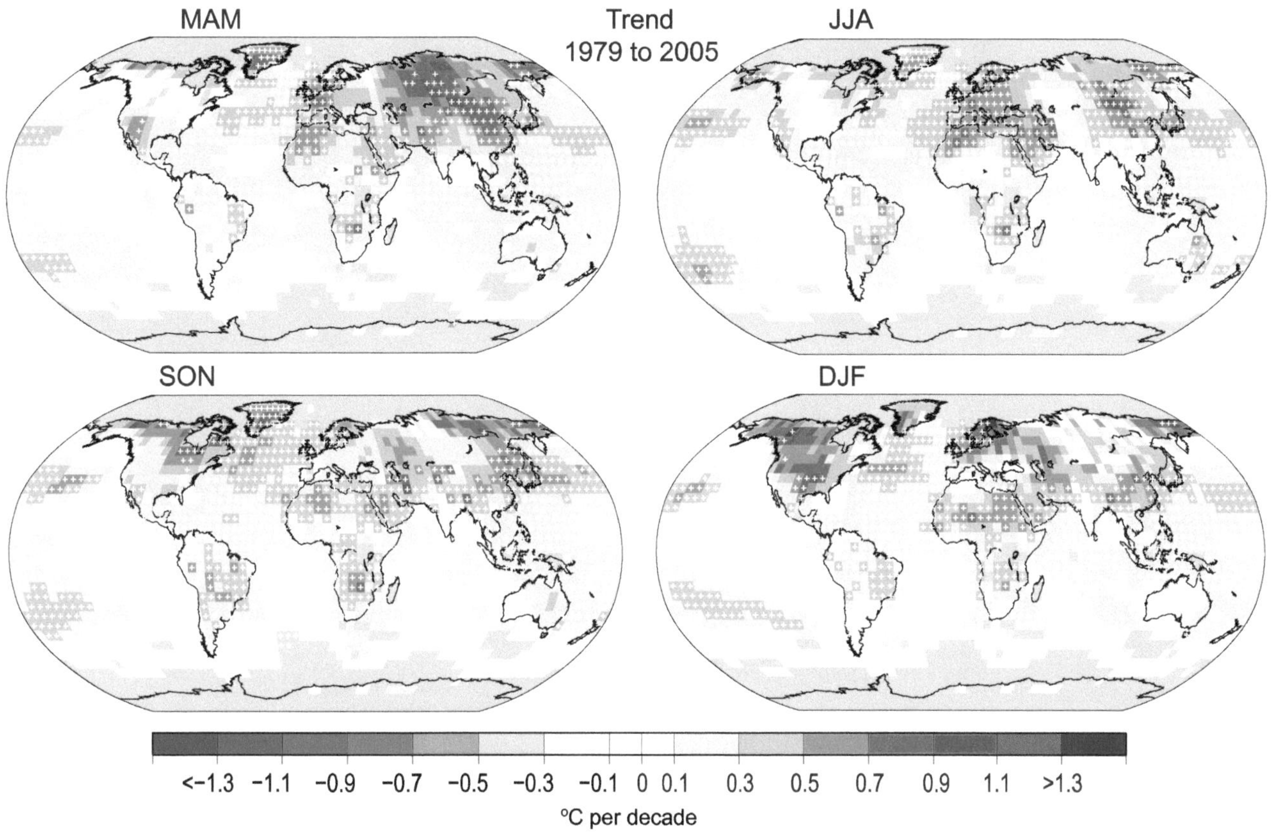


Figure 4.14 Linear trend in seasonal temperature for 1979 to 2005 in units of °C per decade. Areas in grey have insufficient data to produce reliable trends. The dataset used was produced by the National Climate Data Centre (NCDC) using the Global Land and Oceans dataset (Smith and Reynolds 2005). Trends significant at the 5 per cent level are indicated by white + mark. Months are indicated by their initials. (Source: IPCC 2007; Trenbreth et al. 2007.)

during spring months. During summer months, Europe and North Africa show particularly high temperature increases and similar is true for northern regions of North America, Greenland and eastern Asia during the autumn season. North America is also a region of widespread extra-tropical continental warming. One can also find some regions of cooling, for example, the western North Atlantic Ocean. In addition, there has been a decrease in continental diurnal temperature range since around 1950, which coincides with increases in cloud amount and total water vapour.

During the twentieth century, a 2 per cent increase in total cloud amount was observed over the land, and also possibly over the ocean, at least since 1952. The increases in total tropospheric water vapour in the last 25 years are qualitatively consistent with increases in tropospheric temperatures and an enhancing hydrologic cycle, resulting in more extreme and heavier precipitation events in many areas; for example, the middle and high latitudes of the Northern hemisphere show a 5–10 per cent increase and in the tropics a 2–3 per cent increase. In the subtropics, precipitation has decreased by 2–3 per cent. Figure 4.15 shows spatial patterns of the monthly Palmer Drought Severity Index (PDSI) for 1900 to 2002 (IPCC 2007; Trenbreth et al. 2007; Dai et al. 2004). The PDSI is a commonly used measure for determining the severity of drought on long terms (several months). It uses precipitation and temperature data and provides a numerical index with negative numbers indicating drought conditions and positive numbers indicating rainfall. Trenbreth et al. (2007) as part of AR4 of IPCC have analysed long-term trends in precipitation from 1900 to 2005 showing pronounced variations. Areas of eastern North and South America, northern Europe and northern and central Asia are significantly wetter whereas the Sahel, southern Africa, the Mediterranean and southern Asia are becoming drier. Such marked variations are thought to result from the warming of the world's oceans leading to increased water vapour in the atmosphere. The changes in the precipitation patterns are also leading to increases in the occurrences of both droughts and floods in some regions of the world.

Sea ice is expected to become a sensitive indicator of a warming climate (Trenbreth et al. 2007). It is very clear to see the systematic decrease of sea ice extent (nearly 3 per cent from 1973) and thickness in the Arctic especially in spring and summer. This decrease is consistent with an increase in temperature over most of the adjacent land and ocean. The surface air temperature in the Arctic has increased by 1.1 °C over the last 50 years, that is, more than three times as fast as the global mean air temperature (WMO 2003a). The decline of sea ice was strongest in the Eastern hemisphere. In the Antarctic, sea ice extent has shown little trend of changing after a rapid decline in the mid-1970s. A spectacular event was the collapse of Prince Gustav and parts of the Larsen ice shelves in 1995 and 2002. The significant warming over the Antarctic Peninsula of more than 2 °C since the 1940s has led to the southerly migration of the climatic limit of ice shelves and eventually the rapid disintegration of several large ice shelves. The decrease in snow cover and the shortening seasons of lake and river ice relates well to an increase in northern hemispheric land-surface temperatures (IPCC 2001a). Satellite records indicate a decrease in the Northern hemisphere annual snow cover extent by about 10 per cent since 1966. Reduction in snow cover during the mid- to late 1980s was strongly related to temperature increase in snow-covered areas.

The recession of mountain glaciers has often been used as clear and easily understandable evidence of global warming (IPCC 2001a; Koerner and Fisher 2002; Loso et al. 2006; Gordon et al. 2007) because, as a high-altitude ecosystem, they are quite sensitive to global temperature rise. One of the important direct indicators to characterising change in glaciers is the mass balance – the difference between ice and snow accumulating to a glacier and melting from it. Globally averaged values for about 30 glaciers in 10 mountain ranges indicate a trend towards increasingly negative balance, that is, accelerating glacier melting. Worldwide, glaciers have retreated over the twentieth century, although in some regions

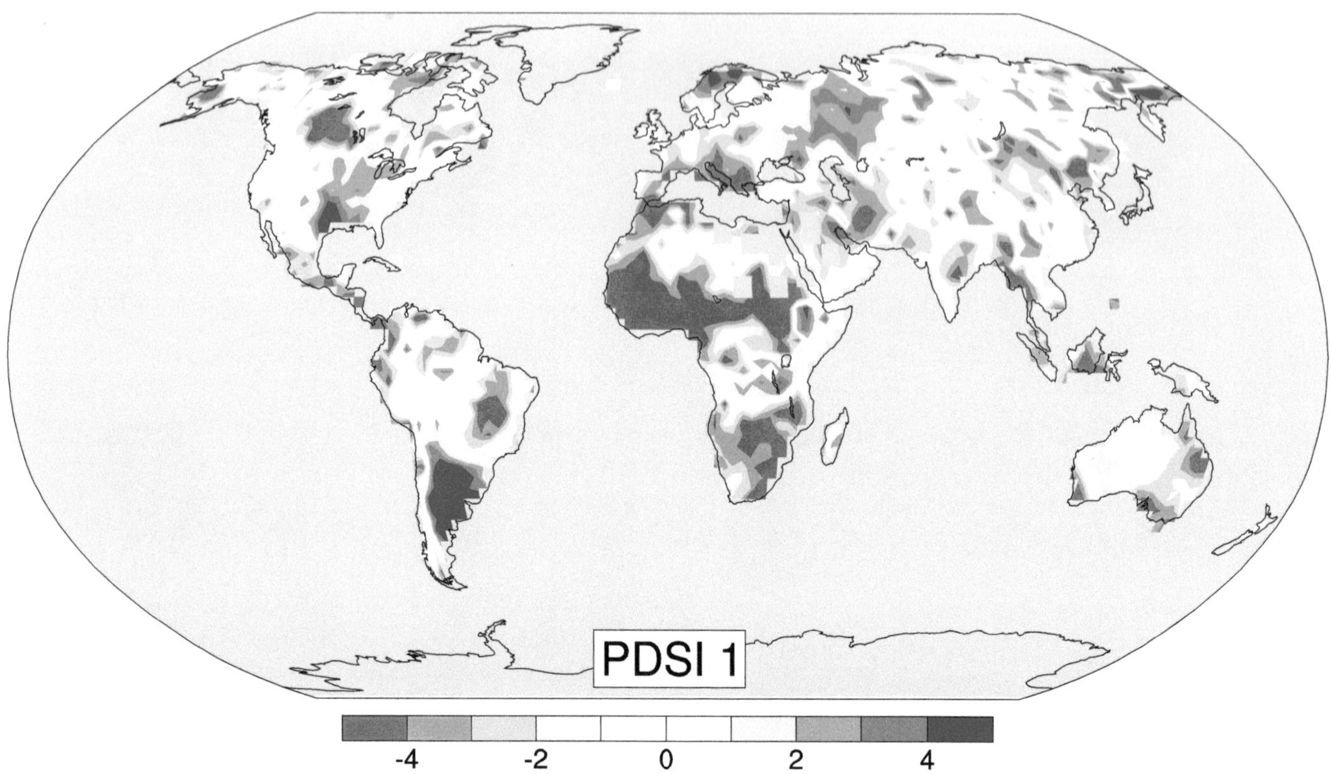


Figure 4.15 Spatial pattern of the monthly Palmer Drought Severity Index (PDSI) for 1900 to 2002. For the recent years the areas in red and orange are drier than average and blue and green areas are wetter than normal. (Source: IPCC 2007; Trenbreth et al. 2007; see also Dai et al. 2004.)

glaciers have increased in length despite the average warming. In low and middle latitudes, the retreat generally started in the mid-nineteenth century.

About 25 per cent of the land surface of the Northern hemisphere is underlain by permafrost. The areas affected by permafrost include Canada, Alaska, Russia and China. Very small changes in surface climate can produce important changes in permafrost temperatures, thus affecting the spatial distribution, thickness and depth of the active layer overlying permafrost. Changes also affect the climate system through the release of huge amounts of carbon stored in shallow layers of permafrost. Recent observations and further analyses since the last IPCC TAR, indicate that permafrost in many regions of the earth is currently warming (WMO 2003a; Osterkamp 2005; Lemke et al. 2007; Marchenko et al. 2007; Anisimov et al. 2007; Swanger and Marchant 2007).

The understanding of the impact of sea-level rise due to climate change is vital as it affects the population directly, especially those living near coasts. Sea-level rise is mainly caused by the thermal expansion of the oceans and increased melting of land ice from mountain glaciers, Greenland and Antartica (Nerem et al. 2006). Ocean heat content has increased since the late 1950s and global average sea level rose between 0.1 and 0.2 metres during the twentieth century. The rate of sea-level rise is expected to increase in this century compared to 1961–2003 period (Bindoff et al. 2007). Figure 4.16 illustrates the sea level over the past, the twentieth century and the projections for the twenty-first century (Bindoff et al. 2007). The data includes the reconstructed past values (Church and White 2006), tidal gauge measurements (Holgate and Woodworth 2004.) and satellite altimetry values since 1992 from TOPEX/Poseidon and Jason satellites (e.g. Leuliette et al. 2004; Cazenave and Nerem 2004). Over the twentieth century an accelerated rate of sea-level rise is observed compared to the past (Church and

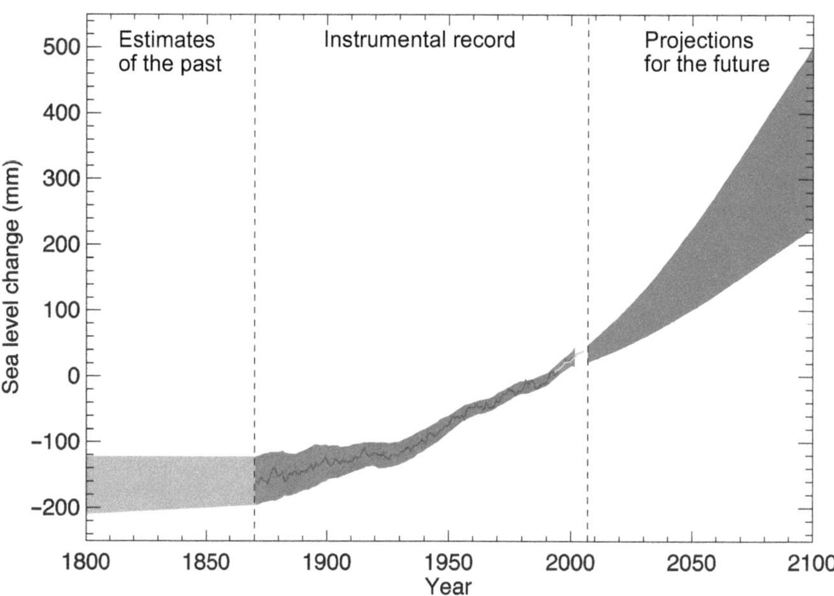


Figure 4.16 Time series of global mean sea level (deviation from the 1980–1999 mean) in the past, twentieth century and the twenty-first century. The grey shading shows the uncertainty in the estimated rate of sea-level change in the past century, the red line represents the global mean sea level from tide gauge measurements and the green line gives the global mean sea level from satellite altimetry. The variations in the measurements are indicated by the red shading and the blue shading represents the range of model projections for the twenty-first century, relative to the 1980 to 1999 mean. (Source: IPCC 2007; Bindoff et al. 2007.)

White 2006). Model predictions for the rest of the twenty-first century continue to show an increase although there is significant variation in the extent of the projected sea-level rise.

## Direct and Indirect Effects of Atmospheric Aerosols

As illustrated previously, atmospheric aerosols are known to influence significantly the radiative budget of the climate system in two distinct ways: (1) the direct effect, where aerosols themselves scatter and absorb solar and thermal infrared radiation; and (2) the indirect effect, where aerosols modify the microphysical and hence the radiative properties (albedo) and amount of clouds (IPCC 2001a). Substantial progress has been achieved in better defining the direct effect of a wider set of aerosols. Advances in observations and in aerosols and radiative models have allowed quantitative estimates of various aerosol components: sulphate aerosols, biomass-burning aerosols, fossil fuel black carbon (or soot), fossil fuel organic carbon aerosols and biomass-burning organic carbon aerosols, as well as an estimate for the range of radiative forcing.

Direct radiative forcing is estimated to be $-0.4$ Wm$^{-2}$ for sulphate, $-0.2$ Wm$^{-2}$ for biomass-burning aerosols, $-0.1$ Wm$^{-2}$ for fossil fuel organic carbon and 0.2 Wm$^{-2}$ for fossil fuel black carbon aerosols. There is much less confidence in the ability to quantify the total aerosol direct effect, and its evolution with time, than for greenhouse gases. For this and other reasons, a simple sum of the positive and negative bars cannot be expected to yield the net effect on the climate system. In simulations of many climate models, the effect of negative radiative forcing produced by anthropogenic sulphate aerosols is to cool the global climate and somewhat offset the warming effect of greenhouse gases (IPCC 1995; IPCC 2001a; Verma et al. 2006; Forster et al. 2007).

There is now more evidence for the indirect effect, but estimates of the indirect radiative forcing by anthropogenic aerosols remain problematic. The total indirect effect is negative ($-1.5$ to $-3$ Wm$^{-2}$), although with very uncertain magnitude. This is caused almost equally by changes in cloud water droplet size and increased cloud liquid water content. Numerous investigators have discussed the impacts of aerosols in the atmosphere (Rostayn and Lohman 2002; Feichter et al. 2004). Their results have shown that the surface temperatures everywhere decreased, with the maximum cooling found in the Northern hemisphere. This led to a change in the sea surface temperature (SST) meridional gradient, thus leading to strengthening of trade winds and weakening of the African monsoon, and eventually droughts in the Sahelian region. Lohman (2002) has studied the impact of anthropogenic soot aerosols on mid-latitude precipitation. The soot aerosols, as ice nuclei, may contribute to rapid growth of ice crystals in super-cooled liquid water, thus leading to increased precipitation. In addition, Roeckner et al. (1999) investigated the impact of the anthropogenic aerosols (mainly sulphate aerosols) on the global hydrological cycle. They have found that the future hydrological cycle would weaken compared to the present climate under the inclusion of the direct and indirect effects of sulphate aerosols. Recent studies such as Tripathi et al. (2007) and Sud and Lee (2007) have also investigated the possible impact of aerosol indirect effects on water clouds.

Whereas the impact of sulphate aerosols on global warming is negative and leads to cooling, the impact of black carbon is positive (e.g. Highwood and Kinnersley 2006; Flanner et al. 2007). Studies on the impact of anthropogenic black carbon on changes in the Asian monsoon and related precipitation patterns began at the turn of the twentieth century. The INDOEX experiment over South Asia and South East Asia indicates that local air heating caused by black carbon can induce enhanced evaporation of cloud droplets and hence affect cloud and precipitation behaviour. Although the local influence of changing aerosol concentration may be significant, the net global effects of all the changes in various tropospheric aerosols, including natural and anthropogenic sources, may be much smaller because many of the individual effects are offsetting, and the presence of clouds can mitigate their influence.

Soil dust aerosols have a significant impact on the earth's radiation budget through their scattering and absorbing of solar and infrared radiation, especially over the Saharan, Arabian and Asian desert regions (Miller and Tegen 1998; Takemura et al. 2002; Mikami et al. 2006; Hara et al. 2006; Shao and Dong 2006). Dust from one region can travel large distances through long-range transport and affect air quality at another region (e.g. Fairlie et al. 2007). Furthermore, the dust can have a harmful influence on human health, especially in East Asia, where dust sources are also close to urban areas (e.g. Lee et al. 2007). However, their interaction with the climate system is more complex than other types of aerosols. They may produce a radiative heating or cooling effect, depending on different conditions, with a large uncertainty range of 0.5 to 0.7 Wm$^{-2}$. The uncertainty of the indirect effect is even larger. It is well known that Saharan dust storms, while episodic, can create significant local radiative influence, with reduction of short-term net radiative fluxes over ocean surfaces by up to 10 Wm$^{-2}$, and half as much over land (Diaz et al. 2001). The Asian dust is emitted from the Chinese and Mongolian arid regions by cold fronts and

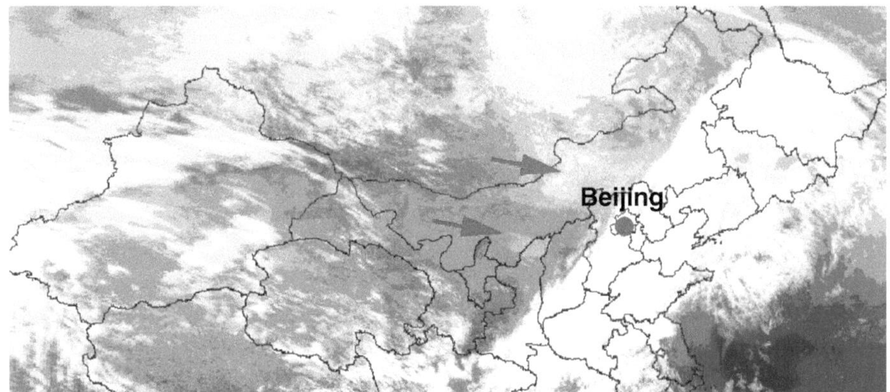


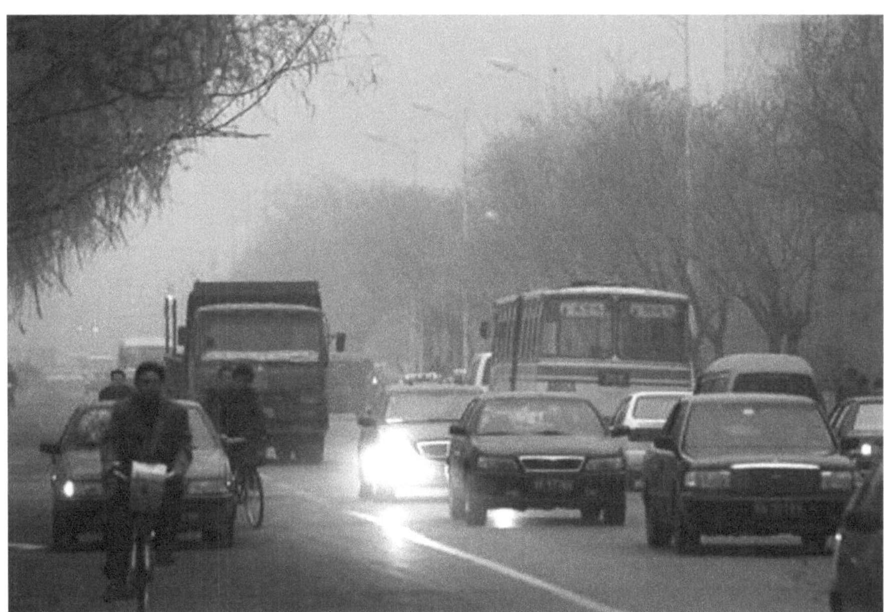

Figure 4.17 Satellite view (upper) of the dust storm event of 19–22 March 2002 over northern China with arrows indicating the movement of the dust and photograph of the dust storm (lower) taken in Beijing on 20 March 2002 (by courtesy of China Meteorological Administration).

extra-tropical cyclone activities, and transported to the Chinese coastal region, Korea and Japan, mainly in the spring. Since the late 1970s, the frequency of dust storms has decreased but with a gradual increase starting from 1997, possibly due to prolonged dry conditions in North China. Figure 4.17 shows the strongest dust storm event of 2000–05 which occurred on 19–22 March 2002. This huge dust storm originated in the Mongolian Gobi desert and the dust was transported over large distances (e.g. Park et al. 2005).

At the western edge of Japan (e.g. Nagasaki), the number of days in which the arrival of Asian dust was observed was 6.1 days per year on an average during the 1990s, but 16 and 15 days in 2000 and 2001, respectively. Another significant dust transport event in April 2001 brought substantial quantities of mineral dust from Asian deserts to the US atmospheric boundary layer, in an amount comparable to the daily emission flux of US sources of $PM_{10}$ (EOS 2003). Effects of Asian dust events provide evidence that the air pollution issue must be viewed in a global context.

## 4.4 Scenarios of Atmospheric Emissions for Climate Change Predictions

Climate models are used with future scenarios of forcing agents (e.g. greenhouse gases and aerosols) as input to make a suite of projected future climate changes that illustrates the possibilities that could lie ahead. With this development of climate change scenarios it is possible to assist in assessment of climate change impacts, adaptation and mitigation. Scenarios of future greenhouse gases (GHGs) are the product of very complex dynamic systems, determined by driving forces such as demographic development, socio-economic development and technological change (Nakicenovic et al. 2000; Meehl et al. 2007). Their future evolution is considered to be highly uncertain.

The IPCC developed long-term emission scenarios in 1990 and 1992, widely used in the analysis of climate change, its impact and options to mitigate climate change. In 1996 the IPCC began the development of a new set of emission scenarios known as the Special Report on Emission Scenarios (SRES) (Nakicenovic et al. 2000), effectively to update and replace the well-known IS92 (Leggett et al. 1992). For the SRES emission scenarios, four different narrative storylines were developed to describe consistently the relationship between forces driving emissions and their evolution, and to add context for scenario quantification. The resulting set of 40 scenarios covers a wide range of the main demographic, economic and technological driving forces of future greenhouse gases and sulphur emissions. Each scenario represents a specific quantification of one of the four storylines. All the scenarios based on the same storyline constitute a scenario family. It is evident that these scenarios encompass a wide range of emissions. Particularly noteworthy are the much lower future $SO_2$ emissions for the six SRES scenarios, compared to the older IS92 scenarios, due to structural changes in the energy system as well as concern about local and regional air pollution.

A wide range of studies have made use of these scenarios. For example, Van Vuuren et al. (2007) have downscaled the scenarios to derive national scale emissions of GHGs. Other studies such as Gaffin et al. (2004) and Bengtsson et al. (2006) have downscaled the SRES scenarios to derive finer scale population datasets. Through the use of this higher resolution data, Bengtsson et al. (2006) have analysed the population densities for the major river basins across world and shown that the largest increases in population (1990–2050) are expected in Indian, African and Middle Eastern basins. SRES scenarios have been used by Märker et al. (2007) to demonstrate how climate change could affect land erosion for a region of Italy. A study by Rounsevell et al. (2006) has examined the land-use changes (e.g. agricultural and urban) that are likely to occur across Europe in the future but also discusses the difficulties in making such predictions. Mirasgedis et al. (2007) have examined the implications of the optimistic (B2) and pessimistic (A2) scenarios on the energy demand of Greece. Although most attention is usually given to $CO_2$ when considering scenarios, it is also important to consider the role of other GHGs in reducing the impact of climate change (Sarofim et al. 2005).

The SRES scenarios have been examined by van Vuuren and O'Neill (2006) and they have concluded that in general they have been consistent with past (1990–2000) and more recent economic, energy and emissions data but do identify some areas which require updating such as A2 projections being too high for population trends, short-term $CO_2$ emissions according to A1 scenario being too high and $SO_2$ emissions being overestimated for some regions. There are other studies in the literature, however, which have been critical of the methodology followed for SRES (e.g. Tol 2007) but the development of IPCC scenarios is continuing (e.g. Hoogwijk 2005; Hanaoka et al. 2006) and new emission scenarios are planned for the fifth round of assessment (http://www.mnp.nl/ipcc/).

For projections of future changes in greenhouse gases and aerosols in the atmosphere, model calculations are used to obtain different concentration trajectories under the illustrative SRES scenarios. By 2100 the $CO_2$ cycle indeed leads to a projection of atmospheric $CO_2$ concentration of 540–970 ppm for the illustrative SRES scenarios (90–250 per cent above the concentration of 280 ppm in 1750). The primary non-$CO_2$ greenhouse gases by the year 2100 vary considerably across the six illustrative SRES scenarios. Future emissions of indirect greenhouse gases ($NO_x$, CO, VOC), together with a change in $CH_4$, are projected to change the global mean abundance of tropospheric hydroxyl radical (OH), by −20 per cent to 6 per cent over the twenty-first century. The large growth in emissions of greenhouse gases and other pollutants, as projected in some of the six illustrative scenarios (A2 and A1FI) for the twenty-first century, will degrade the global environment in ways beyond climate change, with increasing background levels of tropospheric $O_3$, thus threatening the attainment of current air quality standard over most metropolitan and even rural regions, and compromising crop and forest productivity. This problem reaches across boundaries and couples emissions of $NO_x$ on a hemispheric scale. Nearly linear dependence of abundance of aerosols on emissions is projected based on models using present-day meteorology. Sulphate and black carbon aerosols can respond in a non-linear fashion depending on the chemical parameterisation used in the model. Emission of natural aerosols such as sea salt, dust and gas phase precursors of aerosols may increase as a result of change in climate and atmospheric chemistry.

When comparing the estimated total historical anthropogenic radiative forcing from 1765 to 1990 followed by forcing resulting from the six SRES scenarios, it is evident that the range for the SRES scenarios is higher compared to IS92 scenarios, mainly due to the reduced future $SO_2$ emissions and the slightly larger cumulative carbon emissions in SRES scenarios (Nakicenovic et al. 2000; IPCC 2001b). In almost all SRES scenarios, the radiative forcing due to $CO_2$, $CH_4$, $N_2O$ and tropospheric $O_3$ continues to increase, with the fraction of the total radiative forcing due to $CO_2$ projected to increase from slightly more than half to about three-quarters of the total. The radiative forcing due to $O_3$-depleting gases decreases due to the introduction of emission controls as a result of the Montreal Protocol aimed at curbing stratospheric $O_3$ depletion. The direct radiative forcing by aerosols represents changes in radiation budgets in the atmosphere (including its top) and at the surface through their scattering and absorbing solar and infrared radiation. So aerosol concentrations or abundance in the atmosphere in the future will determine their direct radiative forcing. The latter (abundance in the atmosphere), in turn, will be dependent on the future different emissions of aerosols (e.g. scenarios). The total (direct plus indirect) aerosol effects are projected to be smaller in magnitude than those of $CO_2$.

## 4.5 Projection of Global Climate Change, Its Impacts and Atmospheric Composition

Comprehensive climate models can be used to project the climate change or response to the different input scenarios of future forcing agents. Similarly, projection of the future concentrations of emitted $CO_2$ and other greenhouse gases requires an understanding of the biogeochemical processes involved and incorporating these into a numerical carbon cycle model (IPCC 2001a). Since the year 1750 (i.e. the beginning of the Industrial Revolution), the atmospheric concentration of $CO_2$ has increased by 35 per cent due to human activities. $CO_2$ concentrations will continue to grow substantially to the year 2100, compared with the year 2000. If all possible uncertainties are included, then the range of $CO_2$ concentrations in the year 2100 will be about 490 and 1260 ppm (compared to the pre-industrial concentration of about 280 ppm and about 379 ppm in the year 2005).

## Change in Temperature

Under all SRES scenarios, projections show the global average surface temperature will continue to rise during the twenty-first century, at rates that are very likely to be without precedent during the last 10,000 years, based on palaeoclimate data. The range of surface temperature increase for the period 1990–2100 is projected to be 1.4–5.8 °C according to TAR, (IPCC 2001a). It is very likely that nearly all land areas will warm more rapidly than the global average, particularly those at high northern latitudes in the cold season. There are very likely to be more hot days; fewer cold days, cold waves and frost days; and a reduced diurnal temperature range. In a warmer world the hydrological cycle will become more intense, with global average precipitation to increase. More intense precipitation events (hence flooding) are very likely over many areas. Increased summer drying and associated risk of drought is likely over most mid-latitude continental interiors. An increase in temperature globally is likely to lead to greater extremes of drying and heavy rainfall, and a subsequent increase in the risk of droughts and floods during El Niño.

Since TAR, more modelling studies have led to the possibility of using multi-model ensemble results rather than single model predictions to test the impact of the different scenarios for climate change and atmospheric composition (e.g. Dentener et al. 2006; Kunkel and Liang 2005; Meehl et al. 2007). As an illustration Figure 4.18 shows the AR4 (Meehl et al. 2007) multi-model mean surface temperature change for three projected time periods relative to 1980–99. The ensemble predictions are for three scenarios representing low (B1), medium (A1B) and high (A2) emission projections. The ensemble multi-model predictions show higher warming over most land areas for all three scenarios with highest temperature increase for A2 scenario. Model results indicate warming over the ocean is relatively large in the Arctic and the equatorial regions (e.g. the eastern Pacific) and lower warming over the North Atlantic and the Southern Ocean. Ensemble approaches using a single model have also been used for climate studies (e.g. Harris et al. 2006; Hargreaves and Annan 2006) as multi-model runs inevitably are computationally expensive.

## Sea-Level Rise

In a warmer world, the sea level will rise with a range of 0.09–0.88 m for the period 1990–2100, primarily due to thermal expansion and loss of mass from glaciers and ice caps, the rise continuing for hundreds of years, even after stabilization of greenhouse gas concentrations. According to the latest IPCC assessment of sea-level rise (Meehl et al. 2007), predictions from atmospheric-ocean coupled global circulation models (AOGCM) indicate that the average rate of rise during the twenty-first century is very likely to exceed the rate of increase for the period 1961 to 2003 (1.8 ± 0.5 mm/yr). The central estimate of the rate of sea-level rise (taken as the output from an average model) is predicted to be 3.8 mm/y for 2090 to 2099 under the A1B scenario while the equivalent sea-level rise rate for 1993–2003 is 3.1 mm/yr. Meehl et al. (2007) do, however, stress that there is a lack in current scientific understanding of sea-level change leading to uncertainties in the predicted values and caution that this could actually imply an underestimation in the projections. Figure 4.19 shows the projected total global mean sea-level rise for 2090 to 2099 for the six IPCC SRES scenarios.

## Changes in Atmospheric Composition

Measurements and modelling investigations reveal that concentrations and distributions of pollutants can exhibit marked spatial and temporal variabilities (e.g. WMO 2003b; Grewe 2006). These variabilities are caused by complex atmospheric processes and interactions as well as changes in emissions. In addition to transport and mixing processes, chemical transformations can lead to air pollution on urban (see Chapter 2) and regional (see Chapter 3) scales. Globally, emissions of pollutants such as chlorofluorocarbons (CFCs) deplete ozone levels in the stratosphere (Chapter 5)

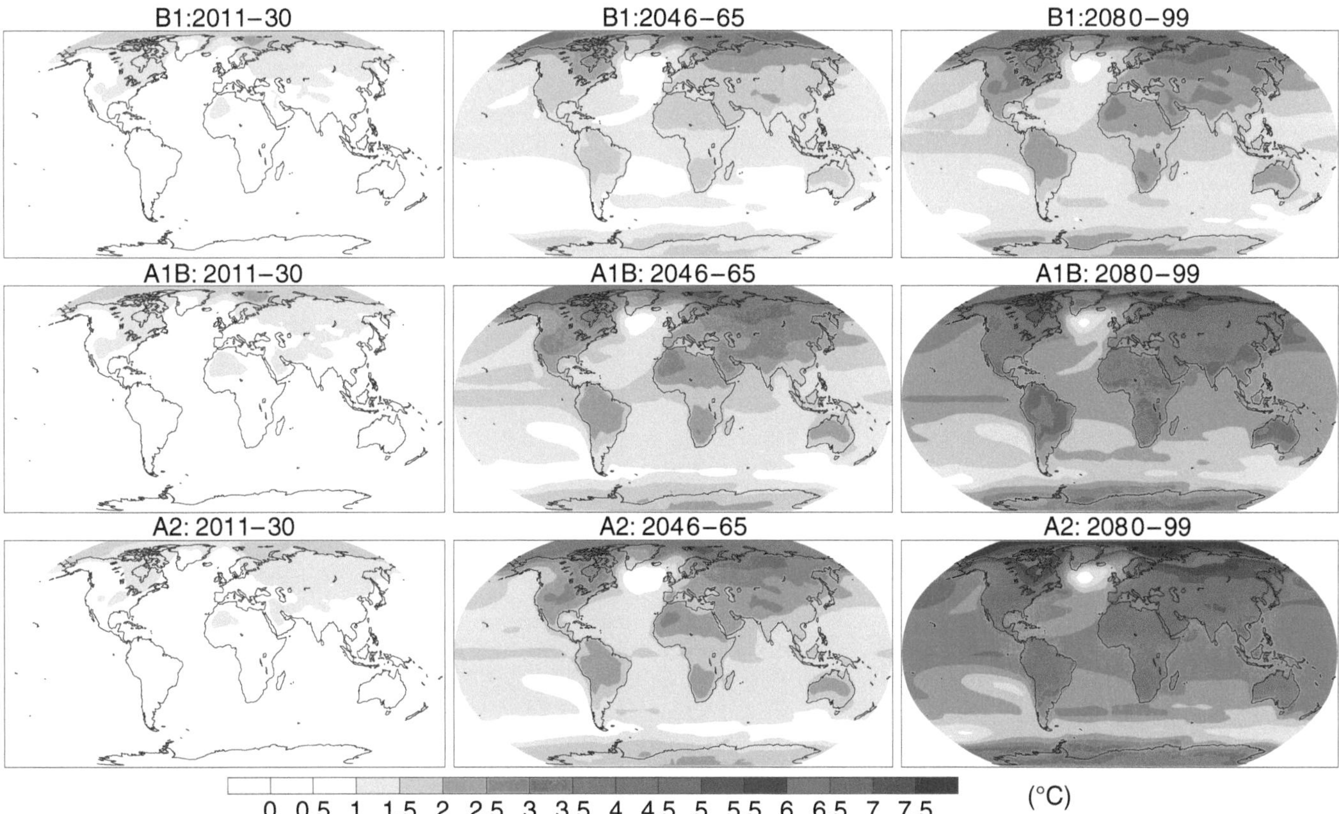


**Figure 4.18** IPCC Fourth Assessment Report multi-model mean of annual mean surface warming (surface air temperature change, °C) for the SRES scenarios B1 (top), A1B (middle) and A2 (bottom), and three time periods, 2011 to 2030 (left), 2046 to 2065 (middle) and 2080 to 2099 (right). Temperature anomalies are relative to the average of the period 1980 to 1999. (Source: IPCC 2007; Meehl et al. 2007.)

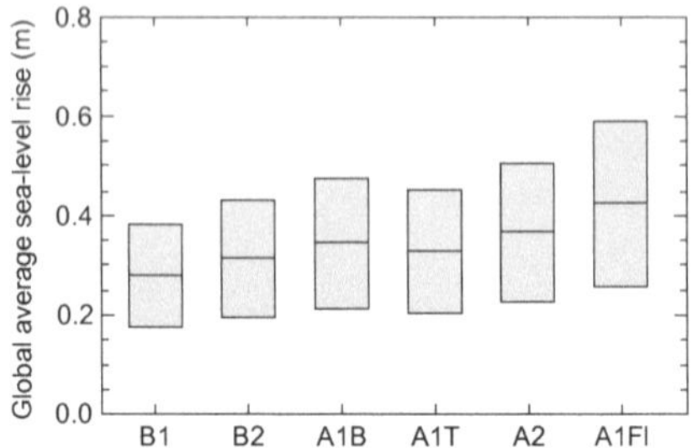


**Figure 4.19** Projections of the total global average sea-level rise in 2090 to 2099 (relative to 1980 to 1999) for the six SRES scenarios as estimated by IPCC AR4. The box plots show the mean values as well as the 5 to 95 per cent ranges (adapted from Meehl et al. 2007).

whereas within the troposphere, some studies are revealing an increase in ozone levels over the 1990s (e.g. Derwent et al. 2004) and longer term (Oltmans et al. 2006; Grewe 2007). Derwent et al. (2006) have reconsidered the possible global ozone background changes for a period of up to 2030 in light of available emission scenarios and highlighted the potential impact on European air quality. The study further demonstrated the need to take such changes into account when formulating regional air quality policies.

Under the current legislation (CLE) scenario, Dentener et al. (2005) have predicted that global ozone levels will rise by about 3–5 ppbv by 2020–30 relative to 1990–2000 over the USA and Europe. The levels are predicted to be higher (8–12ppbv) over the Indian subcontinent, South China and South East Asia. Over the North Pacific and Atlantic Ocean ozone is predicted to increase by 4–6 ppbv under the CLE scenario as a result of increased background ozone and increases in ship emissions. Ozone levels remains largely unchanged over the other world latitude bands. Despite the emission reductions in North America and Europe in the CLE scenario, model predictions do not show a decrease in the ozone levels for the 2020s and the levels may even increase indicating the need for global ozone control strategies (see also Dentener et al. 2006). If an alternative scenario of Maximum technically Feasible Reduction (MFR) is adopted, the TM3 model calculates ozone decreases of about 5 ppbv over most of the Northern hemisphere and up to 10 ppbv in the USA, Middle East and South East Asia for the 2020s compared with the 1990s. A stronger response is predicted by the STOCHEM model to emission reductions with reduced concentrations of ozone in large parts of eastern Europe and Russia. The global spatial variation in the changes in ozone levels between the 2020s and 1990s is shown in Figure 4.20 for the TM3 and STOCHEM models, based on the CLE and the MFR scenarios (Dentener et al. 2005).

Overall the two models show broad agreement but there are significant differences over some regions. These types of differences can occur as a result of the type of mixing schemes used or the treatment adopted for emissions within each model. The study by Dentener et al. (2005) also discusses the importance of controlling the emissions of methane to reduce radiative forcing as well as levels of ozone. Future trends of air quality are considered further in Chapter 7.

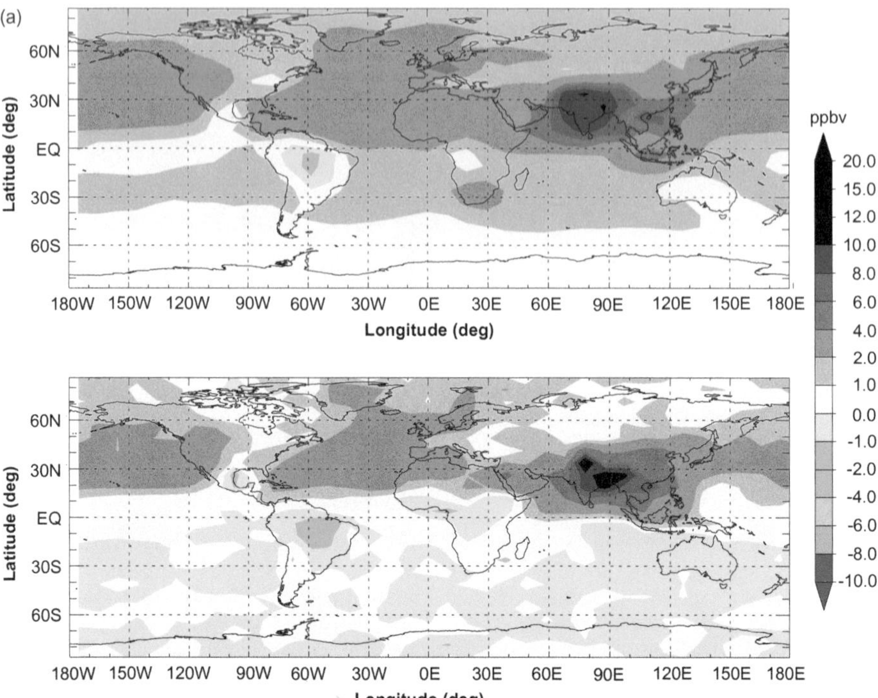


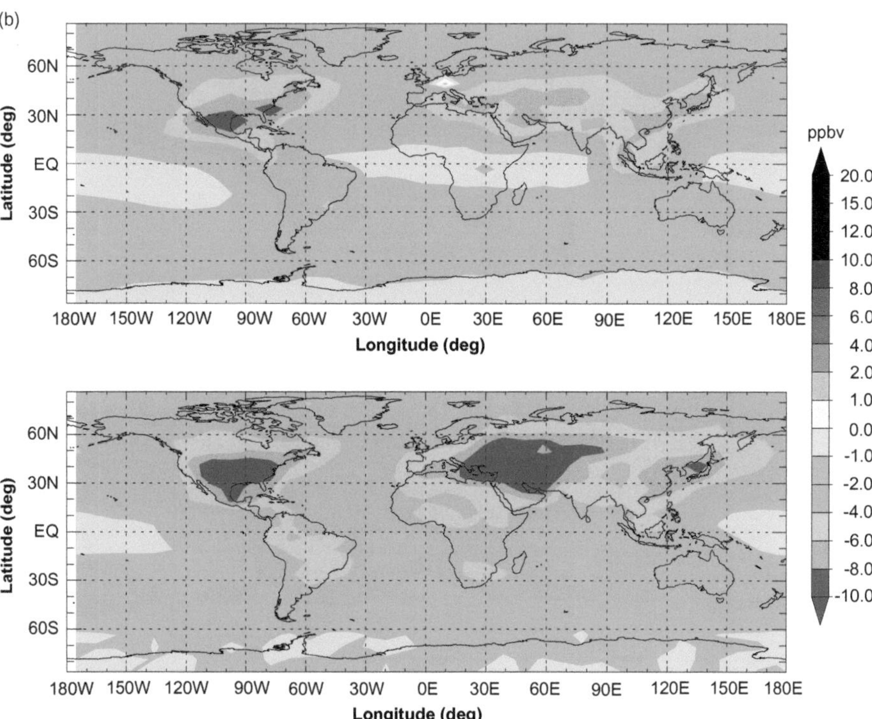


**Figure 4.20** Decadal averaged ozone volume mixing ratio differences [ppbv] between the 2020s and the 1990s for (a) CLE scenario and (b) MFR scenario from the TM3 (upper) and STOCHEM (lower) models. (Source: Dentener et al. 2005.)

As mentioned earlier, aerosols are also of concern regionally and globally in relation to atmospheric composition and climate change. There are a range of modelling studies that have shown the global aerosol loading and the implications on radiative forcing (e.g. Takemura et al. 2005; Liao and Seinfeld 2005; Stier et al. 2006). As an example, Figure 4.21 shows the GISS GCM II' model simulation of total dry aerosol mass ($\mu$g/m$^3$) in the surface layer for the present day and the year 2100 (Liao and Seinfeld 2005). The components of the dry aerosol mass considered are sulphate, nitrate, ammonium, black carbon (BC), primary organic aerosols (POA) and secondary organic aerosols (SOA). For present day conditions the model simulations indicate dry mass concentrations of above 15 $\mu g/m^3$ over parts of Europe, eastern United States, eastern China and over the biomass region in South America. The dry mass concentrations by 2100, when compared to present day values, over parts of Europe, South America and southern Africa are predicted to double or even triple. Aerosol levels over eastern United States are also expected to nearly double but the highest concentrations by 2100 are likely to be observed in eastern China. The study by Liao and Seinfeld (2005) show that heterogeneous reactions (gas-aerosol) can account for a large proportion of the aerosol mass concentrations (mainly nitrate and ammonium aerosols).

## Ecosystems and Health

The change in global climate and the environment as a whole along with the underlying variabilities can have a great variety of impacts on natural ecosystems and socio-economics (e.g. IPCC 2001b; Anderson, P K et al. 2004; Hitz and Smith 2004; Khasnis and Nettleman 2005; Krishnan et al. 2007; Bytnerowicz et al. 2007). Determination of vulnerabilities and possibilities for adaptation has become essential issues to address impacts of global climate change caused by the increase in greenhouse gases and aerosols. It is quite clear that the climate change characterised by changes in temperature and precipitation and sea-level rise can exert important impacts on human health (e.g. weather-related mortality, infectious disease and air pollution respiratory effects), agriculture (e.g. crop yields and irrigation demands), forests (e.g. forest composition, geographic range and change in water quality), water resources (e.g. changes in water supply and change in water quality), coastal zone (e.g. beach erosion, inundation of coastal land and cost of defending coastal communities) and species and natural lands (e.g. loss of habitat and species and shift in ecological zones). There are a large number of ways in which climate change can affect human health in various parts of the globe, often through complex interactions and pathways, see Figure 4.22 (McMichael et al. 1996; Martens 1998; Martens 1999; Haines et al. 2006; Huntingford et al. 2007; IPCC WG II 2007).

Climate is one of the factors that governs the occurrence of many infectious diseases, from the Black Death in the fourteenth-century Europe to modern times, when the spread of Ebola in Africa, cholera in South America and Lyme disease in the United States are affected by changes in temperature, rainfall, sunshine and even ocean currents. It is the interaction among these factors that will, in combination with other non-climatic factors, determine the timing of infectious disease outbreaks. Malaria claims millions of lives every year, mainly in tropical Africa (e.g. Mabasoa et al. 2007), but also in large areas of South America and South East Asia (Martens et al. 1999; van Lieshout et al. 2004; Guerra et al. 2006). Malaria is caused by the malaria parasite, plasmodium, and is spread by the anopheles malaria mosquito, which serves as the vector of the disease. The spread of the disease is thus

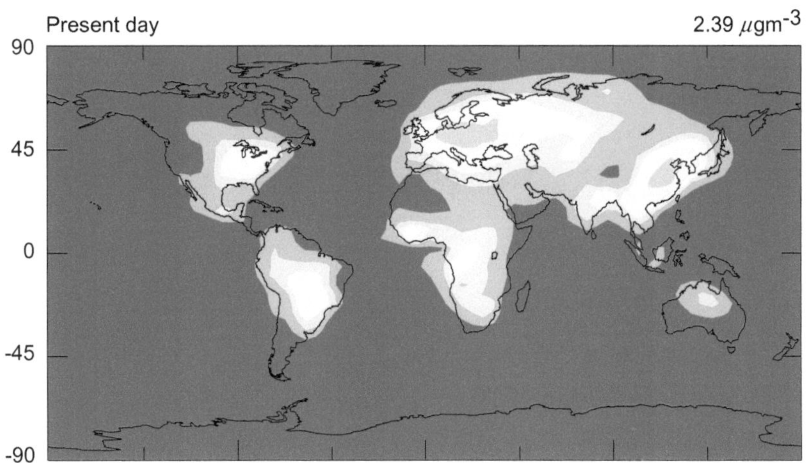


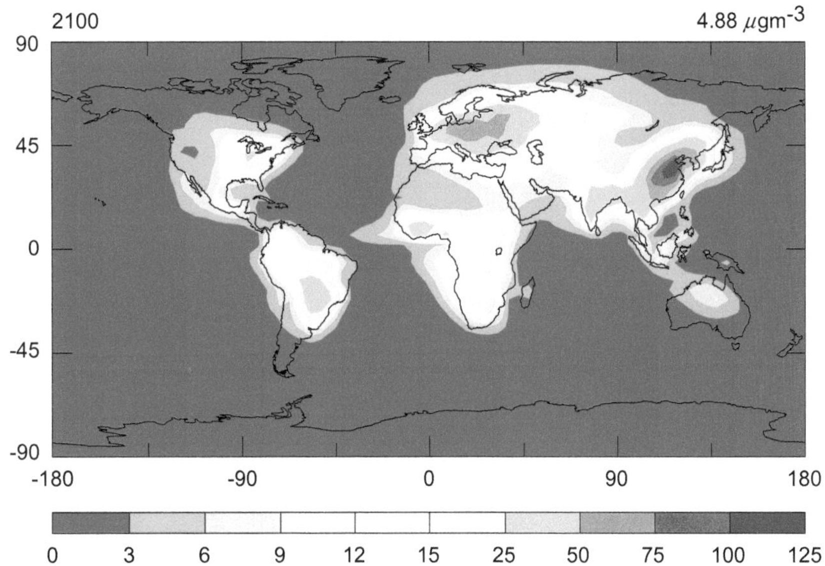


Figure 4.21 Total dry aerosol mass ($\mu g/m^3$) of sulphate, nitrate, ammonium, black carbon (BC), primary organic aerosols (POA) and secondary organic aerosols (SOA) in the surface layer from the baseline simulations for present day (top) and year 2100 (bottom) for IPCC SRES scenario A2 (Source: Liao and Seinfeld 2005). The global mixing ratios are shown on the top corner of each panel.

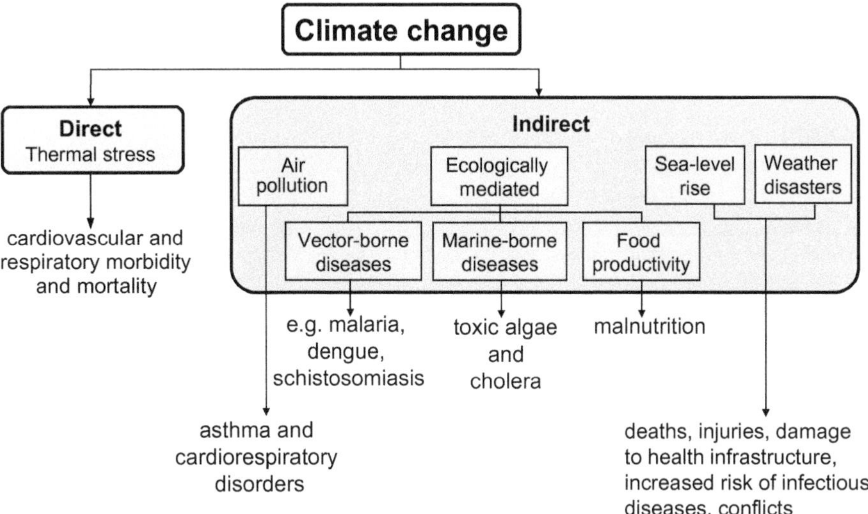


Figure 4.22 Possible effects of climate change on human health (adapted from Martens 2004 personal communication).

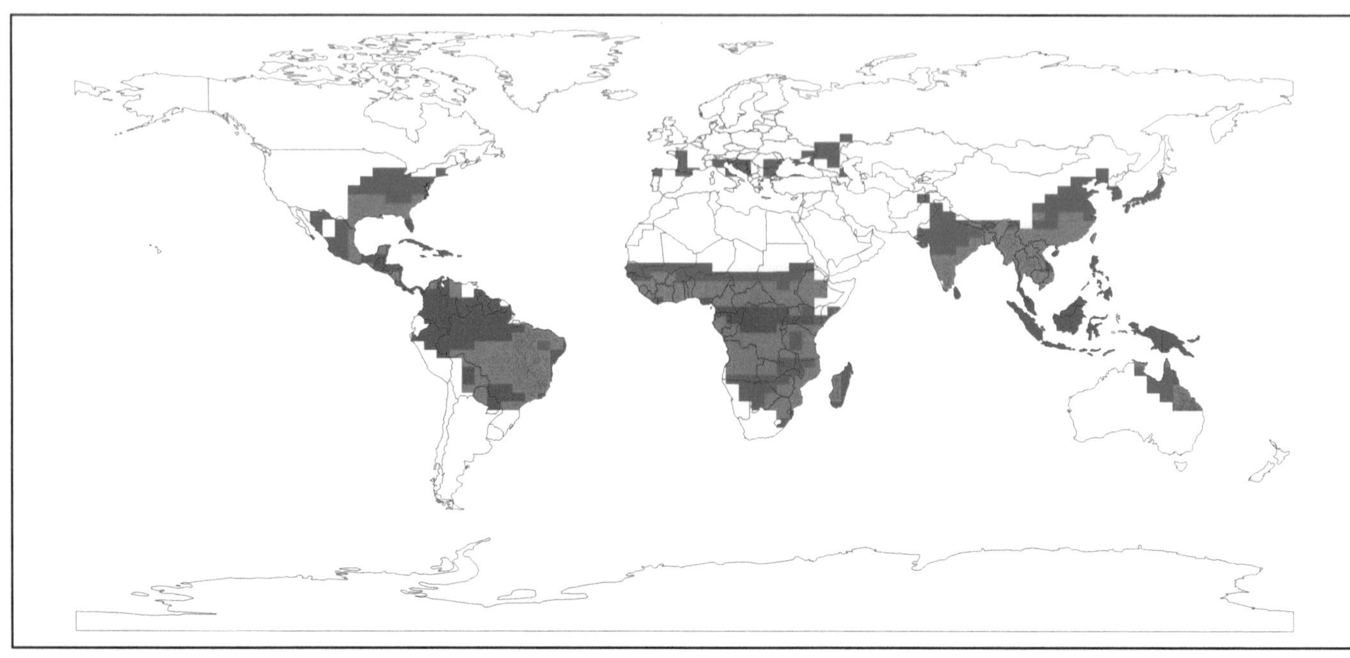

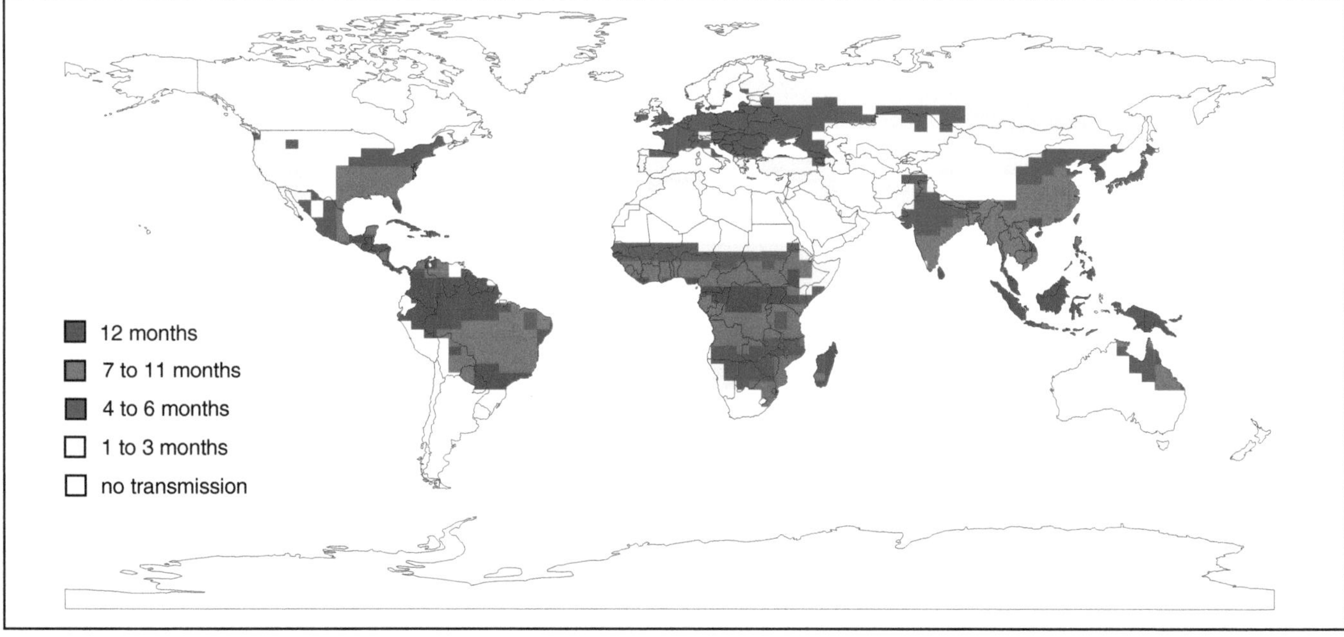


**Figure 4.23** Model projections showing the global distribution of malaria risk for a base line climate (1961–1990) (upper) and 2080s (lower). (Source: Martens 2004 personal communication; Martens et al. 1999; van Lieshout et al. 2004.) The legend represents the number of months of climate suitability for malaria transmission.

limited by conditions that favour the vector and the parasite. The malaria mosquito is most comfortable at about 20–30 °C and a relative humidity of at least 60 per cent. Furthermore, the malaria parasite develops more rapidly inside the mosquito as the temperature rises, and ceases entirely below about 15 °C. Increased rainfall and surface water also provide breeding grounds for the mosquito. Climate change may thus wreak considerable change on the distribution of the disease. Although malaria has disappeared from most wealthy countries, partly due to the use of insecticides and the antimalarial medicine chloroquine, the mosquitoes that transfer the disease are still present. In Europe the potential risk is greatest in countries that surround the Mediterranean region.

Based on TAR (IPCC 2001b), projected climate change will have beneficial as well as adverse effects, but the larger the changes and the greater the rate of change in climate, the more the adverse effects predominate. The adverse effects are of particular concern for developing countries because they usually bring about damaging losses. First, regional changes in climate have already affected and will continue to affect a diverse set of physical and biological systems in many parts of the world. Their rate of change would be expected to increase in the future represented by any of the SRES scenarios. Many physical systems are vulnerable to climate change; for example, the lake-level rise and the continued retreat of glaciers and permafrost. Planned productivity would decrease in most regions of the world for warming beyond a few degrees Celsius. In most tropical and subtropical regions, yields are projected to decrease for almost any increase in temperature.

Ecosystems and species are vulnerable to climate change and other stresses, and some will be irreversibly damaged or lost, including an increased risk of extinction of some vulnerable species. Populations that inhabit small islands and low-lying coastal areas are at particular risk of severe social and economic effects from sea-level rise and storm surges. Projected climate change would exacerbate water shortage and water quality problems in many areas of the world where water is already scarce, but alleviate it in some other areas. Overall climate change is projected to increase threats to human health, particularly in lower-income populations, predominantly within tropical and subtropical countries. For example, model projections show an increased risk of malaria in moderate zones as the climate becomes warmer and more humid (Figure 4.23).

## Uncertainty Issues

As illustrated previously, complex, physically based models are required to make simulations of past and current climate change and projections of future climate change. However, such models cannot yet simulate all aspects of climate and project the future climate change with high confidence, due to the existence of many uncertainties. Among them, key uncertainties that influence the quantification and the details of future projections of climate change are those associated with the SRES scenarios, as well as those associated with the modelling of climate change. In particular, there are uncertainties in the understanding of key feedback processes in the climate system, especially those involving clouds, water vapour, aerosols (including their indirect forcing), sea ice and ocean heat transport. Clouds and their interaction with radiation also represent an essential uncertainty. They not only affect the magnitude of radiative forcing, but also the signs of radiative forcing. Another uncertainty concerns the understanding of the probability distribution associated with temperature and sea-level projections for the range of SRES by developing multiple ensembles of model calculation. It is well known that the climate system is a coupled, non-linear, chaotic system, and therefore the long-term prediction of exact future climate states is not possible. A useful approach to this problem is the prediction of the probability distribution of the system's future possible states by the generation of ensembles of model solutions. Key uncertainties are also reflected in the details of regional climate change because of the limited capabilities of the regional and global models. In addition, there are inconsistencies in results between different models, especially in some regions and when simulating precipitation. A further key uncertainty concerns the mechanisms, quantifications of timescales and likelihoods associated with large-scale abrupt/non-linear changes (e.g. collapse or stagnation of ocean thermohaline circulation (THC) caused by differences in water density which depends on temperature and levels of salinity). In the aspect of impacts of climate change, key uncertainties arise from the lack of reliable local or regional detail in climate change, especially in the projection of extreme events. Regarding adaptation, key uncertainties relate to the inadequate representation by models of local changes, lack of foresight, inadequate knowledge of benefits and costs, possible side effects, including acceptability and speed of implementation, various barriers to adaptation and more limited opportunities and capacities for adaptation in developing countries.

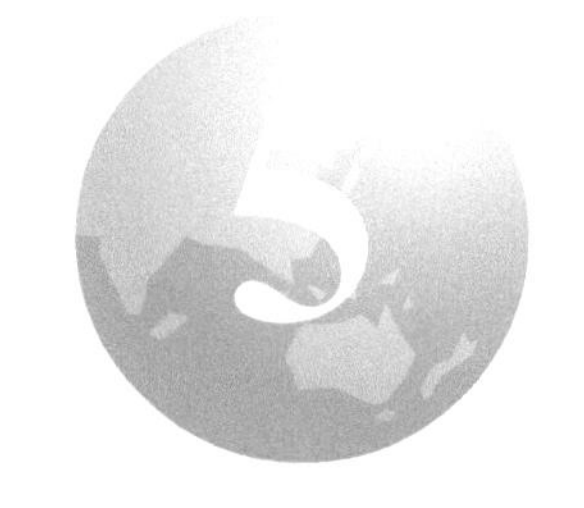

# OZONE DEPLETION

*Richard S Stolarski*



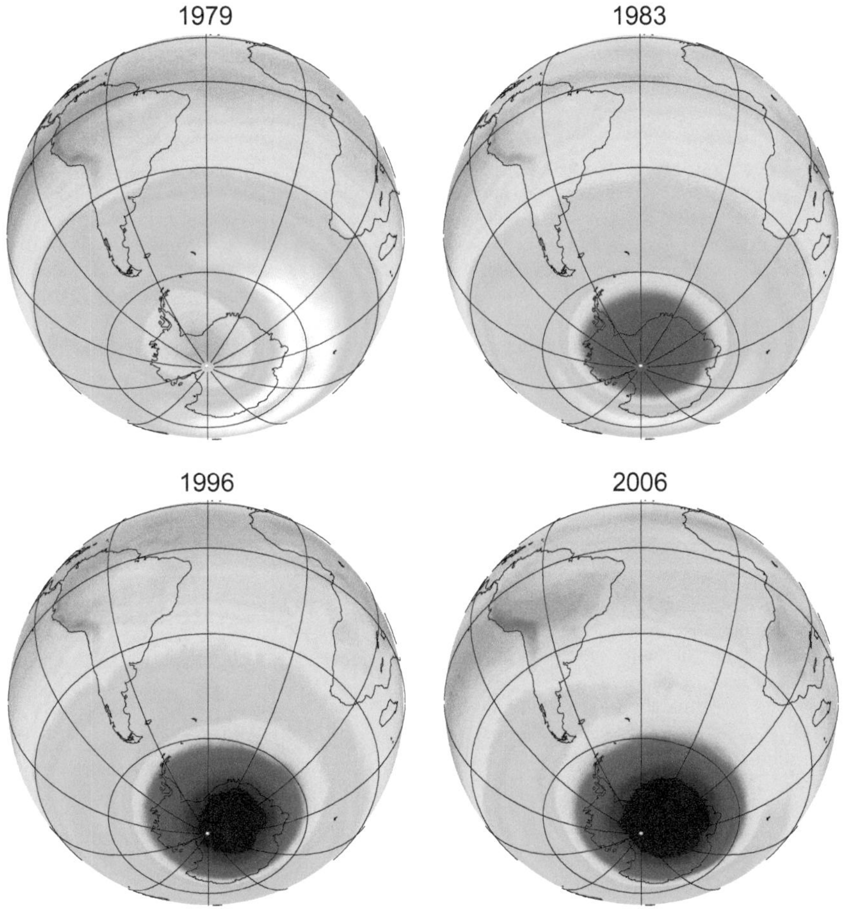


October mean of ozone over the Antarctic measured by satellite instruments for 1979, 1983, 1997, and 2006 showing how the ozone hole has developed since 1979.

Ozone is the triatomic form of oxygen. It is a colourless gas that acts as a highly reactive oxidising agent. It is the primary oxidising irritant in photochemical smog. On the other hand, ozone is used to deodorise air, purify water and treat industrial wastes. Ozone is a strong absorber of ultraviolet radiation. Ozone can be both good and bad: bad to breathe near the surface in the troposphere, but good to shield from ultraviolet (uv) radiation in the stratosphere.

One of the most important properties of ozone is its ability to absorb ultraviolet radiation (discovered by Hartley 1880). The earth is surrounded by a thin layer of ozone that is sufficient to screen us from the ultraviolet radiation from the sun that would otherwise reach the surface, where it would be capable of breaking bonds in biologically important compounds such as DNA (e.g. Björn 2007). The thin layer of ozone in the stratosphere is only about one part in a million of the total molecules that make up our atmosphere. If the entire layer were reduced to surface pressure, it would be only 3 millimetres in thickness. Figure 5.1shows how the air temperature and ozone concentration changes within the troposphere and the stratosphere.

Temperature decreases throughout the region called the troposphere. Temperature is then constant or slowly increasing, forming a permanent inversion layer called the stratosphere. The ozone concentration peaks in a layer-like structure in the stratosphere, where it makes up a few parts per million of all the molecules in that region.

The stratosphere is a region of the upper atmosphere above most of the weather phenomena that we are used to experiencing. Weather, clouds and rain generally occur in the lowest region of the atmosphere called the troposphere. Temperature decreases with increasing altitude as air rises and cools. The tops of high mountains are cold enough to maintain snow cover all the year round (even on mountains in the tropics such as Kilimanjaro). The decrease in temperature ceases when the stratosphere is reached (Figure 5.1). Temperature then begins to increase with altitude, primarily because the ozone in the stratosphere is absorbing the sun's ultraviolet radiation. The stratosphere is a permanent, stable inversion layer. Pollutants injected in the stratosphere will remain there for years. They spread out to fill the entire globe, making the stratosphere a global rather than a regional issue.

The sun emits radiation across the entire electromagnetic spectrum, from x-rays through the visible to microwaves. The distribution of this radiation as a function of wavelength is described approximately by a black body raised to a temperature of about 5500K (approximately 9400 °F). Ozone in our atmosphere acts as a filter for the ultraviolet portion of solar radiation. Ultraviolet radiation contains sufficient energy per photon to break chemical bonds in DNA. The development of an ozone shield appears to have been crucial for the spread of life out of the ocean and onto the land (Berkner and Marshall 1965).

## 5.1 Stratospheric Ozone, Its Abundance and Variability

The global abundance of stratospheric ozone is determined by a balance between its production by solar ultraviolet radiation and its loss by catalytic chemical reactions of the oxides of hydrogen, nitrogen, chlorine and bromine. The balance determines the average amount of ozone present. Winds in the stratosphere blow ozone around and eventually move it from regions where it is produced to regions where it is destroyed by chemical reactions. The winds are variable and the result is a distribution of ozone amounts that have variations with latitude, longitude and season (Dobson et al. 1946). Daily variations in ozone resemble meteorological maps.

The total ozone column is the total amount of ozone in a vertical column overhead, see Figure 5.2. It is measured in a unit called the Dobson Unit (DU), named after Gordon M B Dobson, who designed a spectrophotometer in the 1920s that is still in use around the world today measuring ozone (see e.g. Branstedt et al. 2003; Ziemke et al. 2006). A typical amount of ozone is about 300 DU, which is equivalent to a layer of pure ozone of about 3 mm thickness at standard temperature and pressure. The white areas near the poles during winter indicate no data because the Total Ozone Mapping Spectrometer (TOMS) instrument cannot measure in the darkness.

The sequence in Figure 5.3 shows a similarity to weather maps as the highs and lows travel in an eastward direction over the continent. Note particularly the area around the city of Chicago at the southern end of Lake Michigan. On 1 February the map indicates green to yellow, or about 350 DU, with a front of high ozone approaching from the west. On 2 February the high ozone reaches the Great Lakes and the ozone amount over Chicago is greater than 400 DU. On 3 February the front has passed to the east and Chicago has less than 340 DU (green on the map). On 4 February a smaller secondary maximum of about 370 DU passes over the city.

## 5.2 Chemicals that Destroy Ozone

In the early 1970s, we began to realize that humans produce chemicals that potentially destroy ozone in sufficient amounts to affect the global balance (Crutzen 1971; Johnston 1971). Reactive or soluble industrial compounds may pollute the atmosphere locally, but they are removed from the atmosphere by rainfall. The key to affecting stratospheric ozone is non-reactive compounds like chlorofluorocarbons (CFCs) (Molina and Rowland 1974). In addition to being unreactive, they are insoluble and accumulate in the atmosphere. They eventually drift up into the stratospheric ozone layer where there is uv light to break them apart and release reactive chlorine.

The oxides of chlorine, bromine, nitrogen and hydrogen act as catalysts for the destruction of ozone. Catalysts speed up chemical reactions without being used up. There is normally very little

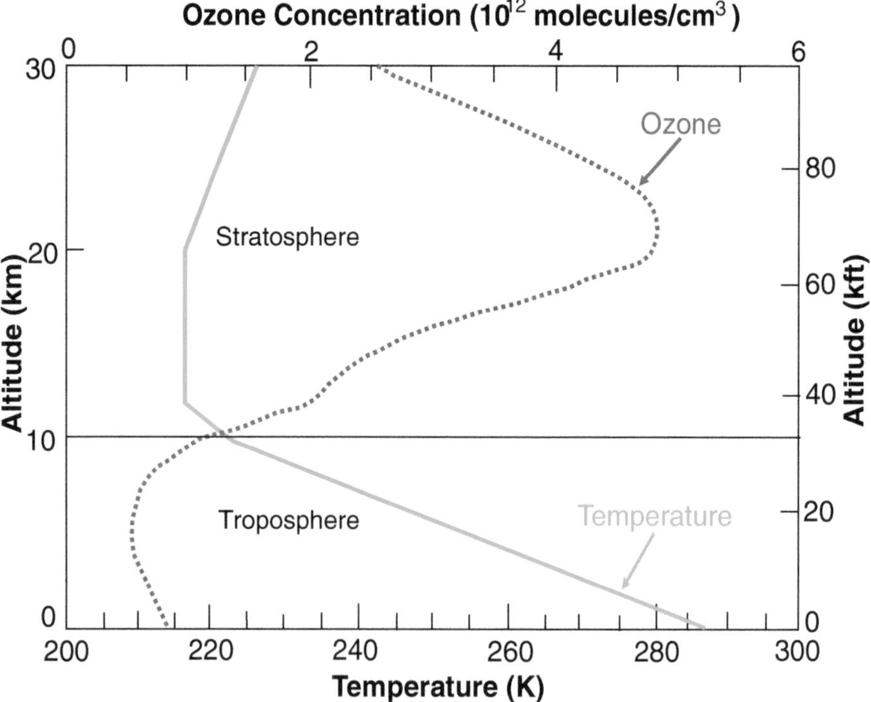


Figure 5.1 Plot of temperature and ozone concentration versus altitude.

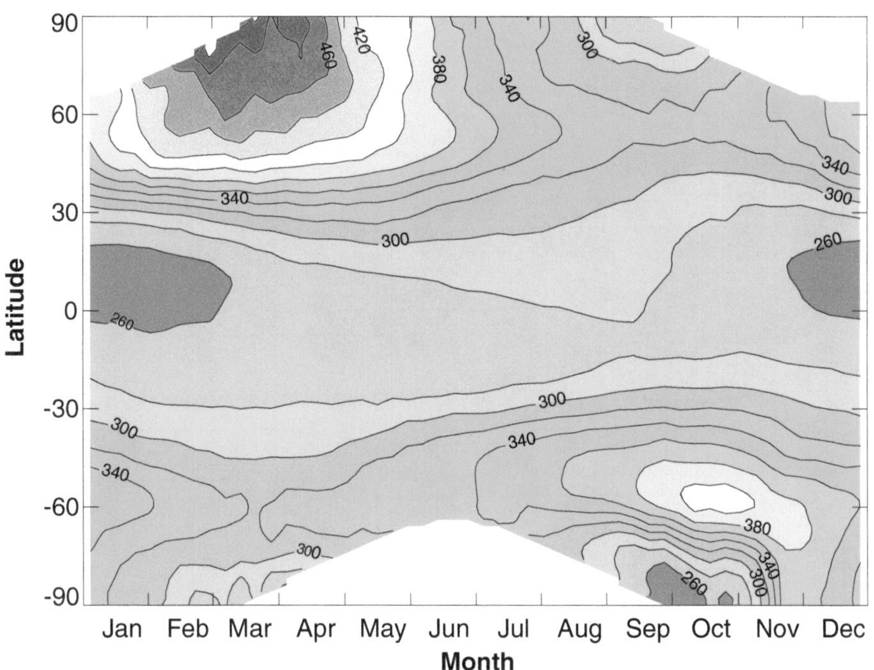


Figure 5.2 Total column amount of ozone measured by the Total Ozone Mapping Spectrometer (TOMS) instrument as a function of latitude and season.

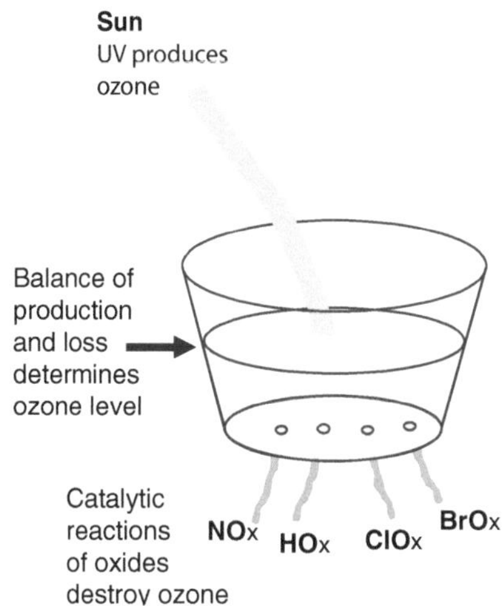


Figure 5.4 Leaky bucket analogy for ozone in the stratosphere.

chlorine in the stratosphere (less than a part per billion). Industrially produced CFCs have more than tripled the amount of chlorine in the stratosphere. The catalytic cycle of chlorine reactions that destroy ozone involve the reaction of chlorine atoms with ozone to form chlorine monoxide (ClO). Chlorine monoxide reacts with atomic oxygen to reform a chlorine atom and molecular oxygen ($O_2$). The net result is that an oxygen atom and an ozone molecule have been recombined to form two oxygen molecules with the chlorine atom available to start the cycle again. During its lifetime in the stratosphere, a chlorine atom can recombine with about ten thousand oxygen molecules. Thus, parts per billion of chlorine can have a significant effect on parts per million of ozone.

A good analogy for understanding the ozone layer is a leaky bucket shown in Figure 5.4. If water is poured into a bucket continuously and that bucket has holes in the bottom, the water level in the bucket will build up until the pressure forces water out of the holes at a rate that just equals the rate at which water is pouring into the bucket. The pouring of water into the bucket represents ozone being created by ultraviolet sunlight. The holes represent catalytic loss of ozone by the reactions of the hydrogen, nitrogen, chlorine and bromine oxides. The level of the water represents the total amount of ozone present in the stratosphere. Ozone depletion can occur when one of the holes is made larger; for instance by adding chlorine from chlorofluorocarbons. The ozone level goes down until the production of ozone is again just balanced by the loss of ozone. Water will flow out at a faster rate and the water level will adjust downward. If the hole is later made smaller, the water level will again rise. It is the same with ozone. If the catalytic loss is decreased, the ozone amount will rise. Ozone is a renewable resource!

CFCs are not the only chemical with a potential for reducing the steady-state amount of ozone. Other potential causes of long-term ozone change (human and natural) include:

- Nitrogen oxides ($NO_x$) from supersonic aircraft (historically the first considered).
- $NO_x$ from nitrogen fertilizers.
- Bromine from methyl bromide used as a fumigant.
- Bromine from Halons used in fire extinguishers.
- Aerosol particles from volcanic eruptions.
- Variations in solar ultraviolet (11-year sunspot cycle and 27-day solar rotation period).
- Atmospheric oscillations; quasi-biennial, El Niño.

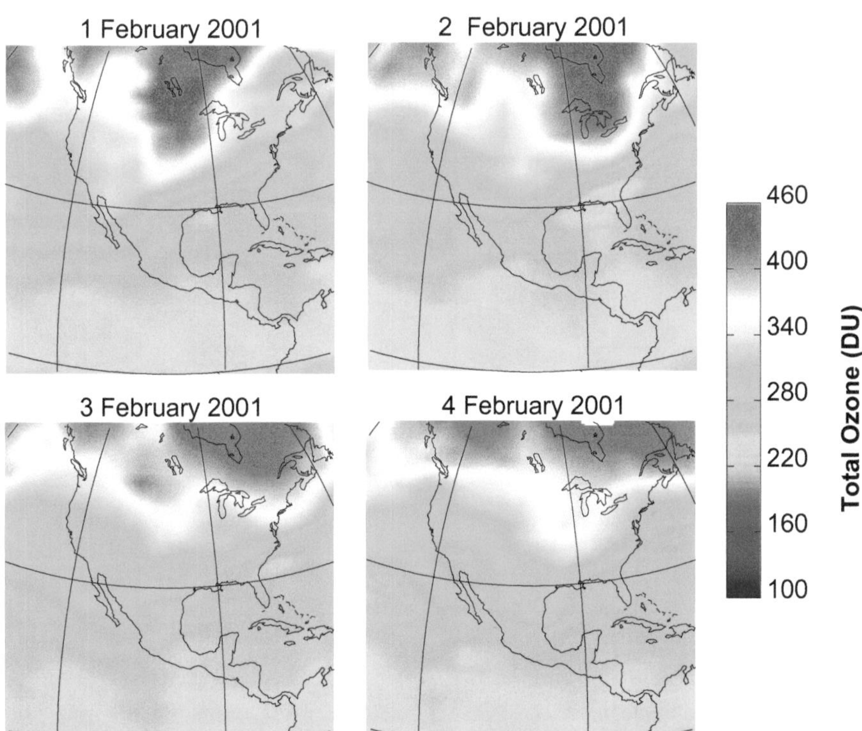


Figure 5.3 A four-day sequence of total ozone column measurements over North America.

## 5.3 The Antarctic Ozone Hole

It was 1985. The debate over the impact of chlorofluorocarbons had been going on for 10 years. The scientific case had been confirmed by many measurements of chlorine-depleting chemicals. Negotiations were in progress to develop what became known as the Montreal Protocol to limit ozone-depleting substances. However, no actual trends had been observed in ozone and calculations indicated that they should not have been seen in the data up to that time. Then, a big surprise: the British Antarctic Survey announced a large change in ozone over their Halley Bay station on the Antarctic ice shelf (Farman et al. 1985). They observed a 40 per cent decrease in ozone during the month of October from the 1970s to 1985. This was confirmed by the Total Ozone Mapping Spectrometer (TOMS) (Stolarski et al. 1986). TOMS was an instrument on the Nimbus 7 satellite, launched in late 1978. It looked downward on the atmosphere from a height of 900 km, to observe reflected solar radiation at six wavelengths. The radiation from some of these wavelengths was absorbed by ozone. By taking the difference between absorbed and unabsorbed wavelengths, the total amount of ozone in an atmospheric column could be deduced. Daily maps made from TOMS data showed that the very low ozone amounts seen at Halley Bay were occurring over much of the Antarctic region (Figure 5.5).

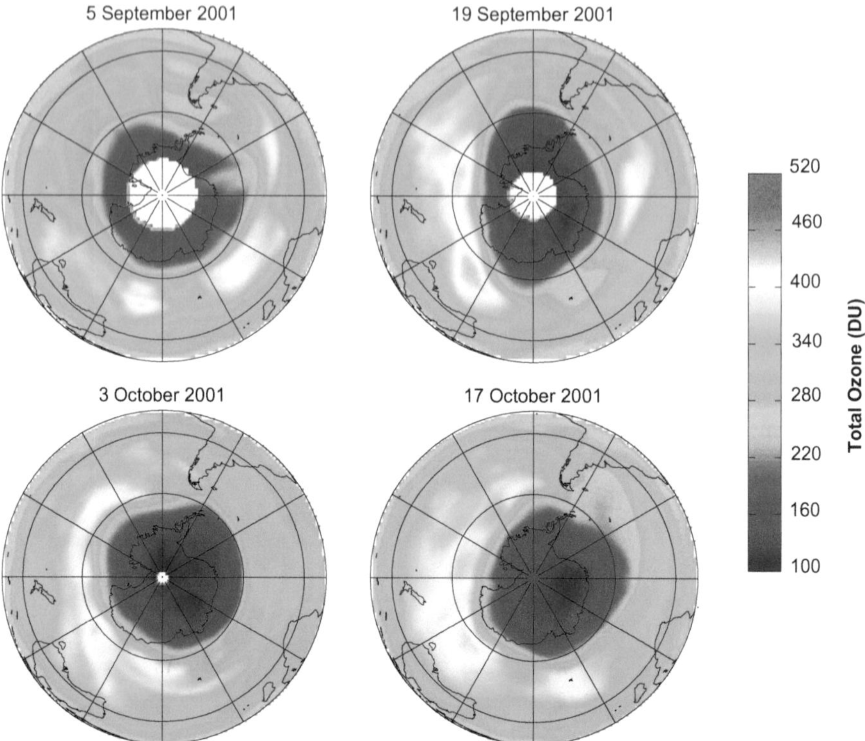


Figure 5.5 Sequence of four maps of ozone measurements made by the Earth Probe TOMS satellite instrument over the Antarctic during the ozone hole period of 2001.

The colours in Figure 5.5 indicate the total column amount of ozone in Dobson Units. Regions of dark blue to purple to black are what is usually termed the ozone hole, where the total column ozone amount is less than 220 DU. The maps are polar orthographic projections, showing the earth as it would look from a great distance above the South Pole. The hole has formed in the upper left map on 5 September. The blue region of less than about 220 DU covers an area slightly greater than the area of the Antarctic continent. The white region near the pole is still in near darkness such that the satellite is unable to make measurements.

Although the ozone hole was a great surprise to the research community, the outline of an explanation for this phenomenon came quickly (Solomon et al. 1986; McElroy et al. 1986; Crutzen and Arnold 1986).

The extreme cold temperatures of the Antarctic stratosphere lead to the presence of polar stratospheric clouds or PSCs (Steele et al. 1983). Reactions on the surfaces of the cloud particles produce chlorine from the inactive reservoirs of hydrogen chloride (HCl) and chlorine nitrate ($ClNO_3$). In most of the stratosphere, the active form, chlorine monoxide (ClO), makes up less than 0.5 per cent of the available chlorine. During the Antarctic winter, the reaction between hydrogen chloride and chlorine nitrate on the surface of PSCs converts most of these compounds to chlorine gas ($Cl_2$). When the sun comes up in the Antarctic spring, the $Cl_2$ is converted by sunlight to ClO. The ClO begins to destroy ozone. By the end of September, we have a full-blown ozone hole (Figure 5.5).

In Figure 5.6, the left globe shows the total concentration of ClO in a vertical column of the stratosphere. The dark blue region surrounding the pole represents peak concentrations of more than one part per billion in the lower part of the stratosphere. The lighter blue regions over the rest of the globe have concentrations of about 1/100th of a part per billion. The right globe shows the total amount of ozone in a vertical column of the stratosphere. This data is for 30 August, when the ozone hole is just beginning to form. The white area around the pole indicates where no data were taken.

Shortly after the ozone hole was discovered in 1985 by Joe Farman and colleagues at the British Antarctic Survey, aircraft missions were organised to fly over Antarctica, measuring key chemical constituents in the stratosphere. These missions established that the key to the formation of the ozone hole was the chemical destruction of ozone by chlorine oxides (e.g. Anderson et al. 1989). Over the next few years these measurements, combined with

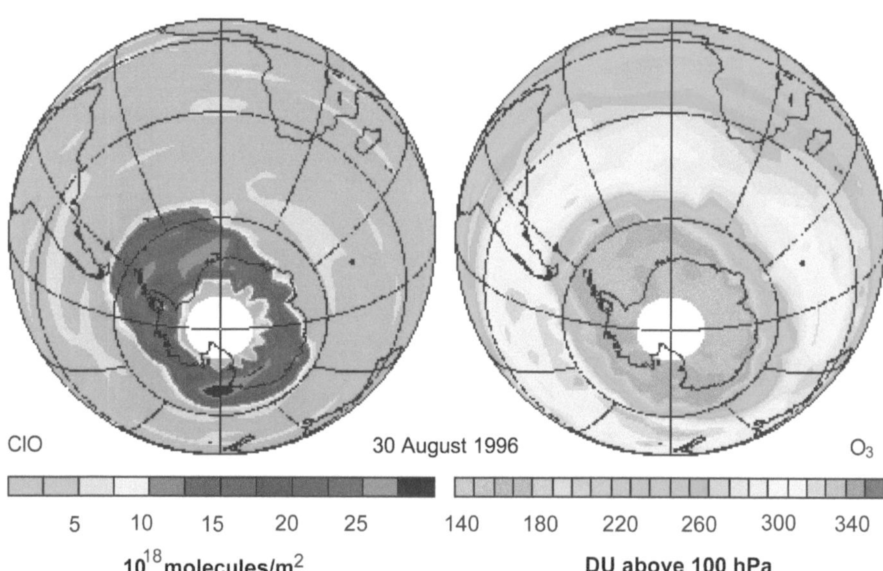


Figure 5.6 Measurements of chlorine monoxide (ClO) and ozone ($O_3$) made by the Microwave Limb Sounder (MLS) experiment on the Upper Atmosphere Research Satellite (UARS).

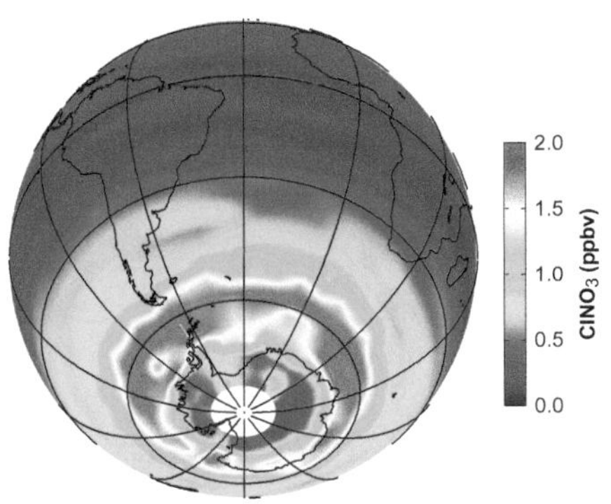


**Figure 5.7** Measurements of chlorine nitrate ($ClNO_3$) during early September 1992 by the Cryogenic Limb Array Etalon Spectrometer (CLAES) on the UARS satellite.

laboratory work on chemical reactions (e.g. Molina and Molina 1987), established a mechanism that could generate an ozone hole similar to that observed. Subsequent satellite measurements mapped the distribution of key chemicals over the entire Antarctic region (e.g. Santee et al. 2003). All of these together present a coherent theory of the formation of the ozone hole.

One aspect of that theory is the conversion of chlorine nitrate to ClO. We saw earlier that ClO was enhanced over the Antarctic. This occurs because $ClNO_3$ reacts with HCl and results in the removal of $ClNO_3$ from the polar region (Figure 5.7).

The measurements shown in Figure 5.7 are in parts per billion at about 20 km altitude. The red region indicates a high amount of $ClNO_3$ surrounding the ozone-hole region. The blue region over the pole indicates removal of $ClNO_3$ (Roche et al. 1994).

Another important factor in the formation of the Antarctic ozone hole is the removal of nitric acid ($HNO_3$). If $HNO_3$ is present when the sun comes up in the Antarctic spring, it will absorb sunlight and form nitrogen dioxide ($NO_2$). The $NO_2$ will react with the ClO to reform $ClONO_2$. There will be some ozone loss, but it will be limited. The PSCs in the Antarctic are actually formed from a combination of $HNO_3$ and water. These particles grow large enough that over the winter they slowly fall out of the stratosphere and take the $HNO_3$ with them. The result is a 'denitrification' of the stratosphere (Figure 5.8). When the sun comes up, there are no available nitrogen oxides to react with ClO. So ClO destroys ozone from late August into early October without interference.

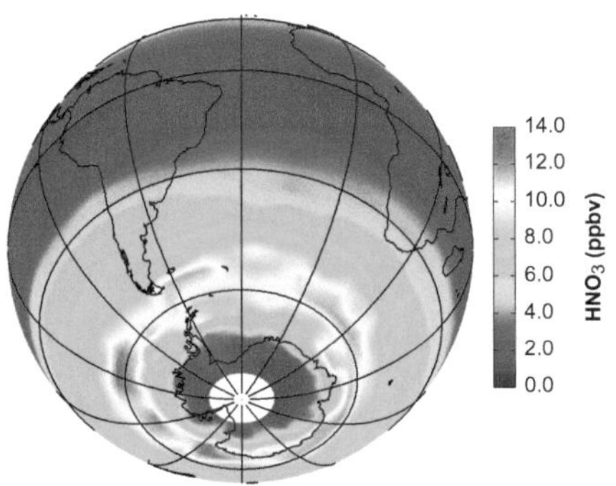


**Figure 5.8** Measurements of nitric acid ($HNO_3$) made by the CLAES instrument on the UARS satellite during early September 1992.

As in the case of Figure 5.7, measurements in Figure 5.8 are shown in parts per billion, for an altitude of about 20 km. The dark blue/purple region surrounding the pole indicates where 'denitrification' has removed virtually all of the available $HNO_3$.

As the amount of chlorine in the stratosphere grew during the 1980s and early 1990s, so did the size and depth of the ozone hole (Figure 5.9).

In 1979 there was a small minimum in total ozone over the polar region, surrounded by a crescent-shaped maximum. Throughout the 1980s the minimum got deeper, while the surrounding maximum also decreased in the amount of total ozone. Finally, from the 1990s to the present, the hole has stayed relatively constant in size and depth. Slow recovery is expected over the coming decades as chlorine is slowly removed from the stratosphere.

## 5.4 Arctic and Global Ozone

The Arctic region shows some similarities with the Antarctic, but also significant differences. An ozone hole does not occur in the Arctic, but significant ozone loss occurs each spring as chlorine and bromine plus sunlight destroy ozone. In the Arctic autumn the total column of ozone has a ring of maximum amount surrounding a shallow minimum over the pole, much as it does in the Antarctic autumn, As the winter progresses, stratospheric motions move ozone downward over the Arctic, resulting in a maximum over the pole. This build-up is stronger than that over the Antarctic. In the spring, ClO has also built up in the Arctic. When the sun comes up, the ClO begins to destroy ozone. However, the $HNO_3$ has not been significantly removed in the Arctic and $ClNO_3$ begins to be reformed, removing the ClO and limiting ozone depletion. Thus, springtime ozone in the Arctic has a maximum over the pole and some regions of ozone loss (Figure 5.10).

The major difference between the Arctic and Antarctic is that the Arctic has a more disturbed stratospheric circulation. This is because flow near the surface in the Arctic is disturbed by mountains and land-sea contrast. This causes wave motions that propagate upward into the stratosphere. These disturbances distort the vortex that isolates polar chemistry and tends to weaken that vortex. One result is more downward motion and accumulation of ozone, and warming of the atmosphere that tends to limit PSC formation. The polar disturbances are seen in the ozone field (as illustrated earlier in Figure 5.5).

The measurements shown in Figure 5.10 are made by observing ultraviolet light from the sun reflected off the earth to the satellite. The high patterns over the Arctic are associated with weather systems and move from day to day.

Because of the disturbed meteorology of the Arctic, the year-to-year variation of springtime ozone over that region is highly variable. A particularly interesting year was 1997, when the vortex was colder and more stable than most years. This resulted in a symmetric ozone loss over the pole that led to an ozone minimum in March of that year that looked like a miniature version of the Antarctic ozone hole (Figure 5.11).

One method to summarise the overall decrease in ozone is to calculate linear trends over the last two decades. This is done using statistical time-series models (see e.g. Stolarski et al. 1991). These models attempt to determine the factors leading to ozone change by determining the best fit of a linear combination of terms representing each of the major influences on ozone. These include a seasonal cycle, an 11-year solar sunspot cycle, the effects of volcanic aerosols, an internal atmospheric oscillation called the quasi-biennial oscillation (QBO) and a long-term trend. The long-term trend has its largest negative value in the Antarctic spring, as expected. It has a smaller negative value in the Arctic spring. Over the equator there is a slight positive trend that is not statistically significant (Figure 5.12).

The trends represented in Figure 5.12 are shown as a function of season and latitude. At each latitude they have been averaged around the globe. The largest negative linear trend occurs over the Antarctic in the spring, between September and November. These trends reach -30 per cent per decade, or a 60 per cent decrease over the two decades represented in the data. The trends are negative everywhere except near the equator, where they are slightly positive during northern spring and slightly negative during northern fall.

The linear trends discussed above have been averaged around a latitude circle. But there are variations around that latitude circle that are interesting. The trends for the month of March over the northern middle and high latitudes are generally negative. Their variation with specific location (Figure 5.13) ranges from strongly negative to barely positive. These variations are related to the variability of the underlying ozone amounts themselves. As the ozone amount varies over the 20-year record, a linear trend fit through that data will have some dependence on whether the ozone is high or low near the beginning or the end of the record. This leads to an uncertainty in the derived trend. Some studies indicate that the patterns of ozone trend mirror patterns in basic atmospheric oscillations, such as that called the North Atlantic Oscillation. When the data are averaged around the latitude circle, much of the variability is averaged over and the trends can be determined with less uncertainty.

The blue regions north of 30 °N indicate negative ozone trends of up to −8 per cent per decade. Green regions interspersed indicate smaller trends that are near zero, with a maximum of just over 2 per cent per decade.

Trends over the Southern hemisphere are larger and less variable. They are centred around the South Pole in the Antarctic spring (Figure 5.14).

Trends are given in percentages per decade, with the largest negative trend being -36 per cent per decade in the middle of the ozone hole. Trends around the equator are slightly positive

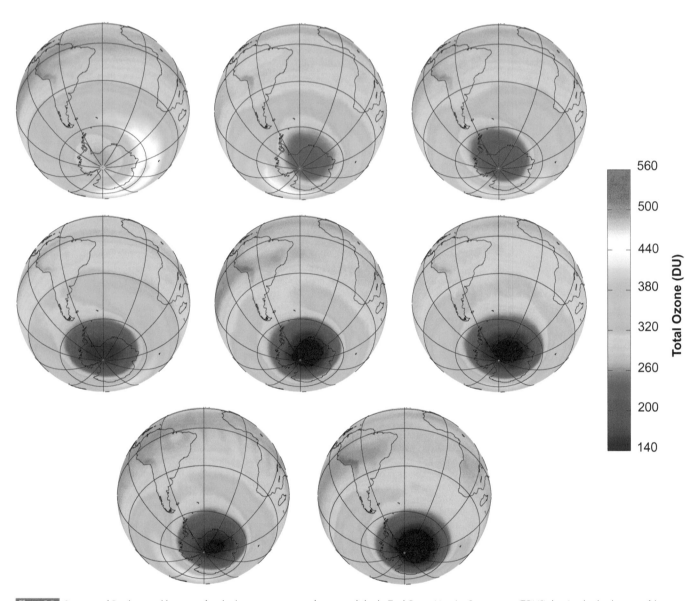

**Figure 5.9** Sequence of October monthly means of total column measurements of ozone made by the Total Ozone Mapping Spectrometer (TOMS) showing the development of the ozone hole over the southern polar region (from the top left, the sequence is 1979, 1982, 1984, 1989, 1997, 2001, 2003,2006). Data for 2006 are from OMI (Ozone Monitoring Instrument) on the Aura satellite.

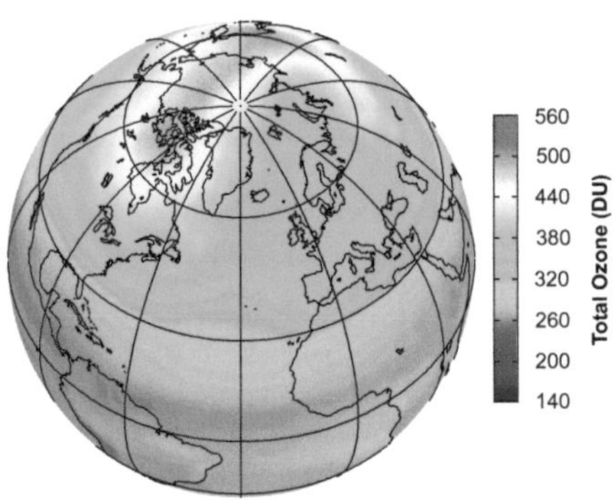


**Figure 5.10** Measurements of the total column amount of ozone for 25 March 1991 made by the Total Ozone Mapping Spectrometer (TOMS).

(approximately 0–2 per cent per decade). Note the change in scale from Figure 5.13.

## 5.5 Volcanic Eruptions and Stratospheric Ozone

Of the many causes of ozone variation, the effect of volcanic eruptions is particularly interesting. There are several different kinds of volcanic eruptions, which can be classified as explosive and non-explosive. Non-explosive eruptions are like those of Kilauea crater in Hawaii. A lot of lava flows from them and gases are released. They occasionally spew lava and sparks a few hundred feet in the air. All this occurs near the surface and has little effect on the stratosphere. Another example is Mount Erebus in Antarctica. Although the stratosphere dips to relatively low altitudes over Antarctica, the eruptions of Mount Erebus put little debris into the stratosphere.

Explosive eruptions can inject material directly to the stratosphere. These occur sporadically. The most important recent eruption was that of Mount Pinatubo in the Philippines in June 1991. The eruption sent more than 5 billion cubic metres of ash and debris to altitudes in excess of 30 km. The larger pieces of debris fell rapidly back to earth, with no lasting impact on the stratosphere. The ash and smaller particles drifted downward and also had little impact. The key to volcanoes affecting the stratosphere is the gas that is emitted.

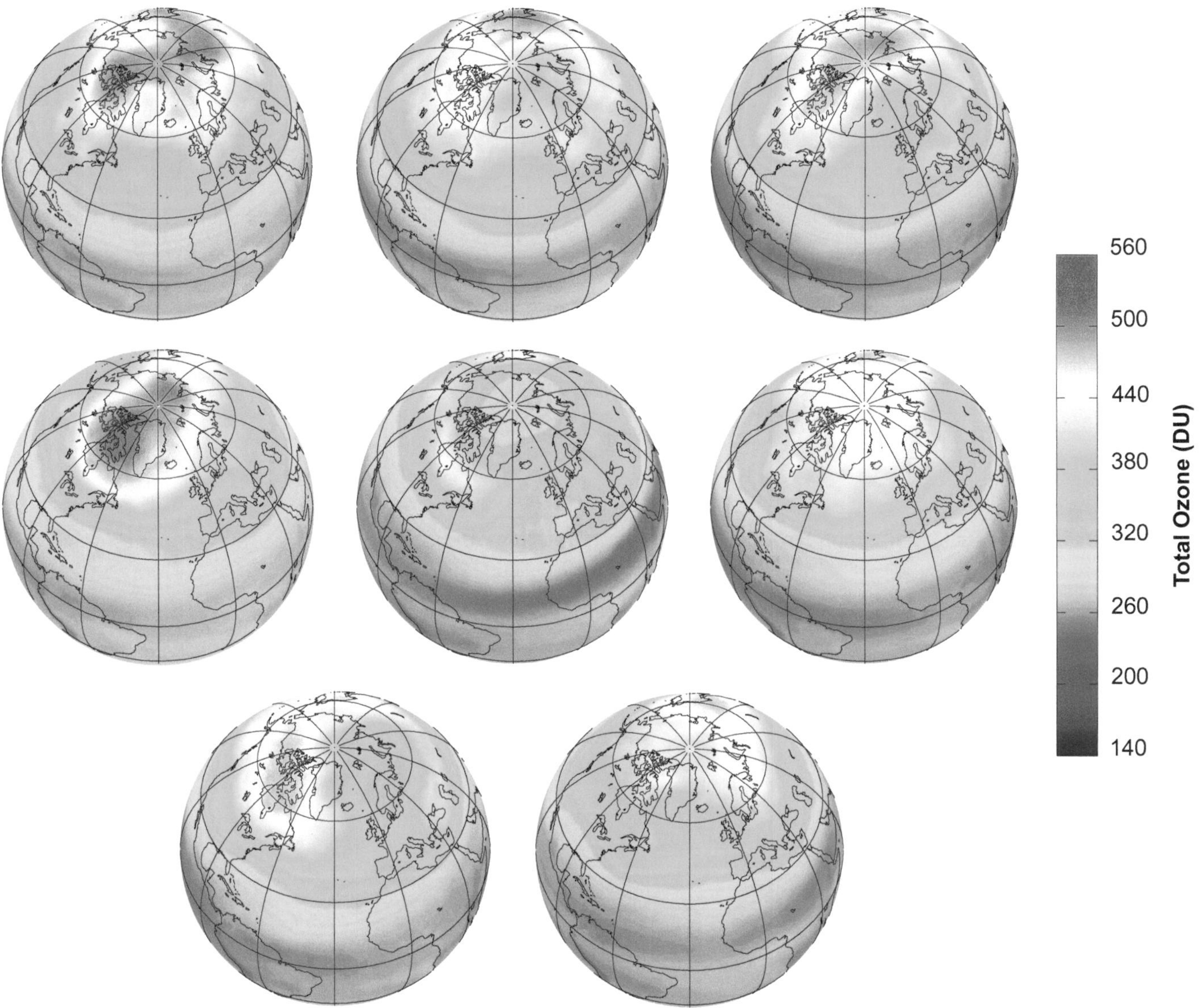


Figure 5.11 Sequence of March monthly mean total ozone measurements made by the Total Ozone Mapping Spectrometer (TOMS) over the northern polar region (from the top left, 1979, 1982, 1984, 1989, 1997, 2001, 2003, 2006). Data for 2006 are from OMI (Ozone Monitoring Instrument) on the Aura satellite).

The most important gas emitted from explosive volcanoes for the stratosphere is sulphur dioxide ($SO_2$) which can be observed with satellite probes (e.g. Khokhar et al. 2005). The gas does not fall out of the stratosphere. Turbulent motions of the air keep it suspended while winds blow it around the globe (Figure 5.15). Sulphur dioxide injected into the stratosphere will slowly react with hydrogen oxides and water to form sulphuric acid ($H_2SO_4$). Sulphuric acid and water then begin to form small particles of sulphate. These particles are small enough that they remain suspended by turbulent motions in the stratosphere. They are spread throughout the globe (Figure 5.16) and are removed when downward motion brings the air back into the troposphere where particles can be dissolved in rain and removed.

The upper left panel shown in Figure 5.15 was the first one available after the launch of UARS, three months after the eruption. The high concentrations of $SO_2$ are indicated in red forming a band around the equator. By the last panel, two months later, the concentrations have been reduced to near zero as the $SO_2$ is converted into stratospheric aerosols.

The upper left panel of Figure 5.16 shows a typical clean stratosphere before the Mount Pinatubo eruption (Trepte et al. 1993). The upper right panel shows the aerosols shortly after the eruption. The lower left panel shows the spread of these aerosols created from $SO_2$ oxidation several months after the eruption. The lower right panel shows the residual aerosols more than two years after the eruption.

The surfaces of small sulphate particles catalyse reactions that convert the oxides of nitrogen to nitric acid. Removal of nitrogen oxides prevents the formation of chlorine nitrate, thus allowing ClO to destroy ozone more effectively. Thus the effect of the Pinatubo eruption is to decrease ozone by several per cent. This decrease in ozone then recovers as the sulphate particles are removed from the stratosphere in the next couple of years.

Mount Pinatubo was not the only explosive volcanic eruption in recent decades. In April 1982, El Chichón in Mexico erupted material well up into the stratosphere. Mount St Helens in Washington State erupted in May 1980. This was an explosive eruption with little effect on the stratosphere. Mount St Helens erupted somewhat sideways so that its ash and gas were not lifted to great heights. The record of stratospheric aerosols can be measured from satellites by observing the extinction of sunlight as it passes through the limb of the atmosphere.

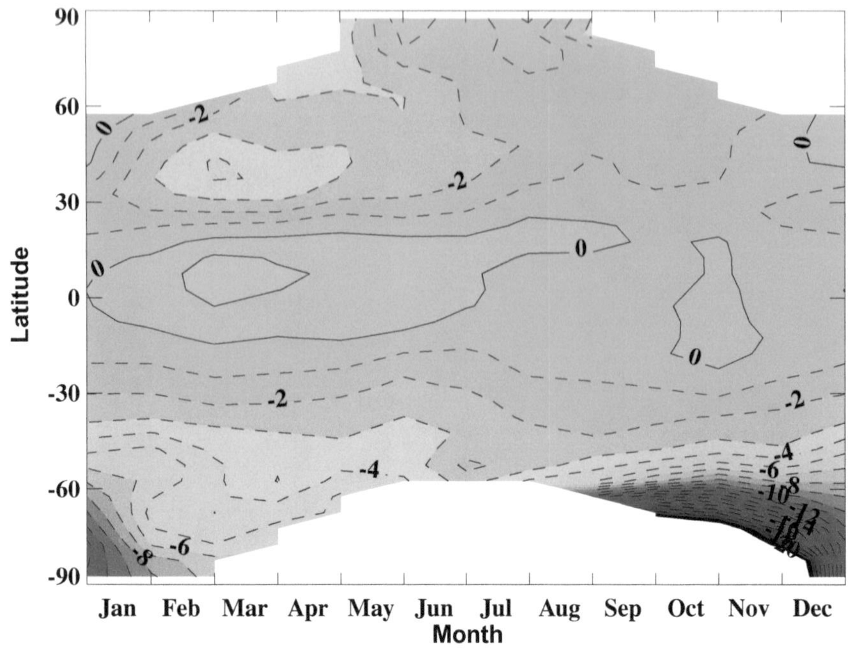


Figure 5.12 Linear trends in percentage per decade, calculated from total ozone data between 1979 and 1999.

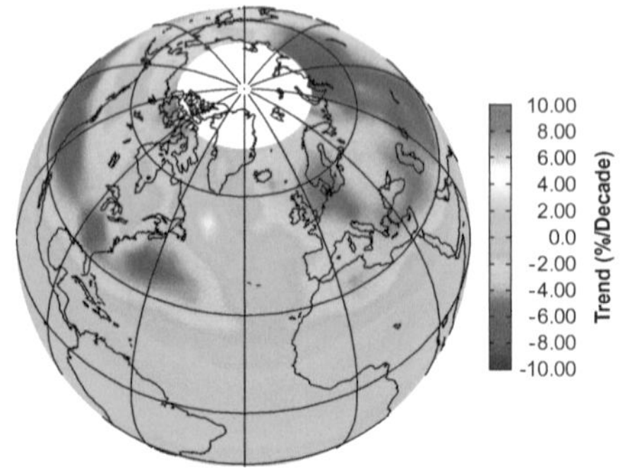


Figure 5.13 Linear trends for the month of March calculated using TOMS data from 1979 to 1999.

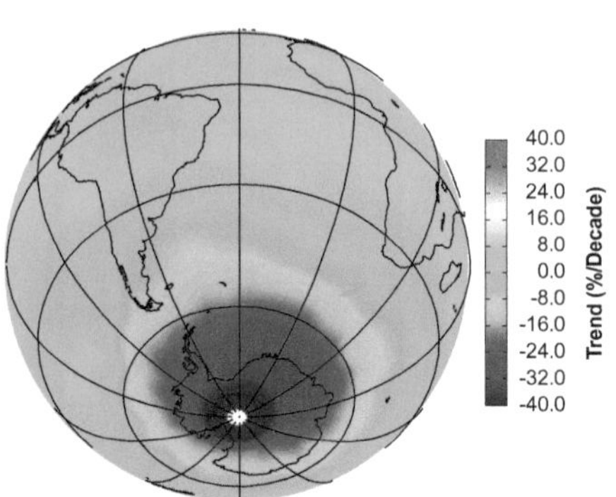


Figure 5.14 Linear trends for the month of October calculated using TOMS data from 1979 to 1999.

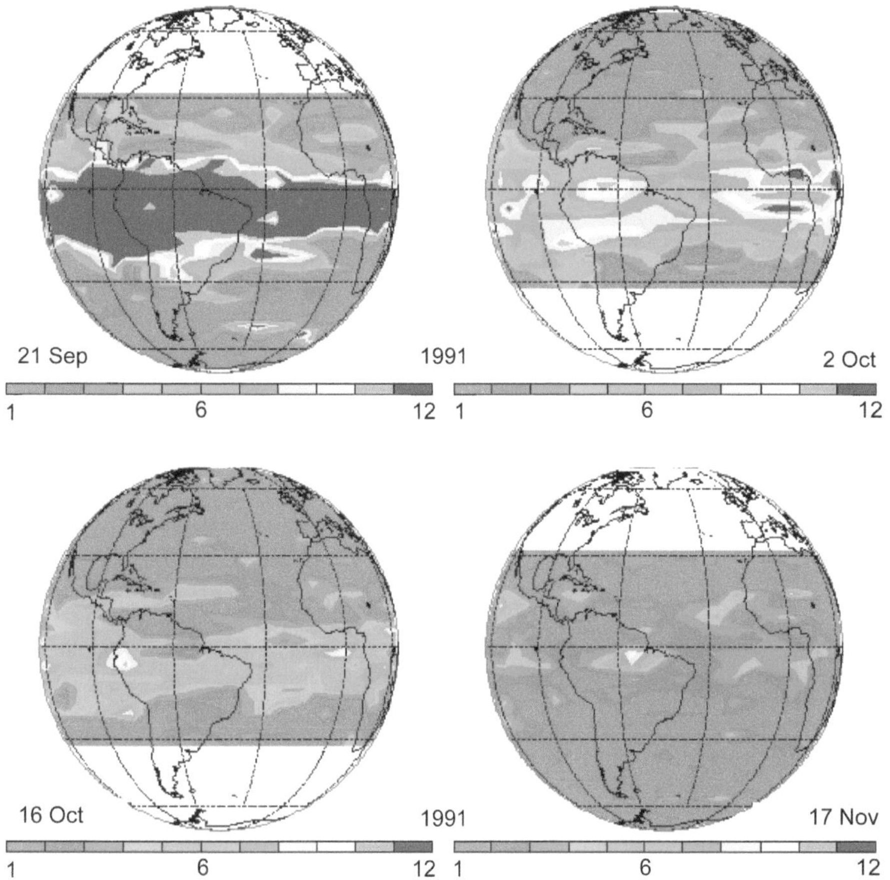


Figure 5.15 Sulphur dioxide ($SO_2$) amount measured by the MLS instrument on the UARS satellite several months after the eruption of Mount Pinatubo.

The previous large explosive eruption before El Chichón was Mount Agung on the island of Bali in Indonesia in 1963. Before that there had been very few explosive eruptions in the previous 50 years. During the period from 1780 to 1840 there were many explosive eruptions, including that of Tambora. The aerosols from Tambora blocked the sun and resulted in no real summer occurrence in Europe in 1816. The effects of these earlier eruptions on ozone may have caused an increase by the removal of nitrogen oxides that catalytically destroy ozone. Chlorine concentrations would have been small and not a factor.

## 5.6 Where Are We Going With Stratospheric Ozone?

We have conducted a large-scale global experiment over the last several decades by releasing chlorine- and bromine-bearing compounds into the atmosphere. We raised the stratospheric chlorine loading from less than one part per billion to about three and a half parts per billion. Ozone levels have declined and the Antarctic ozone hole has formed. In response, more than a hundred countries throughout the world have worked

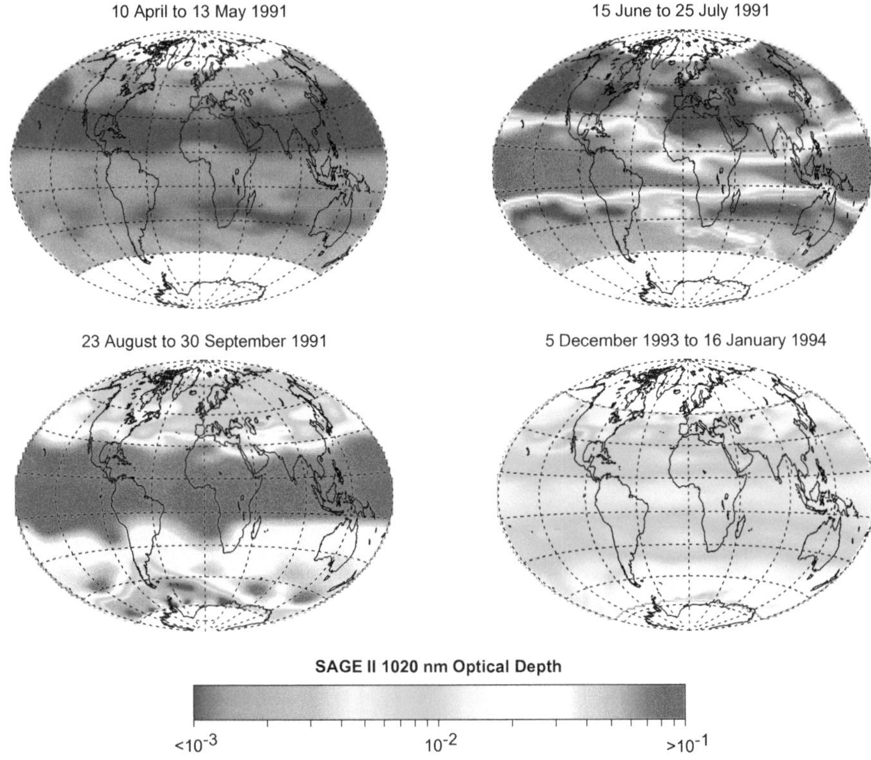


**Figure 5.16** Total optical depth of stratospheric aerosols as measured by the Stratospheric Aerosol and Gas Experiment II (SAGE II).

together to formulate the Montreal Protocol to limit ozone-depleting substances (Benedick 1991). There have been subsequent amendments to strengthen the protocol. Chlorine in the stratosphere has levelled out and is about to begin its slow decline in response to the provisions of the protocol. We have embarked on the ambitious experiment to return chlorine and ozone to their pre-ozone-hole levels (Andersen and Sarma 2002).

Will ozone recover to be the same as before the increase in chlorine? Will the ozone hole disappear? We are quite confident that ozone will increase as chlorine decreases in the stratosphere. However, the recovery is occurring within a climate system that is variable and being driven slowly towards a warmer surface and a cooler stratosphere by increased greenhouse gases. Methane ($CH_4$) is increasing and putting more water vapour into the stratosphere. Could cooler Antarctic temperatures with more water vapour lead to increased clouds and cause the ozone hole to persist despite lower amounts of chlorine? We are not sure. Ozone is an absorber of the sun's ultraviolet radiation. This absorption leads to heating of the stratosphere, affecting the wind systems. The entire system exists in interlocking feedback loops so that changes in ozone interact with other changes in climate. Research in the coming years should begin to unravel the chemistry–climate connection.

# ENVIRONMENTAL AND HEALTH IMPACTS OF AIR POLLUTION

*Mike Ashmore, Wim de Vries, Jean-Paul Hettelingh, Kevin Hicks, Maximilian Posch, Gert Jan Reinds, Fred Tonneijck, Leendert van Bree and Han van Dobben*



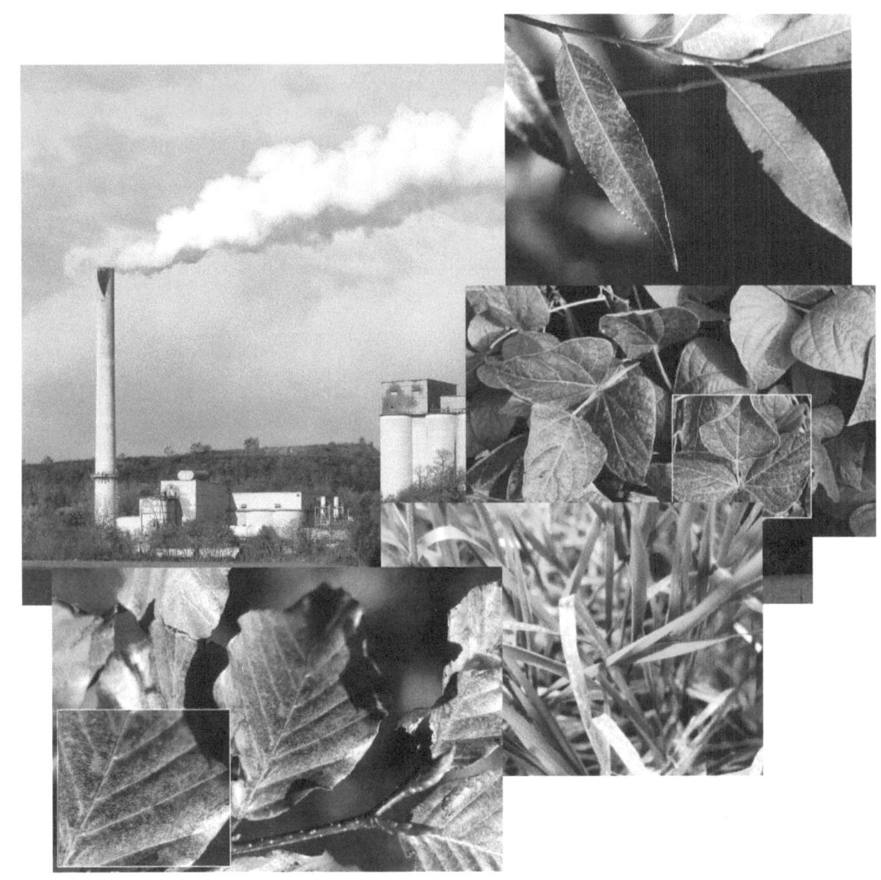

Examples of damage to plants resulting from air pollution. (Source: UNECE 2002.)

Air pollution is known to have a range of effects, including those on human health, crop production, soil acidification, visibility and corrosion of materials. This Chapter focuses on the two major impacts of air pollution that have most strongly influenced the development of policies to reduce emissions: those on the natural environment and on human health.

In broad terms, the major impacts of air pollution on the natural environment can be placed into three categories, representing different spatial scales:

- Local impacts of major industrial or urban sources, for example, instances of damage to ecosystems and crop production close to emission sources. Historically, the biggest impacts have been through the direct effects of sulphur dioxide and particles – either around large point sources such as power stations and smelters, or in urban areas with domestic coal burning – and the accumulation of toxic metals in soils around smelters. However, a range of other pollutants from specific local sources can have direct impacts on vegetation.
- Regional impacts of ozone, which is a significant global air pollutant in terms of impacts on vegetation, since high concentrations are found in rural areas.
- Regional impacts of long-range transport and deposition of sulphur and nitrogen, which have effects on soil acidity, nutrient availability and water chemistry, and hence on ecosystem composition and function.

The Chapter first considers direct effects of air pollution on vegetation and the visible symptoms of damage that can result, illustrating the spatial variation in damage by reference to national and local studies in the Netherlands. Impacts of sulphur and nitrogen deposition on soils, forest health and biodiversity on a European scale are then discussed, with particular emphasis on the development of methods of risk assessment (through the critical load approach) which have led to international agreements on measures to reduce pollutant emissions. A global perspective is also provided, with brief case studies of the impacts of local pollution sources, ozone and nitrogen deposition. The health impacts arising from exposure to pollutants such as ozone and particulate matter are then examined before reviewing pollution abatement strategies and the resulting health benefits.

## 6.1 Direct Effects of Air Pollution on Plants

It has been known for centuries that ambient air pollution can affect plants adversely. Many cases of visible plant injury have been recorded in the vicinity of point sources and near industrial areas. Sulphur dioxide from coal combustion and smelters, and hydrogen fluoride from superphosphate and glass factories, are among the main air pollutants that can reach phytotoxic levels at local scales. Sulphur dioxide and hydrogen fluoride, together with compounds such as nitrogen oxides and particulates, belong to the category of primary pollutants since they are emitted directly into the atmosphere.

New air pollution problems arose with the first observations of visible symptoms on agricultural crops in the Los Angeles area in the 1940s. Richards et al. (1958) showed that ozone was a phytotoxic constituent of this air pollution complex, while Stephens et al. (1961) discovered that the typical bronzing of leaves was caused by peroxyacetyl nitrate (PAN). Pollutants such as ozone and peroxyacetyl nitrates are formed in the atmosphere as a result of chemical reactions of hydrocarbons and nitrogen oxides under the influence of sunlight. Ozone is the most important constituent of photochemical air pollution and has now become the most important phytotoxic pollutant throughout the industrialized world, causing leaf injury and yield losses in many arable and horticultural crops (Krupa and Manning 1988).

Effects of air pollutants can occur at different organizational levels, ranging from plant cells and organs to plant communities. Besides primary effects, air pollutants can also cause secondary effects by predisposing plants to drought, frost and pathogens. This section gives a brief overview of the direct effects of gaseous air pollutants on plants and attempts to highlight the various aspects that are relevant for a first understanding of the relationships between ambient exposures and plant responses. Indirect effects through deposition of pollutants on soils and waters not discussed in Section 6.2. More extensive information is provided by Flagler (1998) and Yunus and Iqbal (1996).

### Plant Response to Gaseous Air Pollution

Plant responses to air pollution range from clearly visible injury to subtle changes at the biochemical or physiological level. The type and magnitude of these responses to a given air pollutant depend on exposure characteristics, external growth conditions and plant properties (Figure 6.1; see also Guderian et al. 1985). Exposure concentration and duration are basic to an understanding of pollutant effects on vegetation and to the development of air quality criteria and standards. Short-term exposures to high concentrations generally result in visible acute injury. Chronic exposures to relatively low concentrations can cause physiological alterations that can result in growth and yield reductions or a reduction in reproductive capacity. These physiological alterations can occur without visible symptoms.

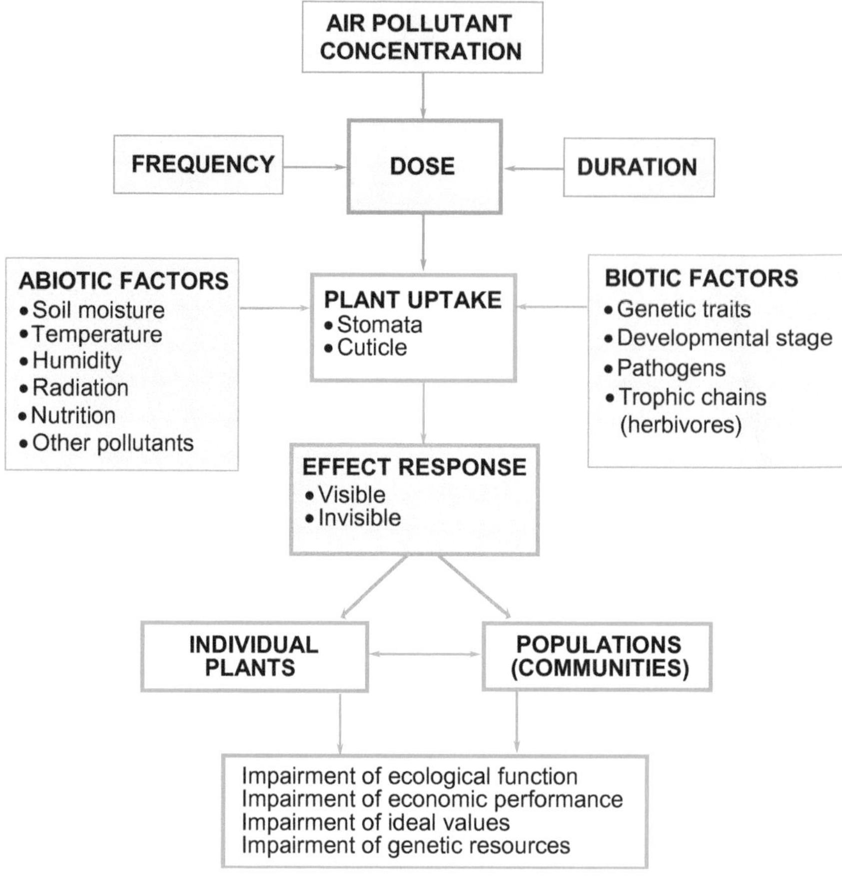


Figure 6.1 Plant responses to air pollutants and the factors involved (adapted from Guderian et al. 1985).

Gaseous air pollutants are primarily taken up via the stomata of plant leaves since the waxy cuticle of the leaf surfaces generally restricts diffusion. Gaseous uptake depends on factors such as total leaf area, stomatal density and stomatal resistance. This resistance is strongly influenced by internal and external growth factors. The cuticular pathway cannot be neglected, however, especially for volatile organic compounds that are soluble in the wax phase of the cuticle (Schönherr and Riederer 1989).

External growth factors modify plant responses by influencing the uptake of air pollution and the physiology of the plants. These factors relate to climate and soil conditions, and the presence of other air pollutants, pathogens and pests. Stomatal resistance to pollutant uptake is regulated by the stomatal aperture, which is influenced by environmental factors such as water deficit, $CO_2$ concentration and light intensity. Plants growing on soil with low water content, for example, are likely to decrease their stomatal aperture, thus reducing the uptake of pollutants via the stomatal pores (Jones 1992).

Biotic factors outside the plant, as well as characteristics of the plant itself, may influence pollutant uptake and plant response. Differences in genetic constitution form the basis for differential sensitivities of plant species and varieties. The developmental stage of the plant will influence the type and degree of plant uptake and reaction to a given air pollutant. Biotic stress factors, such as insect infestation, and viral, bacterial or fungal pathogens, may interact with air pollutants, usually weakening the plant and increasing its susceptibility to injury.

Pollution-induced injury may lead to:

- Impairment of ecological functioning, that is, changes in species composition, expansion of eroded areas, inhibition of water and climatic stabilization and damage to consumers (e.g. fluorosis in cattle and sheep).
- Impairment of economic performance, that is, reduced crop yield, economic loss and visible damage to ornamentals.
- Impairment of ideal values, that is, reduction of scientific and aesthetic values.
- Impairment of genetic resources, that is, reduced genetic variability, and reduced abundance of plant species and genotypes.

## Mode of Action

Injury and damage result from biochemical and physiological reactions within the leaves once the pollutants have entered the intercellular spaces through the stomata. Following the sorption of pollutants on to the wet cell surfaces, liquid phase reactions, including diffusion and eventual reactions with scavenging systems, control further pollutant movements within the leaves. Thus, the amount of pollution that reaches the target sites is influenced by factors such as solubility, absorption rate, transport, metabolism and detoxification processes.

The initial phytotoxic event results from air pollution-induced changes in the structure or function of leaf cells. These target sites are different for the various pollutants. For example, the primary sites for ozone reactions are the cellular membranes resulting in leakage of cell contents into the intercellular spaces. Following uptake, sulphur dioxide is dissolved in the cell walls to form bisulphite and sulphite, which inhibit enzyme activity, resulting in accumulation of oxidation products of phenolic compounds and death of cells. Nitrogen oxides taken up by plants are dissolved and dissociate to form nitrite, nitrate and protons. Protons lead to increasing cellular acidity and when the nitrogen compounds cannot be sufficiently reduced to amino acids and proteins, the highly toxic nitrite ion may accumulate. Exposure to hydrogen fluoride results in the accumulation of fluoride in the leaf margins where it forms complexes with metal ions at the active sites of some enzymes. Plants try to re-establish their normal metabolic states after pollution-induced perturbations at the target sites by repair and compensatory processes.

## Symptomatology and Relative Sensitivity of Plants

Visible injury in plants is often the first sign of enhanced levels of atmospheric pollution. All air pollutants known to visibly affect plants cause a range of injury symptoms. These symptoms have been classified as being either acute or chronic. Acute injury can result when plants are exposed to high concentrations, usually for short durations. The type of acute injury depends on the nature of the pollutant and the plant species and includes symptoms such as necrotic spots (e.g. on tobacco leaves exposed to ambient ozone) and necrosis between the veins or at the leaf tips and margins (e.g. in gladiolus exposed to ambient hydrogen fluoride). Chronic injury can result when plants are exposed to low, sub-lethal concentrations of ambient pollution for an extended time period and generally include symptoms such as yellowing and pigmentation of leaves and needles, and premature senescence.

The symptoms observed on field-grown plants following exposure to air pollutants are often not specific, but can be caused by entirely different pollutants or by other so-called mimicking factors. Thus, several aspects should be considered when attempting to determine the type of pollutant that caused the observed injury. Injury by primary pollutants such as sulphur dioxide and hydrogen fluoride generally occurs at a local scale around emission sources, whereas injury by the photochemical oxidants can be observed at regional scales. There are many listings (see for example, Taylor et al. 1998; Flagler 1998; Bell and Treshow 2002) of the relative sensitivity of plant species to ambient air pollution. These listings are generally based on the extent of visible foliar injury that is observed on plants after pollutant exposure. A brief selection of crops, trees and native herbaceous species that are well known for their high sensitivity to particular pollutants in terms of foliar injury is listed in Table 6.1.

## Biomonitoring

It is well known that certain plant species or cultivars respond to air pollutants at concentrations much lower than those that elicit responses in humans and animals (Manning and Feder 1980). Sensitive plant species, which respond with rather specific and visible symptoms, have been used as indicator plants to monitor air pollution-induced effects in many countries. Other plant species can readily accumulate specific air pollutants without symptoms becoming visible. If the compounds accumulate in the plant, material can be analysed easily and these plants may be used as accumulators (Posthumus 1982). The concepts of biomonitoring have been summarised extensively by Falla et al. (2000).

The idea of biomonitoring goes back to the nineteenth century, when Nylander (1866) used the abundance of lichens as a measure for air pollution effects. For the purpose of surveying ambient air quality, the highly ozone-sensitive tobacco cultivar Bel W3 was used for the first time in 1958 in Los Angeles (Heck 1966). Bioindication and biomonitoring with plants to detect air pollution-induced effects also have a long tradition in western European countries. Biomonitoring with plants is relatively cheap compared to chemical measurements, and can be applied for demonstration of pollution-induced injury and recognition of its causes, delimitation of exposed areas, risk assessments for various types of vegetation and surveillance of permissible ambient concentrations (Guderian et al. 1985).

Using plants from natural sites *in situ* is considered passive biomonitoring. Depending on the sensitivity of the selected plant species, measured responses concern leaf injury or accumulation of deposited substances. This method is used frequently for source identification of primary air pollutants or monitoring networks at small scales. Specific plant species with known sensitivity can also be used for active biomonitoring. In this case, the methods for plant cultivation and exposure are fully standardised from planting to harvest.

Sensitive plant species that have been used frequently to determine ozone-induced effects in large-scale biomonitoring networks include tobacco

**Table 6.1** Selection of plant species that are relatively sensitive to common air pollutants.

| | Plant Species | |
|---|---|---|
| **Ozone** | Ash | Grape |
| | Bean | Hybrid poplar |
| | Black cherry | Milkweed |
| | Brown knapweed | Nettle |
| | Clover | Tobacco |
| | Eastern white pine | Tulip poplar |
| **Sulphur dioxide** | Barley | Larch |
| | Beech | Lucerne |
| | Birch | Pine |
| | Clover | Poplar |
| | Common plantain | Wheat |
| **Fluorides** | Apricot | Goatweed |
| | Douglas fir | Peach |
| | Freesia | Pine |
| | Gladiolus | Tulip |
| **Nitrogen oxides** | Bean | Norway spruce |
| | Juniper | Scots pine |
| | Lettuce | Tobacco |
| **Ammonia** | Clover | Pine |
| | Mustard | Sunflower |
| **Ethylene** | Marigold | Potato |
| | Petunia | Tomato |

(cultivar Bel W3), various clover species and Phaseolus bean for ozone. For example, plants of tobacco Bel W3 were used in a nationwide monitoring network in the Netherlands during the 1970s and 1980s. Results from this network (Figure 6.2) clearly showed that visible injury in this plant differed between regions and was generally more severe in the west than the east of the country. In another biomonitoring programme, bean plants were exposed to ambient air at four rural sites in the Netherlands during the growing seasons of 1994 to 1996 (Tonneijck and van Dijk 2002). Ozone-induced injury in these plants varied between sites and years (Figure 6.3).

Various clover species have been used to detect effects of ambient sulphur dioxide, and ornamental species such as the gladiolus and tulip have been used to monitor phytotoxic levels of hydrogen fluoride. Marigold and petunia are relatively sensitive to ethylene in terms of flower formation and growth, and Figure 6.4 illustrates the results from using these plants to determine the areas of risk of impacts of ethylene around a complex of polyethylene manufacturing plants (Tonneijck et al. 2003). Plant performance was severely reduced close to the sources, but the number of flowers was not affected adversely beyond 400–500 m.

## Effects on Plant Performance

Chronic exposures to air pollution can result in growth and yield reductions, loss of viable seeds and decreased vitality. It is widely recognized that ozone has become the most important pollutant in the northern hemisphere, causing direct effects such as foliar injury and yield losses in many agronomic and horticultural crops. Van Dingenen et al. (2009) estimated that annual yield losses globally for four major staple crops were $14–26 billion, with about 40 per cent of the loss occurring in India and China. This may be an underestimate, as it does not consider evidence (Emberson et al. 2009) that Asian grown wheat and rice may be more sensitive than the North America dose-response relationships used by van Dingenen et al. (2009). Ambient ozone levels are also considered sufficiently high reduce the production of forest trees and the biodiversity of ecosystems. Direct effects on

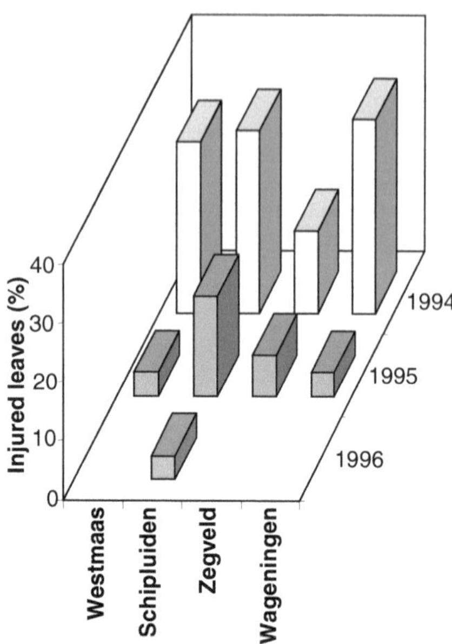


Figure 6.3 Ozone injury (percentage of leaves injured) in bean after exposure to ambient air at four rural sites in the Netherlands in the growing seasons of 1994 to 1996 (data from Tonneijck and van Dijk 2002).

plant performance of pollutants other than ozone have also been documented, but these effects generally do not occur at large scales.

## Critical Levels

The fact that levels of air pollutants within the industrialized world are sufficiently high to cause a decrease of plant performance in many countries has led to interest in defining threshold concentrations for adverse effects which can be used in policy evaluation. In Europe, critical levels for the main air pollutants have been defined within the United Nations Economic Commission for Europe (UNECE) and subsequently adopted by the World Health Organization (WHO). These critical levels were defined as the concentrations of pollutants in the atmosphere above which direct adverse effects on receptors, such as plants, ecosystems or materials, may occur according to present knowledge (UNECE 1988). At the time of writing, critical levels are generally defined in $\mu g/m^3$ or parts per billion (ppb) for a specific duration of exposure, since concentration and duration are the main variables to describe pollutant exposures. Current information on critical levels used within UNECE for the long-term phytotoxic effects of ozone, sulphur dioxide, nitrogen oxides and ammonia is presented in Table 6.2.

There is still a significant problem in defining the ambient exposure in terms of plant response. This response is a function of the pollutant that is absorbed and reaches the target site inside the leaf rather than a function of ambient exposure characteristics. The uptake rate of pollutants, however, is generally not known since this includes measurements of gas exchange properties such as stomatal conductance. The flux concept has now been adopted within UNECE as an alternative

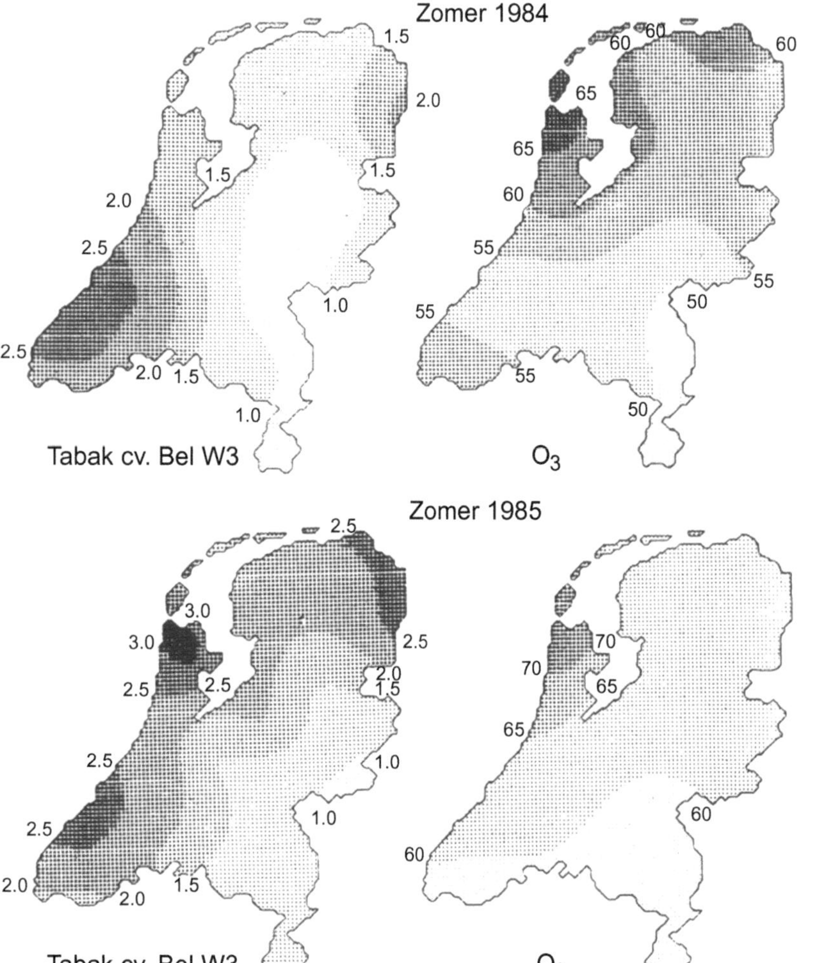


Figure 6.2 Spatial distribution of mean values of 24-hour average ozone concentrations (right, $\mu g/m^3$) and injury intensities in tobacco Bel W3 (left) in the Netherlands for the summers of 1984 and 1985. Darker areas indicate more severe visible injury or higher ozone concentrations.

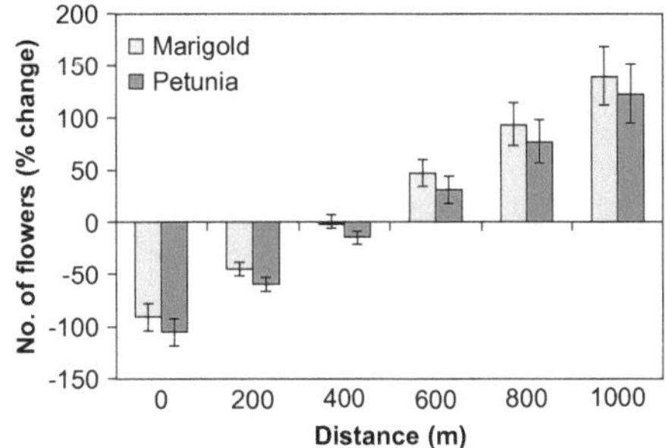


Figure 6.4 Mean number of flowers (percentage change relative to unexposed control) in marigold and petunia at different distances from a complex of polyethylene manufacturing plants for 1977 to 1983. The vertical bars represent ± standard error (SE) (data from Tonneijck et al. 2003).

approach to improve the assessment of ozone impacts on forest trees and crops (CLRTAP 2010). This approach models the flux of ozone into the leaf rather than the external exposure and accounts to some extent for the influence of climatic and developmental factors (Ashmore 2002). Use of a flux-based risk assessment leads to a very different spatial distribution of risks of ozone impacts across Europe compared with use of AOT40 (Simpson et al. 2007; Mills et al. 2011).

## 6.2 Impacts of Sulphur and Nitrogen Deposition

In the 1970s, acidification of lakes and streams in Europe and North America increased public awareness of the risk of air pollutants in general and sulphur emissions in particular. The effects of acidification, nitrogen deposition and ozone on terrestrial ecosystems, and in particular in causing forest decline, started to become a concern in the 1980s. In 1979, the Convention on Long-Range Transboundary Air Pollution (LRTAP), under the United Nations Economic Commission of Europe (UNECE), was established to address regional-scale air pollution problems in Europe and North America. The focus was first on protocols on emission reductions that were negotiated based on technical and economic information, and environmental impacts themselves were not considered explicitly. In the 1990s, the attention of air pollution increasingly focused on eutrophying effects of deposition of nitrogen compounds. In that period, mathematical models were used more and more under the LRTAP Convention to quantify and compare both investment costs and environmental impacts of policy alternatives. Under the Convention, European-scale assessments of the environmental impacts of sulphur (S) and nitrogen (N) have become the most advanced in the world, and therefore this section focuses on Europe only.

The growing use of modelling also calls for backing by empirical findings. The verification of modelled impacts of atmospheric deposition of sulphur and nitrogen compounds in Europe, based on recorded trends, has become important for the support of air pollution abatement policies in Europe. An extensive effects programme under the LRTAP Convention oversees information on *recorded* impacts as well as the *modelled* risk of impacts on European ecosystems. The decrease of acidifying emissions over past decades is reflected in biogeochemical recovery of surface waters (Stoddard et al. 1999) and forests (e.g. decreased S content of tree needles; Lorenz et al. 2003). However, modelling for both freshwaters (e.g. Larssen et al. 2010) and forest soils (e.g. Reinds et al. 2009) indicates that further emission reductions will be needed to achieve complete recovery.

To allow scenario analyses of the impacts of emission reduction policies, the LRTAP Convention's effects programme has developed methods to compute deposition thresholds – so-called *critical loads* – for long-term effects on forest soils. Acidifying and eutrophying deposition below these critical loads does not lead to damage according to current scientific knowledge. Maps of critical loads were used to support protocols for the reduction of acidification and later also of eutrophication (Hettelingh et al. 1995, 2001). The first such protocol was signed in 1994, addressing a single pollutant (sulphur) and a single effect (acidification). In 1999 a more complex protocol was signed. This protocol addresses the reduction of emissions of sulphur, nitrogen oxides, ammonia and volatile organic compounds (VOCs) simultaneously, while considering multiple effects, that is, acidification (by sulphur and nitrogen), eutrophication and the formation of tropospheric ozone (by nitrogen oxides and VOCs). The results of this method (which is described in more detail below) were then used in integrated assessment models such as RAINS (Schöpp et al. 1999). RAINS identifies emission reduction alternatives which limit the exceedance (see Posch et al. 2001) of pollutant deposition over critical loads. Finally, European maps of exceedances provide the location of ecosystems at risk of acidification and eutrophication.

In this section, an attempt is made to relate *in situ* information, including *recorded* trends, on the one hand, and *modelled regional* trends of ecosystems at risk of air pollution effects in European ecosystems, on the other, focusing on the effects of nitrogen.

### Effects and Trends of Nitrogen and Sulphur Inputs on Forest Ecosystems

In 1994 a Pan-European Programme for Intensive and Continuous Monitoring of Forest Ecosystems was established to gain a better understanding of the effects of air pollution and other stress factors on forests. At present, 862 permanent observation plots for intensive monitoring of forest ecosystems have been selected. The Intensive Monitoring Programme includes the assessment of crown condition, increment and the chemical composition of foliage and soil on all plots, with atmospheric deposition, meteorological parameters, soil solution chemistry and ground vegetation composition monitored at selected plots.

In this section, field evidence of impacts of elevated atmospheric sulphur (S) and nitrogen (N) inputs in these intensive monitoring plots is given. Results focus specifically on the possibility of deriving critical loads, concentrating on the effects of elevated N inputs, that is:

- Elevated N leaching (N saturation of forests).
- Release of Al and accumulation of $NH_4$ in soil that may disturb nutrient uptake.
- Elevated N contents and N/base cations ratios in foliage that may cause stress due to drought, frost, pests, diseases and nutritional imbalances.

Although N is not the only substance inducing effects on forest ecosystems, it plays an important role in the multiple stresses that forests experience, and therefore N is at the centre of this evaluation.

#### *Elevated leaching of nitrogen*

A first indication of adverse impacts of N inputs in forest ecosystems is elevated leaching of N (or $NO_3$, which dominates N leaching) that may cause acidification of groundwater and surface water. At more than a hundred intensive monitoring plots across Europe, the input and output of different N compounds ($NH_4$ and $NO_3$) has been derived. Results of N leaching plotted against N deposition show that the leaching of N is generally negligible below a total N deposition of 10 kg ha$^{-1}$ yr$^{-1}$ (Figure 6.5).

At N inputs between 10 and 20 kg ha$^{-1}$ yr$^{-1}$, leaching of N is generally elevated, although lower than the input, indicating N retention at the plots. At N inputs above 20 kg ha$^{-1}$ yr$^{-1}$, N leaching is also mostly elevated, and in seven plots it is near or even above N deposition (Figure 6.5). The latter situation indicates a clear disturbance in the N cycle in response to the elevated N input. In summary, these data indicate a critical N load to avoid elevated N leaching of 10 kg ha$^{-1}$ yr$^{-1}$.

Table 6.2 Long-term critical levels for the phytotoxic effects of various air pollutants (CLRTAP 2010).

| Air Pollutant | Type of Vegetation | Concentration/Exposure | Duration |
|---|---|---|---|
| **Ozone (AOT40)**[a] | Agricultural crops | 3 ppm.h | 3 months |
| | Horticultural crops | 3.5 ppm.h | 3 months/growing season |
| | Natural vegetation dominated by annuals | 3 ppm.h | 3 months/growing season |
| | Natural vegetation dominated by perennials | 5 ppm.h | 6 months |
| | Forest trees | 5 ppm.h | 6 months/growing season |
| **Sulphur dioxide** | Agricultural crops | 30 $\mu g/m^3$ | Annual mean |
| | Forests and semi-natural vegetation | 20 $\mu gm^3$ | Annual and winter mean |
| | Lichens | 10 $\mu g/m^3$ | Annual mean |
| **Nitrogen oxides** | All vegetation | 30 $\mu g/m^3$ | Annual mean |
| **Ammonia** | Lichens and bryophytes | 1 $\mu g/m^3$ | Annual mean |
| | Higher plants | 3 $\mu g/m^3$ | Annual mean |

[a] Exposures to ozone (AOT40) are expressed as accumulated exposure over a threshold of 40 ppb.

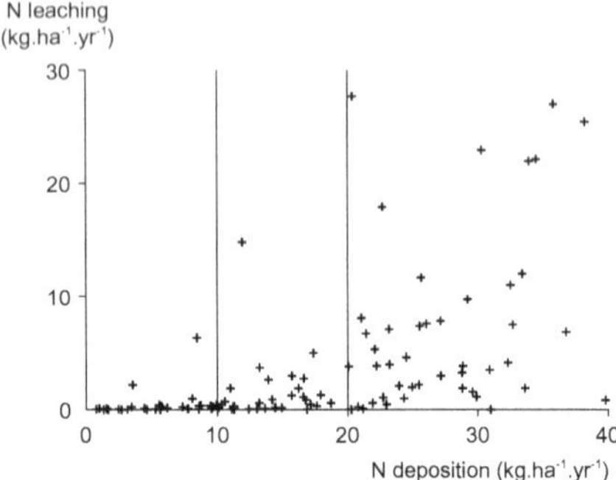


**Figure 6.5** Scatter plots of total N leaching against total N deposition at more than a hundred intensive monitoring plots in Europe (de Vries et al. 2001).

The geographic variation of N leaching and N retention (deposition minus leaching) over the investigated plots is shown in Figures 6.6 (a) and (b). High N leaching fluxes (>1000 $mol_c$ $ha^{-1}$ $yr^{-1}$) do occur in Belgium and central Germany, where the input of N (specifically of $NH_4$) is also high. In northern Europe and France, N leaching fluxes are low (<200 $mol_c$ $ha^{-1}$ $yr^{-1}$). However, the geographic variation of N leaching is large (specifically in Germany), indicating that both N deposition and soil characteristics influence N leaching. Sites with a net release of N are found in Belgium and north-western Germany, an area that has received high N deposition over a prolonged period of time. The high N retention in south-eastern Germany is remarkable; according to present calculations, these sites still retain a lot of N, despite relatively high N deposition. This may be explained by the centuries of intensive removal of litter from these poor soils until the 1950s, leading to a deficit in the N budget that still exists.

### *Soil acidification and ammonium accumulation*

In acid soils, the atmospheric deposition of S and N compounds leads to elevated aluminium (Al) concentrations, in response to elevated soil concentrations of sulphate ($SO_4$) and nitrate ($NO_3$), and also to accumulation of ammonium ($NH_4$) in situations where nitrification is strongly inhibited. This may cause nutrient imbalances, since the uptake of base cation nutrients, namely calcium (Ca), magnesium (Mg) and potassium (K), is reduced by increased levels of dissolved Al and $NH_4$ (Boxman et al. 1988). This effect may be aggravated in systems of low N status, where an elevated input of N will increase forest growth, thus causing an increased demand for base cations. Observations of increased tree growth of European forests in recent decades may be an effect of increased N inputs (e.g. Spiecker et al. 1996; de Vries et al. 2009).

Results obtained for the concentrations of Al, $NH_4$ and base cations in the soil solution of the intensive monitoring plots show a clear increase in Al concentration, and in the ratio of Al to base cations, going from the organic layer to the mineral soil, whereas the reverse is true for the $NH_4$ concentration and the ratio of $NH_4$ to potassium (K). Insight into the possible impact of acid deposition on Al release and of N deposition on $NH_4$ accumulation is given in Figures 6.7 (a) and (b). The release of Al in response to elevated $SO_4$ and $NO_3$ concentrations in subsoils (20–80 cm) with a low pH (below 4.5) is shown in Figure 6.7a. In these soils more than 80 per cent of the variation in Al concentration is explained by the variation in sulphate ($SO_4$) and nitrate ($NO_3$) concentrations, which in turn are strongly related to the deposition of S and N, respectively. Although $SO_4$ is important in releasing Al, results showed that $NO_3$ concentrations were mostly higher, reflecting the increasing role of N in soil acidification.

The $NH_4$/K ratio in the mineral topsoil in response to elevated N deposition is shown in Figure 6.7b. The critical $NH_4$/K ratio of 5, mentioned in the literature (e.g. Roelofs et al. 1985; Boxman et al. 1988), is only exceeded once in the topsoil, at an N input near 30 kg $ha^{-1}$ $yr^{-1}$. Results indicate that below an N deposition of approximately 10 kg $ha^{-1}$ $yr^{-1}$, the $NH_4$/K ratios are hardly elevated, whereas they do increase above this value.

The geographic variation in the leaching fluxes of Al and base cations (BC) (taken as the sum of Ca, Mg and K) is presented in Figures 6.8 (a) and (b). BC leaching is relatively high in areas with a high N or S deposition, such as Belgium, north-western Germany and the area around the German-Czech border, because of high BC release. Extremely high leaching fluxes for BC (above 7000–8000 $mol_c$ $ha^{-1}$ $yr^{-1}$) occur at near-neutral or calcareous sites in central Europe, where the leaching of Ca is high due to natural decalcification. Results show that the critical molar Al/BC ratios of 0.5–1.5 (Sverdrup and Warfvinge 1993) are regularly exceeded. Very high leaching fluxes of Al mainly occur in western and central Europe (Belgium and parts of Germany and the Czech Republic), indicating the occurrence of an acid soil releasing mainly Al in response to the high input and leaching of $SO_4$. Sites with the highest $SO_4$ release are located in central Europe, where the strongest reduction in $SO_4$ deposition has taken place over the last decade (de Vries et al. 2001).

### *Nutritional imbalances*

An excess input of N may increase the N content in foliage, which in turn may cause an increased sensitivity to climatic factors, such as frost and

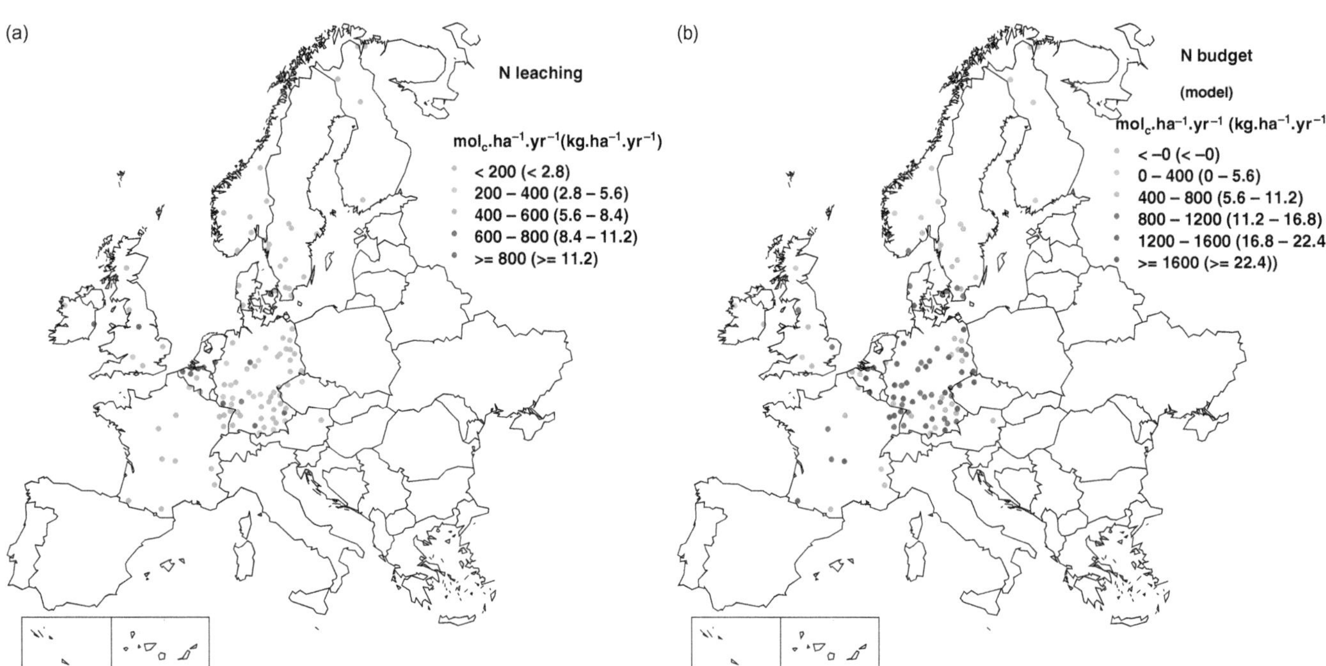


**Figure 6.6** Geographical variation in (a) leaching fluxes ($mol_c$ $ha^{-1}$ $yr^{-1}$) and (b) budgets of N at the investigated intensive monitoring plots throughout Europe (de Vries et al. 2003b).

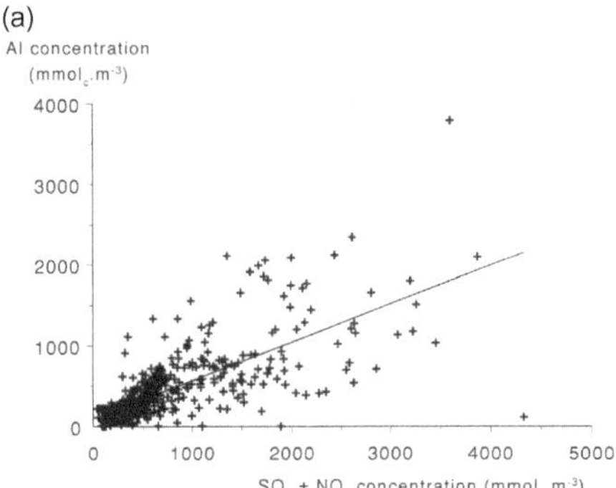


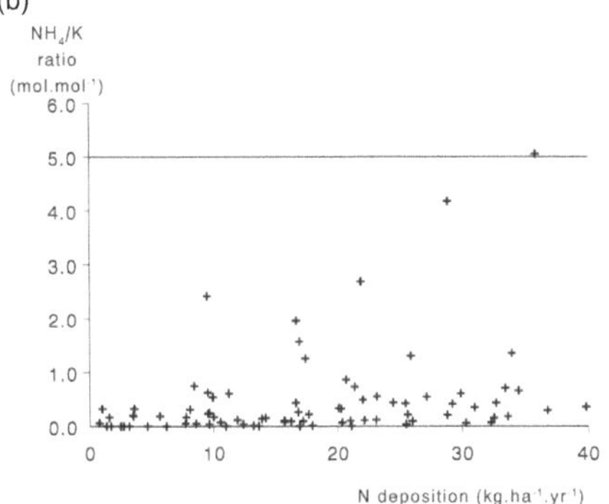


**Figure 6.7** Scatter plots of (a) the concentration of total Al against total $SO_4$+$NO_3$ in the subsoil of intensive monitoring plots with a pH < 4.5 (the solid line represents a regression line being equal to: Al=−95 + 0.74 ($SO_4$+ $NO_3$) ($R^2$=0.86)) and (b) the $NH_4$/K ratio in the mineral topsoil against the total N deposition (de Vries et al. 2000, 2003b).

drought (e.g. Aronsson 1980), and diseases and plagues, such as attacks by fungi (e.g. van Dijk et al. 1992; Flückiger and Braun 1998). In this context, a critical N content of 1.8 per cent in needles has often been mentioned in the literature. More information on the impacts of nitrogen deposition on nutrient imbalances in forests is given in de Vries et al. (2003b). The relationship between N contents in first-year needles of Scots pine and total N deposition at 68 intensive monitoring plots in Europe (Figure 6.9) indicates a critical N load of 20 kg ha$^{-1}$ yr$^{-1}$ for this effect. Above this input level, N contents in foliage may exceed the critical N content of 1.8 per cent related to drought and frost stress.

### *Trends in sulphur and nitrogen deposition*

Changes in N and S deposition have been derived from a comparison of annual throughfall fluxes assessed at some 120 plots in the 1980s and at more than 300 plots in 1996 and 1997. The first set of data consists of a literature compilation, whereas the latter data set is based on a Europe-wide monitoring programme in forests using stands with similar forest types (pine, spruce or broadleaves) located within a distance of 10 km of each other. Results for a total of 53 plots showed a clear decrease, specifically for the $SO_4$ input, but also for the total N input in throughfall (Figures 6.10 (a) and (b)). The decrease in N inputs was due to a strong decrease of $NO_3$, whereas values of $NH_4$ remained relatively constant.

### *Trends in soil solution concentrations*

The large decrease in S and N deposition and the strong response of acid sandy soils to these inputs, in terms of Al and BC release, implies that considerable changes in soil solution chemistry are to be expected in these soils. To illustrate the effect, cumulative frequency distributions of the dissolved concentrations of $SO_4$, $NO_3$ and Al in 124 non-calcareous forest soils in the Netherlands for the years 1990, 1995 and 2000 are presented in Figures 6.11 (a)–(d). The results illustrate the much larger decrease in dissolved $SO_4$ concentration compared to the $NO_3$ concentration, and the strong relationships between the decrease in $SO_4$ and Al concentrations and consequently the Al/Ca ratio.

The reversibility of acidification of soils and waters is supported by field experiments. The most illustrious example is the RAIN project (Reversing Acidification in Norway), where a 860 m$^2$ head water catchment has been covered by a transparent roof to exclude ambient acid precipitation and where rain with natural levels of seawater salts is sprayed out underneath the roof. After two and a half years, concentrations of $SO_4$, $NO_3$ and $NH_4$ in runoff were lowered by more than 50 per cent, compensated by a decrease in BC concentrations (45 per cent) and an increase in alkalinity (55 per cent). Similar results were observed for ion concentrations in the soil solution underneath two roofed sites in the Netherlands (Boxman et al. 1995).

## Effects of Sulphur and Nitrogen Deposition on Biodiversity

Generally recognized effects of acidification on biodiversity are confined to lakes and streams in areas with acidic bedrock (Bronmark and Hansson 2002), to moorland pools on sandy soils (Roelofs et al. 1996) and to epiphytic lichens (van Herk 2001).

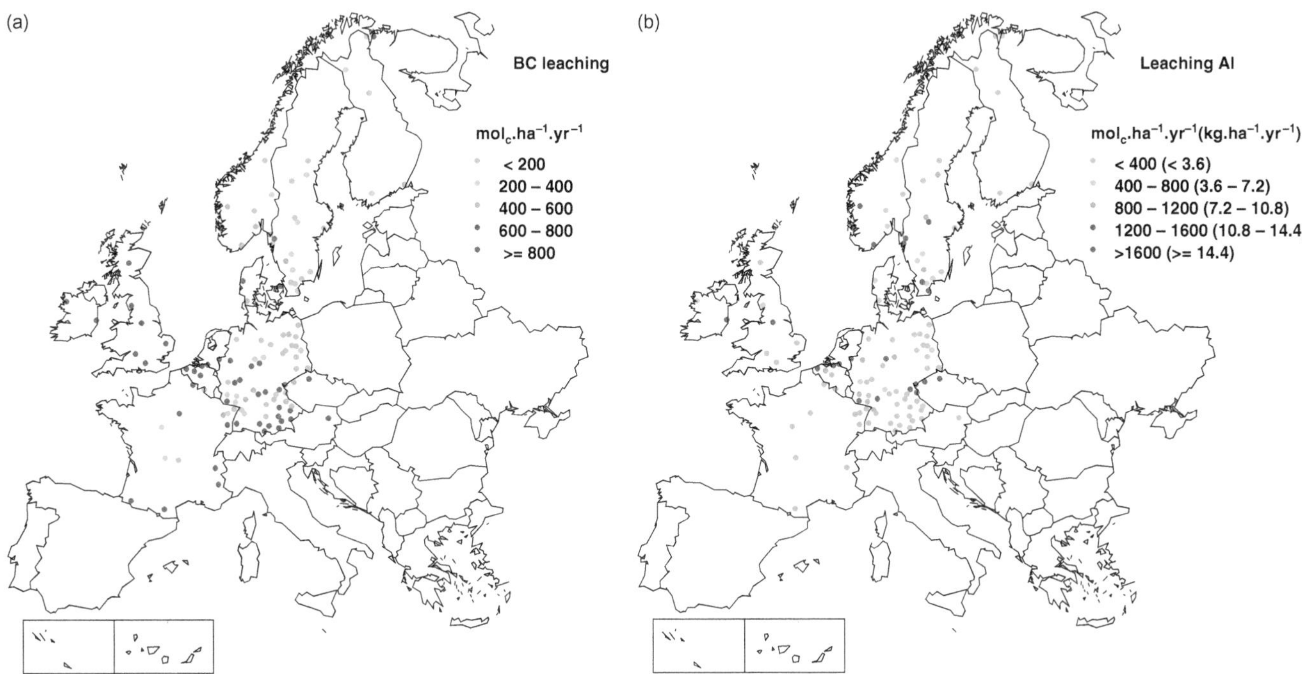


**Figure 6.8** Geographical variation in leaching fluxes (mol$_c$ ha$^{-1}$ yr$^{-1}$) of (a) base cations (BC) Ca+Mg+K and (b) Al at the investigated intensive monitoring plots throughout Europe (deVries et al. 2001).

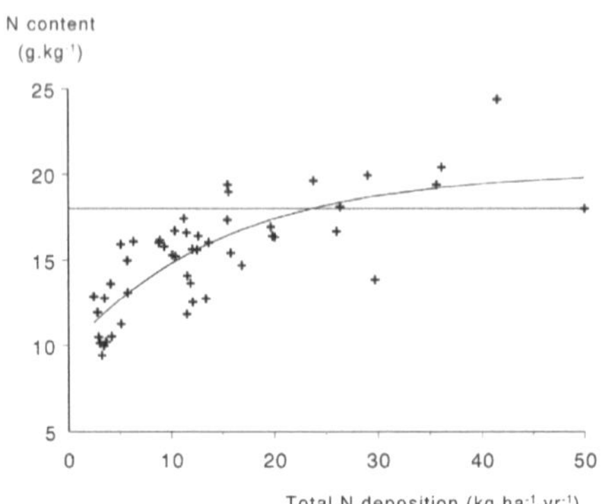


Figure 6.9 Relationship between N contents in first-year needles of Scots pine and total N deposition at 68 plots in Europe (de Vries et al. 2003b).

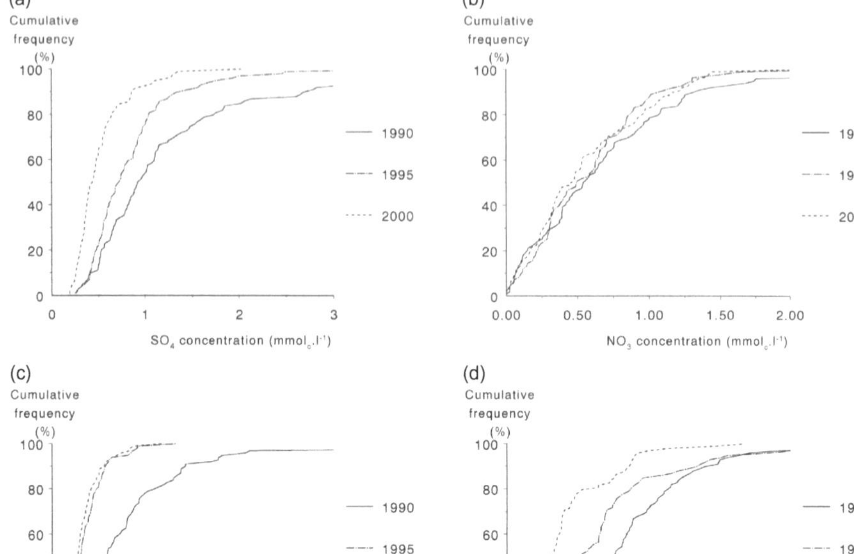


Figure 6.11 Frequency distributions of the dissolved concentrations of (a) $SO_4$, (b) $NO_3$ (c) Al, and (d) the ratio of Al to Ca,in 124 non-calcareous forest soils in the Netherlands for the years 1990, 1995 and 2000.

Such effects are well described and generally entail a loss of species, sometimes accompanied by a strong dominance of one or a few acid-resistant species. Some authors also ascribe the decline of the diversity of grasslands on poor, sandy soil to acidification (de Graaf et al. 1997). However, in many of these cases other factors besides acidification seem to be responsible for the reported decline. The decline of epiphytic lichens is probably the ecological effect of air pollution that is best documented over a long time (van Dobben 1996; van Herk et al. 2002). Lichens mainly respond to direct toxicity of $SO_2$ and, to a lesser extent, to acidification, eutrophication and toxicity of $NO_x$ (van Dobben and Ter Braak 1999; van Herk 2001). The dramatic changes that took place in the epiphytic lichen flora of north-western Europe in the last two decades of the twentieth century and the early years of the twenty-first can be ascribed to a combination of decreasing $SO_2$ concentration, increasing $NH_3$ concentration and climate change (van Herk et al. 2002).

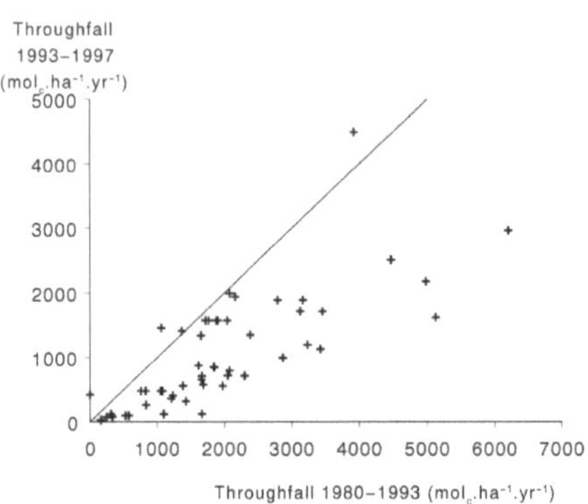


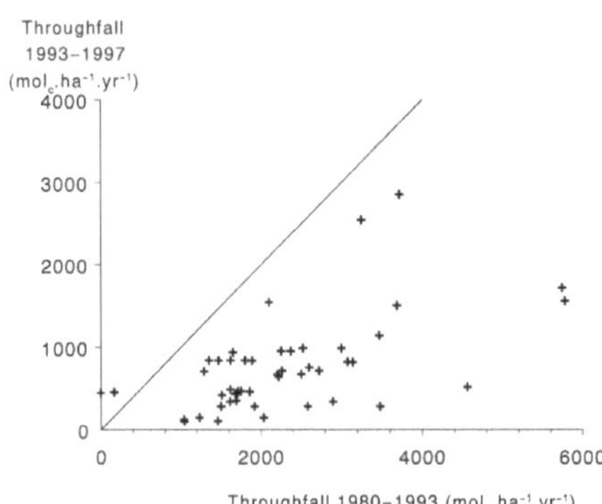


Figure 6.10 Comparison of throughfall of total N and $SO_4$ measured at 53 plots located within 10 km of each other in the eighties (1980–93) and nineties (1993–97) (the solid line represents the 1:1 line) (de Vries et al. 2003a).

In Europe, the effect of N deposition is now considered the most important effect of air pollution on biodiversity. There are three main reasons for this: (1) the atmospheric concentration of $SO_2$ has dramatically decreased over this period in most parts of Europe; (2) the expected large-scale forest dieback did not occur; and (3) effects of N deposition and resulting eutrophication have appeared to be much more widespread than effects of S deposition and resulting acidification. Effects of nitrogen deposition are now recognized in nearly all natural ecosystems with low nutrient levels in Europe; these include aquatic habitats, forests, grasslands (including tundra, montane and Mediterranean grasslands), wetlands (mire, bog and fen), heathlands, and coastal and marine habitats (Bobbink et al. 2010). In such nutrient-limited systems, nitrogen is generally the most important growth-limiting element, and species are adapted to a nitrogen-deficient environment. If the availability of nitrogen increases, other species that use the available nitrogen more efficiently may outcompete the less productive species that are adapted to nitrogen deficiency.

Since the recognition of nitrogen deposition as a driver of loss of biodiversity in Europe, a number of expert workshops have taken place to reach agreement on critical loads of nitrogen for various ecosystems (Bobbink et al. 2010). The harmful effects considered in defining values of critical loads may be chemical changes in soils and waters which might cause direct or indirect effects on organisms, or changes in individual organisms, populations or ecosystems (Nilsson and Grennfelt 1988). The studies that have been carried out on a European scale (see below) concentrate on chemical changes in soil or water that are hypothesised to be 'harmful' to organisms. When the focus is on the organisms themselves or on ecosystems, the setting of critical load values becomes more complicated because of the intrinsic variability of natural systems, and expert knowledge and empirical or observational studies play an important role. At present, two approaches exist for setting critical loads of nitrogen deposition in Europe; the empirical one and the modelling one. The empirical approach completely relies on experiments and observation of the effects of nitrogen deposition on vegetation, whereas the modelling approach uses observations or expert knowledge to determine critical limits for the vegetation (e.g. in terms of pH or N availability), and then uses a model to translate these critical limits into critical loads at steady-state.

Empirical critical loads for ecosystems are extensively discussed by Bobbink et al. (2003). In their approach, long-term (i.e. more than one year) experimental effects of nitrogen addition to existing vegetation play a central role. Such addition experiments may be carried out either in the field, or in the laboratory using artificial or transplanted plant communities. Because of the time-and labour-intensive nature of such studies, results are only available for a limited number of broadly defined ecosystems. In some cases, experimental results are supplemented by observational studies (e.g. time series under a known increase in deposition). In this approach, the critical load is the highest addition of nitrogen that does not lead to adverse physiological changes (at the individual level) or loss of species (at the ecosystem level).

The modelling of critical loads is based on the principle that a certain chemical threshold for effects (e.g. N availability, N leaching or loss of acid neutralising capacity (ANC)) is defined, and a model is used to determine the N (or N and S) deposition that results in this threshold value at steady state. The chemical thresholds can be made ecosystem-dependent, that is, as a function of the known environmental demands of a given vegetation type. This leads to critical load values for narrowly defined ecosystem types. However, for most ecosystems hard data on their environmental limits are lacking, and combinations of field observations and expert knowledge have to be used instead. Van Dobben et al. (2006) made a detailed analysis of the sources of uncertainty and their effect on the final critical load per ecosystem type, and concluded that: (1) the uncertainty in the simulated 'overall' critical loads (i.e. including all terrestrial vegetation types) is low, and well in agreement with empirical studies (namely, in the range 15–25 kg $ha^{-1}$ $y^{-1}$); (2) the uncertainty in the simulated critical loads per vegetation type is also low, but there is little agreement with values per vegetation type from empirical studies; and (3) the uncertainty in the simulated critical loads for discrete sites is extremely high. Table 6.3 gives a comparison of empirical critical loads for Europe agreed on in an expert workshop (Achermann and Bobbink 2003), and simulated critical loads for the Netherlands, by European Nature Information System (EUNIS) class (Davies and Moss 2002).

Table 6.3 shows that there is a fair agreement between the empirical and the simulation approach. In general, the empirical critical loads tend to be somewhat lower and to have narrower ranges than the simulated ones. The empirical ranges are the result of an interpretation of a large number of studies, and this interpretation is usually based on a precautionary principle, that is, it tends to search the lower end of all reported no-effect levels. On the other hand, the simulated critical loads are determined as an average over all vegetation types that belong to a given ecosystem, under average environmental conditions for that ecosystem. Some of the differences may also be due to modelling errors (e.g. for 'raised and blanket bogs'), where the range in empirical critical loads is judged 'reliable' and is far lower than the simulated one. This difference may be due to an underestimation of mineralisation in organic soils.

## Critical Loads for N and Acidity and their Exceedances over Europe

Critical loads are used in a policy context to assess the relative benefits of different emission reduction alternatives. To do this, critical loads for European ecosystems (predominantly forest soils) are compared to acidifying and eutrophying deposition rates. When critical loads are not exceeded, the ecosystem is assumed to be protected. This section focuses on the modelled trends of critical load exceedances across Europe that result from the sulphur and nitrogen emissions agreed in the Gothenburg Protocol.

### *The computation of critical loads*

The critical loads consist of four variables, which were submitted by the parties under the LRTAP Convention and were used to support the 1999 Gothenburg Protocol (Hettelingh et al. 2001). These variables are the basis for the maps used in a comparison between effect modules of the European integrated assessment modelling effort:

- The maximum allowable deposition of S, $CL_{max}(S)$, that is, the highest deposition of S which does not lead to 'harmful effects' in the case of zero nitrogen deposition.
- The minimum critical load of nitrogen.
- The maximum 'harmless' acidifying deposition of N, $CL_{max}(N)$, in the case of zero sulphur deposition.
- The critical load of nutrient N, $CL_{nut}(N)$, preventing eutrophication.

Critical loads have been computed for forest soils and other ecosystems in Europe. Thus, ecosystem-dependent combinations of sulphur and nitrogen deposition can be determined which do not cause 'harm' to the ecosystem. For policy support, these variables have allowed for the first time the assessment of acidification and eutrophication effects *together* (i.e. the effects of simultaneously reducing emissions of sulphur and both oxidised and reduced nitrogen).

Figure 6.12 shows maps of critical loads applied to each of the approximately 1.3 million ecosystem points distinguished by parties under the Convention (Hettelingh et al. 2007). The fifth and fiftieth percentile (median) maps (top and bottom, respectively) of $CL_{max}(S)$ (left) and $CL_{nut}(N)$ (right) reflect values in grid cells at which 95 and 50 per cent of the ecosystems are protected. In these maps, critical loads of different ecosystems have been combined into one map on a 50×50 $km^2$ grid cell resolution. Comparison of the fifth and fiftieth percentile maps shows that low (including 200–700 eq $ha^{-1}$ $yr^{-1}$) values for $CL_{max}(S)$ are required to protect 95 per cent of the ecosystems in north and central-west Europe, while the protection of 50 per cent of the ecosystems continues to require low critical loads in northern Europe in particular. The difference between the fifth and fiftieth percentiles of $CL_{nut}(N)$ also illustrates the occurrence of low values in other areas of Europe, including Spain and southern Italy.

### *The assessment of areas where S and N critical loads are exceeded*

Maps of deposition of sulphur and nitrogen compounds are provided under the LRTAP Convention by the Cooperative Programme for Monitoring and Evaluation of the Long-Range Transport of Air Pollutants in Europe (EMEP), using a Eulerian model for the dispersion computations (EMEP 2006). Ecosystem-specific deposition patterns computed from this model are compared to the critical loads as mapped in Figure 6.12, by using a method described by Posch et al. (2001). Modelled deposition fields from 1980 to 2010 allow trends to be assessed of the modelled risk of acidification and eutrophication to European ecosystems (i.e. where critical loads are exceeded). Figures 6.13 and 6.14 show the trend of ecosystem protection for acidification and eutrophication, respectively. Comparison between the relatively large areas where less than 10 per cent of ecosystems (red shading) were protected in 1980 with those predicted for 2010, shows a tendency towards increased protection, in particular for acidification (Figure 6.13). For eutrophication the areas at risk remain widespread (Figure 6.14).

The significant decrease in acidified areas between 1980 and 2010 (Figure 6.13) of forest soils has not yet been recorded in the field (see above). The discrepancy between the modelled and measured result is partly due to lack of knowledge of the relationship between a change in deposition and a change in biogeochemistry. Such time lags can be simulated using dynamic models (Posch et al. 2003a, 2003b); however, the verification *in situ* requires long time series of recorded changes in soil condition.

Of all reported ecological effects of atmospheric deposition in Europe, the effect on lake and stream acidification is most clear-cut. The large-scale dieback of fish populations in sensitive areas has been linked to deposition of acidifying compounds with a high degree of certainty. However, the focus of this case study of spatial risk assessment is on

**Table 6.3** Comparison of simulated and empirical critical loads (kg N $ha^{-1}$ $yr^{-1}$). Empirical data are taken from Bobbink et al. (2003); ##=reliable, #=quite reliable. Correspondence between simulated and empirical critical loads is given in the last column: < simulated range below empirical range, > simulated range above empirical range, = ranges overlap.

| Ecosystem Type (EUNIS Class) | Empirical Critical Load | Reliability | Simulated Critical Load | Simulated Compared to Empirical |
|---|---|---|---|---|
| Ground vegetation (Temperate and boreal forests) | 10–15 | # | 8–41 | = |
| Dry heaths | 10–20 | ## | 4–31 | = |
| Sub-Atlantic semi-dry calcareous grassland | 15–25 | ## | 15–31 | = |
| Non-Mediterranean dry acid and neutral closed grassland | 10–20 | # | 10–31 | = |
| Heath (Juncus) meadows and humid (Nardus stricta) swards | 10–20 | # | 4–33 | = |
| Raised and blanket bogs | 5–10 | ## | 26–33 | > |
| Poor fens | 10–20 | # | 5–30 | = |
| Permanent oligotrophic waters; soft-water lakes | 5–10 | ## | 21–22 | > |
| Coastal stable dune grasslands | 10–20 | # | 15–24 | = |

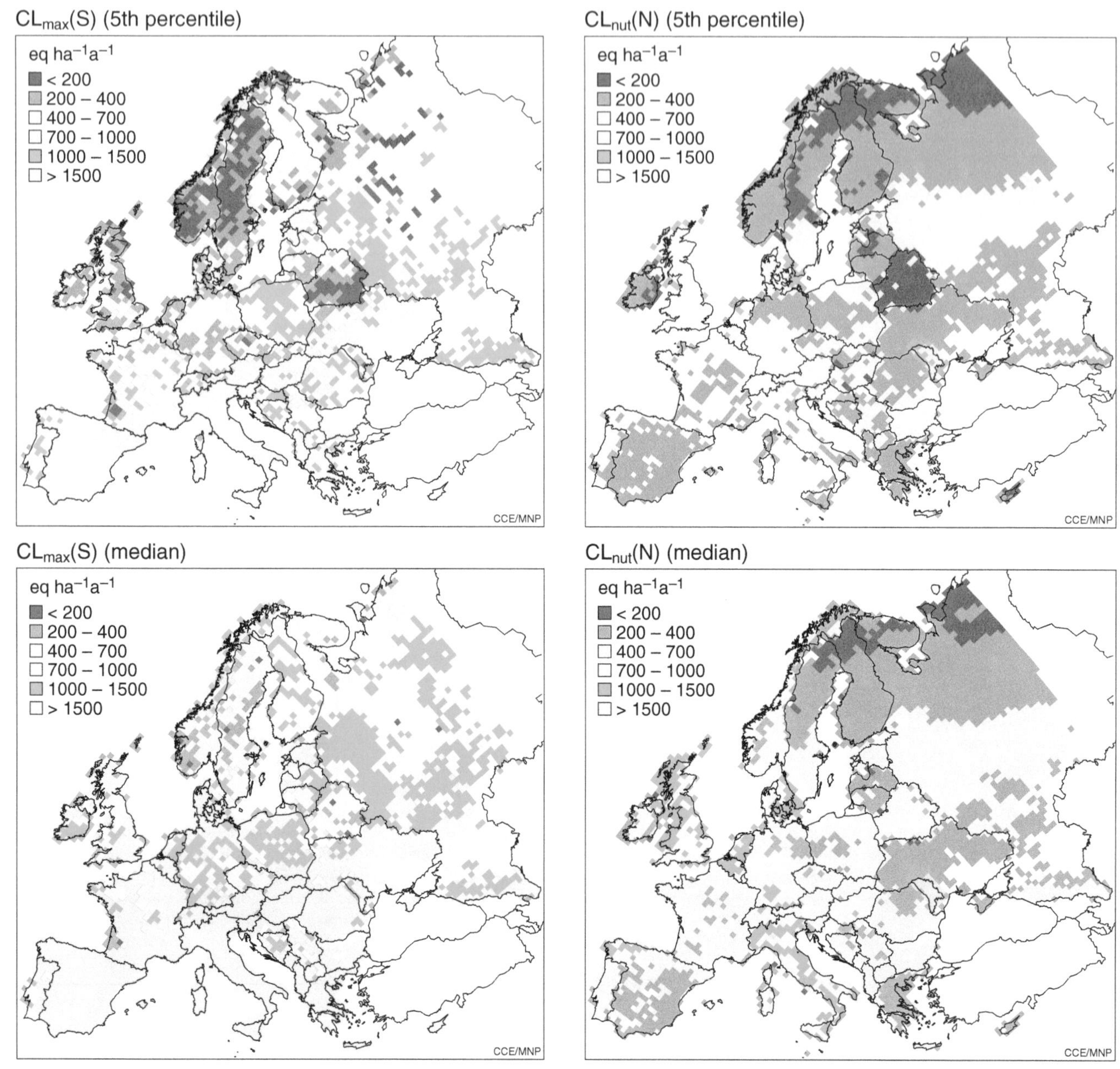


**Figure 6.12** The fifth percentile of the maximum critical loads (eq $ha^{-1}$ $yr^{-1}$) of sulphur (top left) and of nutrient nitrogen (top right). The corresponding fiftieth percentiles (medians) are shown at the bottom. The maps present these quantities on the EMEP 50x50 $km^2$ grid. (Source: Hettelingh et al. 2007.)

terrestrial ecosystems, for which effects are much more subtle. Research has concentrated on forest ecosystems because of the expected large-scale forest dieback, and many effects have been reported. The most important of these effects are: changes in soil chemistry; leaching of base cations and nitrogen; and changes in nutrient contents in leaves and needles, all of which have been reported to have adverse effects on trees, mostly in laboratory experiments. However, up to now large-scale forest dieback has only occurred in the most extreme situations, such as the 'Black Triangle' in the border area between Germany, Poland and the Czech Republic. Therefore, critical loads have to be considered as 'risk indicators' rather than as hard no-effect levels as far as forest trees are concerned.

The effects of N deposition on natural vegetation seem to be much more clear-cut than those on forest trees. There is a fair agreement between critical loads estimated by different methods, and these critical loads are also corroborated by field observations. Ecological effects of deposition are not confined to forests and lakes, but have been reported from nearly all nutrient-limited ecosystems. Also, effects have not only been reported for vegetation, but also for other groups like mushrooms or insects. Ecological effects of nitrogen enrichment are extensive (Bobbink et al. 2010, Stevens et al. 2010), and are more widespread than those of acidification, probably because nitrogen is often the main driver of ecosystem structure and composition. Usually, nitrogen is the element that is most growth limiting, and a change in its availability will lead to a change in the competitive relationships between organisms, and thus to a change in species composition.

The critical loads determined by a modelling approach, such as the Simple Mass Balance (SMB) model, are mostly lower than the empirical critical loads for vegetation. This may have several causes: SMB is a steady-state model, which does not take account of changes in storage over time; and SMB parameters are not based on biological criteria but rather on acceptable amounts of leaching. Therefore, the critical loads determined by SMB should be viewed as risk indicators, and further research will have to elucidate the relationship between biological

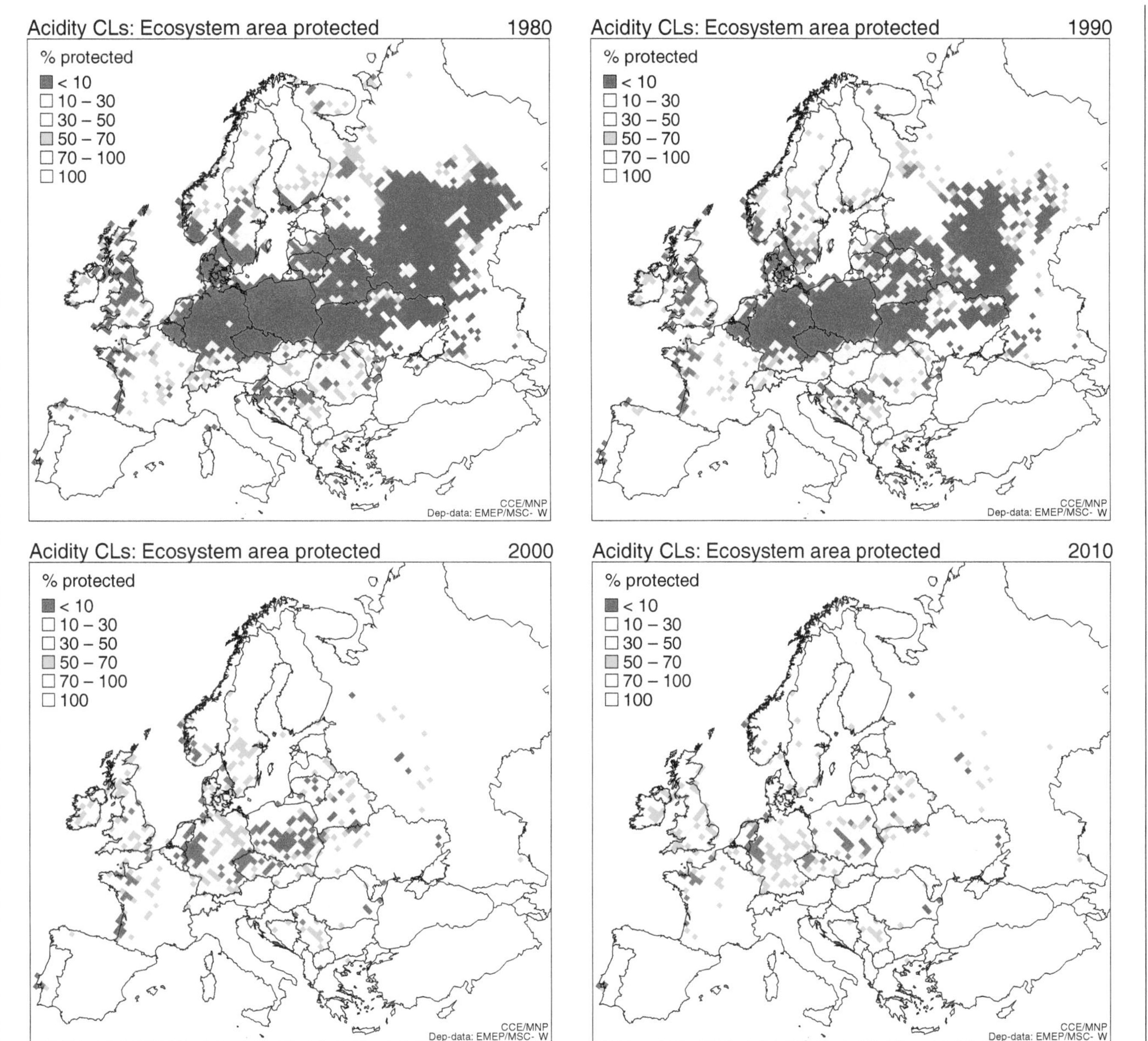


Figure 6.13 Percentage of ecosystem protection from 1980 (top left) to 2010 (bottom right) using critical loads of acidity. The maps show a marked decrease of areas where less than 10 per cent of ecosystems are protected, due to the reduction of deposition of acidifying compounds between 1980 and 2010. Acid deposition in 2010 is simulated using emissions that are prescribed in the 1999 Gothenburg Protocol.

criteria and the amounts of leaching that are judged acceptable.

The use of critical loads as a risk indicator in integrated assessment for the support of European policies in the field of air pollution has been successful. It has supported various international emission control protocols of increasing complexity, that is, extending from the assessment of a single pollutant and single effects to multiple pollutants and multiple effects. The critical load concept has increased the cost-effectiveness of air pollution control policies. The reason is that in addition to economic and technical consequences (costs) of policy alternatives, impacts (benefits) can also be compared. In the near future the concept of multiple pollutants and multiple effects under the Convention may be stretched to include climate change. This will enable more opportunities for cost-efficient policies by taking full account of linkages between climate change and air pollution policies.

## 6.3 Global Perspectives of Air Pollution Impacts on Vegetation

The impacts of air pollution on vegetation in western Europe, North America and Japan are well established. In these regions of the world, control of emissions has improved greatly over recent decades, and the concentrations and deposition of the major air pollutants are tending to decrease. However, in many developing countries, accelerating urbanisation and industrialisation have resulted in large increases in emissions from transport, energy and industry. The effect of these increased exposures to phytotoxic pollutants is often uncertain, and very little field or experimental work has been carried out to assess the scale of impacts; hence the detailed assessments for Europe, which are described in Sections 6.1 and 6.2, are not possible for many parts of the world.

The information used in this section is partly based on the work of Emberson et al. (2003), who commissioned 'state of knowledge' reviews on the

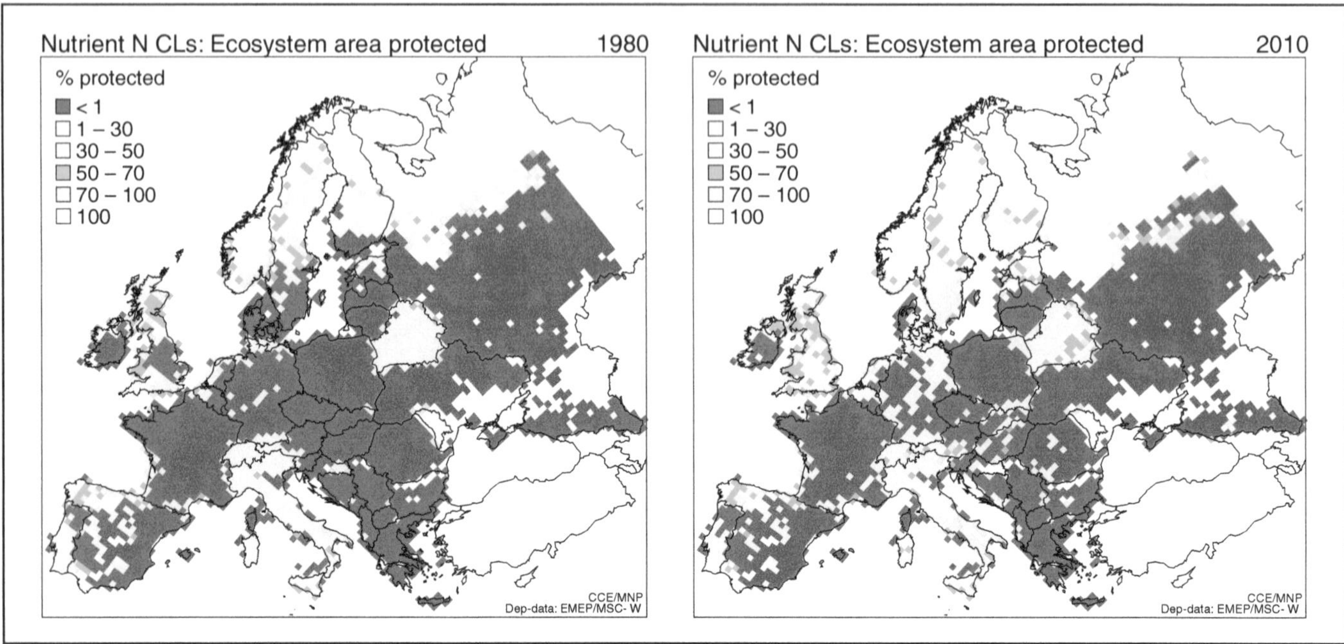


**Figure 6.14** Percentage of ecosystem protection in 1980 (left) and 2010 (right) using critical loads of nutrient N. The maps show that areas where less than 1 per cent of ecosys tems are protected continue to occur broadly in Europe due to nitrogen deposition in 2010. N deposition in 2010 is simulated using emissions that are prescribed in the 1999 Gothenburg Protocol.

impacts of air pollution on crops and forests for 12 different countries. We do not claim that the information is comprehensive – rather the maps aim to be illustrative of the range of problems and issues that have been identified. In all likelihood there are many other instances of reduced crop yields, visible damage and forest decline that are simply not recognized because of limited awareness of air pollution as a rural, as well as an urban, problem. It is also important to note that current projections are that global impacts of ozone and nitrogen deposition will increase over the first three decades of the twenty-first century, due to continued increases in global emissions of nitrogen oxides (Ashmore 2005; Phoenix et al. 2006; Royal Society 2008).

Figure 6.15 provides examples of damage to vegetation caused by urban and industrial emissions of pollutants, such as sulphur dioxide, particulates and fluorides. Figure 6.16 shows examples of locations where visible injury to vegetation, characteristic of ozone, has been reported.

## Global Impacts of Nitrogen Deposition on Biodiversity

There is now evidence that current levels of nitrogen deposition in Europe are having significant effects on the species composition of a range of nutrient-limited habitats (see also Section 6.2). Much of the earth's biodiversity is found in semi-natural and natural ecosystems, and many plant species from these habitat types are adapted to low nitrogen availability (Vitousek and Howarth 1991) and may therefore be at risk from elevated nitrogen deposition. Nitrogen deposition is now increasing in many regions beyond Europe and North America, where it was first recognized as a problem, as developing economies drive emission increases from intensive agricultural practices and fossil fuel combustion (Galloway and Cowling 2002).

Figure 6.17 overlays maps for global nitrogen deposition in the mid-1990s (Galloway et al. 2004) with 34 global biodiversity hotspots for conservation priorities (Myers et al. 2000; Mittermeier et al. 2005). It highlights seven regions where total nitrogen deposition is modelled to exceed 10 kg N $ha^{-1}$ $yr^{-1}$ in at least 10 per cent of the hotspot. European experience suggests that there is a potential threat to biodiversity for plant communities receiving deposition above 10 kg $ha^{-1}$ $yr^{-1}$ (Bobbink et al. 1998). The Western Ghats and Sri Lanka, Indo-Burma, the Atlantic Forests of Brazil and the mountains of south-west China are the four hotspots with the greatest risk, with over 30 per cent of the area estimated to receive more than 10 kg N $ha^{-1}$ $yr^{-1}$. Evidence of any atmospheric nitrogen deposition effects in these regions is currently lacking, but studies in these regions have shown the importance of nitrogen availability on species composition, inter alia, for dry forest in India, mangroves in Malaysia and the floristically rich Cape Province of South Africa (Lamb and Klaussner 1998). Nitrogen deposition in many of these hotspots is expected to increase by 2050, thus increasing the risk of significant species loss (Phoenix et al. 2006, Bobbink et al. 2010).

Although the hotspot approach does not cover all sensitive ecosystem types, some of which may be equally valuable but relatively species poor, it does provide a useful focus for increased awareness of the issue and a platform for action to protect vulnerable ecosystems. The threat to biodiversity may not be as large as that for land clearance, or indeed climate change effects, but it certainly should not be ignored or underestimated.

# 6.4 Health Effects of Air Pollution

Air pollutants appear to pose a serious threat to human health and may result in life shortening. Although the relative risks for ambient air pollution are small, the impact is considerable because of the large number of people affected and the existence of sub-populations at increased risk (Brunekreef and Holgate 2002). Asthmatics and patients with cardiovascular and chronic lung diseases seem more susceptible to air pollution-related illness compared to the average population. Short-term and long-term exposure to ambient air pollution may result in a variety of health effects, the occurrence of which in the population follows a more or less pyramid-like structure, as visualised originally by the American Thoracic Society (Figure 6.18), constituting an important environmental disease burden.

A wide variety of gases and particles in ambient air have been directly or indirectly linked to adverse effects on human health. Ground level ozone ($O_3$), nitrogen dioxide ($NO_2$) and particulate matter with an aerodynamic size of 10 $\mu$m or less ($PM_{10}$) or with an aerodynamic size of 2.5 $\mu$m or less ($PM_{2.5}$) are major and ubiquitous air pollutants. Exposure to their ambient levels appears to be associated with a variety of adverse health effects, ranging from respiratory symptoms and complaints to enhanced morbidity and premature mortality from cardiac and respiratory

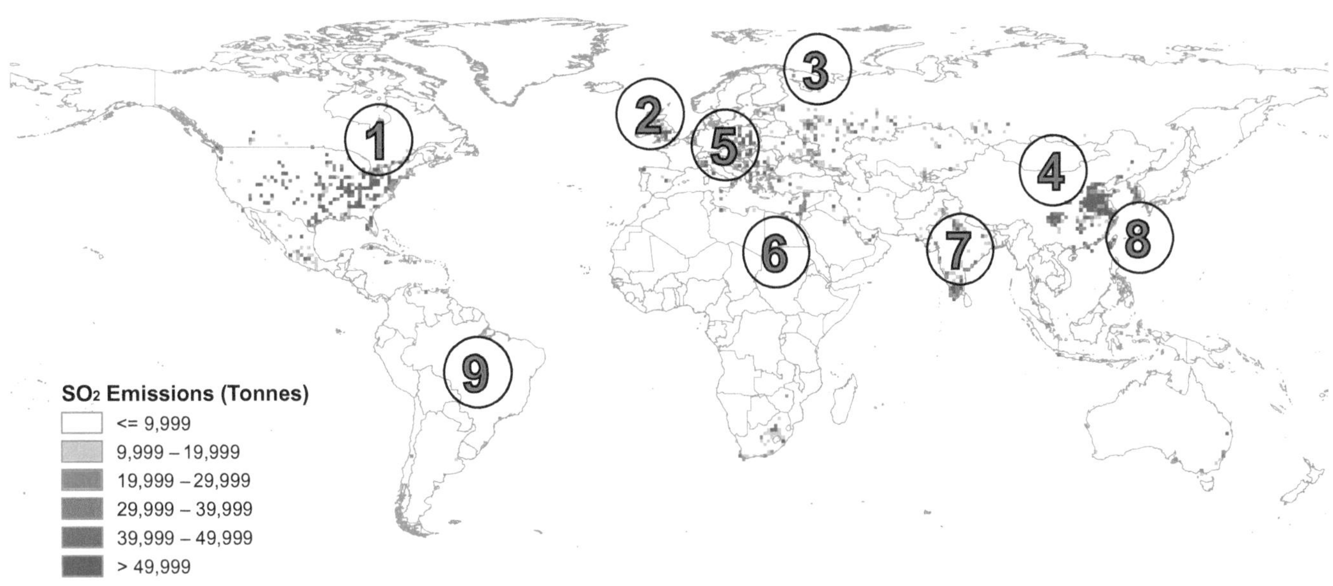


**1. Smelting, Sudbury, Canada**
Areas devoid of any vegetation occurred up to 8 km from this smelter in the 1970s, with species numbers and growth reduced up to 20–30 km away, probably as a result of combined effects of $SO_2$ and heavy metal emissions (Freedman and Hutchinson 1980).

**2. Urban coal burning, Leeds, UK**
Smoke and $SO_2$ emissions from industry and domestic coal burning up to the early 1960s were associated with ecological impacts in English cities such as Leeds, where early studies of impacts of urban air pollution were made (Cohen and Rushton 1925).

**3. Nickel smelting, Kola Peninsula**
Nickel smelting in this Arctic area has emitted $SO_2$ and metals since the 1950s. Extensive death of plants is reported up to 20–40 km from smelters, with visible damage to mosses, lichens and pine needles observed over a greater area (Kashulina et al. 2002)

**4. Urban pollution, Chongqing, China**
Chongqing is a highly industrialized city in China, with high levels of $SO_2$ and dust in the 1990s. Fruit and vegetable yields are greatly reduced in the city, e.g lettuce, radish and brassica yields are reduced by over 75 per cent (Zheng and Shimizu 2003).

**5. Black Triangle, Central Europe**
Extensive forest damage in the Black Triangle, covering parts of Poland, Germany and Czechoslovakia, was linked to emissions from power stations burning lignite during the 1960s–1980s; 1 million ha. of Norway spruce was severely injured (Godzik 1984).

**6. Industrial zones, Cairo**
Field studies in the 1980s with species such as clover, barley and lettuce showed large yield reductions, visible injury and accumulation of potentially toxic metals in crops grown close to major industrial zones near Cairo (Abdel-Latif 2003).

**7. Power plant, Uttar Pradesh, India**
Field studies around industrial sources in India have shown large effects on crop yield. Yield losses of 50 per cent for wheat, pea and beans, attributed to the effects of $SO_2$, were found within 3 km of a 1500 MW coal-fired power station (Agrawal 2003).

**8. Fluoride emissions, Taiwan**
Studies around ceramic factories and brickworks in Taiwan have shown typical symptoms of visible injury due to high concentrations of hydrogen fluoride on sensitive species such as aubergine, banana, betel nut and eucalyptus (Sheu and Liu 2003).

**9. Atlantic rainforest, Cubatão, Brazil**
Extensive damage to forests around the industrial city of Cubatão covered 60 $km^2$ in the 1980s.There were secondary effects on nutrient and water cycling, and soil stability, leading to landslides threatening populated areas (Domingos and Klumpp 2003).

Figure 6.15 Examples from across the world of significant plant damage caused by local industrial or urban emissions of pollutants such as sulphur dioxide, particulates and fluorides. The global background is of emissions of sulphur from major point sources around the globe for the year 2000.

causes. Other components, like sulphur dioxide ($SO_2$), lead and carbon monoxide (CO), are also important pollutants, but these will not be treated in this section; nor will air pollutants which have a more local impact, such as benzene.

In general, urban air quality is improving in Europe and North America, but it remains a serious problem, or is worsening, in other parts of the world. For example, Cohen et al. (2005) estimated that the countries of East and South Asia contribute about two-thirds of the world's premature deaths due to indoor and outdoor particulate matter. However, rather than provide a global perspective, for which comparable data from different locations are very difficult to find, this section will focus on Europe, where a large fraction of the urban population is still exposed to air pollution levels that pose a serious health risk. The proportion of Europe exposed to air pollution episodes above limit values fell during the 1990s for $SO_2$ and $NO_2$ (Figure 6.19), but ozone and PM show no detectable improvement from 1997 to 2010, and they remain important problems. These pollutants are considered in more detail below.

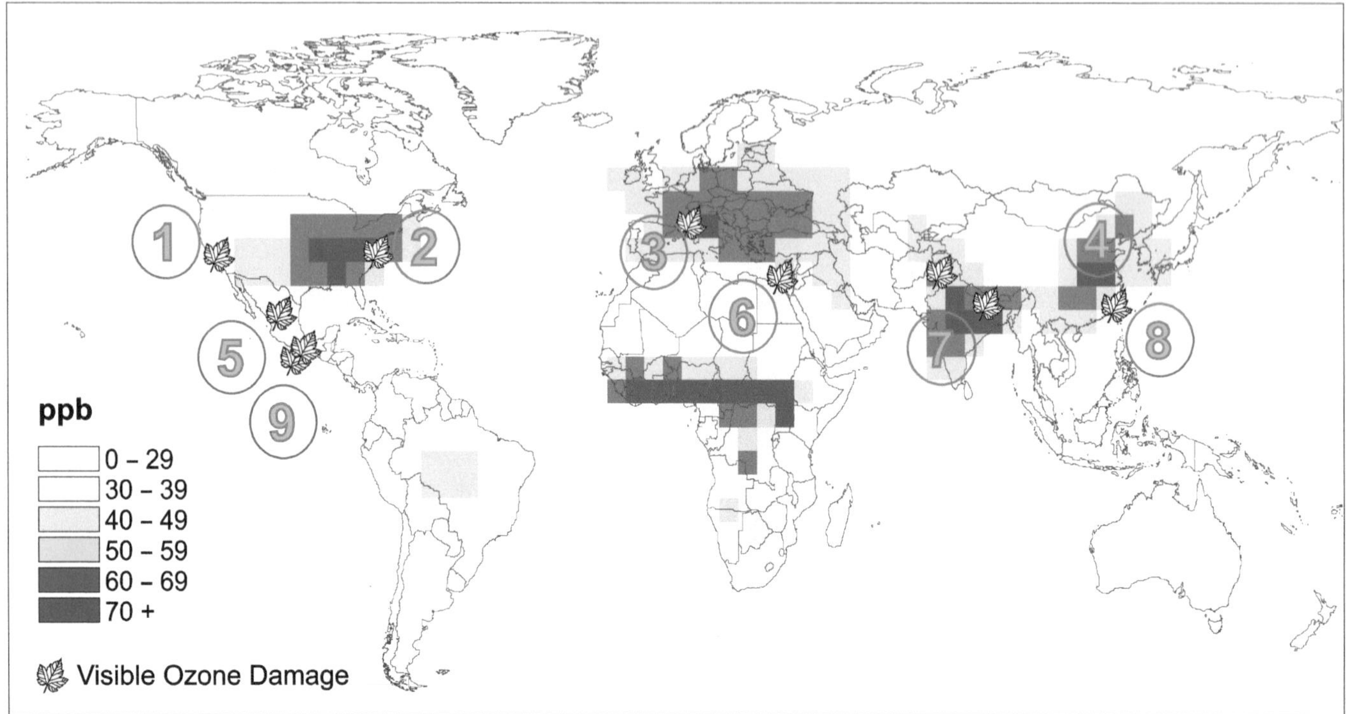


**1. San Bernadino Forest, California**
Declines of ponderosa and Jeffrey pines in southern California were associated with high ozone levels in the 1960s and 1970s. Early senescence and reduced growth led to increased attack by bark beetles and complete death of trees (Miller and McBride 1999).

**2. Impacts on crop yield in the USA**
A national coordinated programme during the 1980s assessed experimentally the impacts of ozone on major US crops. On the basis of these experiments, the annual economic impact was estimated as 3 billion US dollars in the 1980s (Tingey et al. 1993).

**3. Visible injury to Mediterranean crops**
Visible ozone injury, in some cases leading to major commercial losses, is often reported in Mediterranean areas of Europe on crops such as bean, courgette, grape, lettuce, onion, peach, peas, spinach, tobacco, tomato and watermelon (Fumigalli et al. 2001).

**4. Future crop yield in China**
Global models suggest that rural ozone levels in China will increase greatly by 2030. US dose-response relationships suggest that national yields of crops such as wheat and soybean would then be significantly reduced (Wang and Mauzerall 2004).

**5. Sacred fir forests, Mexico**
Sacred fir has shown widespread symptoms of decline and large growth reductions in the mountains around Mexico City. The visible symptoms are characteristic of ozone, which is found in high concentrations in the affected areas (De Bauer 2003).

**6. Vegetables around Alexandria**
Experiments with a protectant chemical show that ozone causes visible injury and reduces the yield of local varieties of radish and turnip at a rural site outside the city of Alexandria, Egypt where there are elevated ozone concentrations (Hassan et al. 1995).

**7. Crop yield in the Punjab**
Studies close to Lahore show that filtering ambient air increases yields of local wheat, rice, soybean, chickpea and mung bean cultivars by 25–50 per cent; it is thought that these large effects are primarily due to ozone (Wahid et al. 1995; Wahid 2003).

**8. Horticultural crops in Taiwan**
Visible ozone injury on tobacco was reported in the 1970s in Taiwan. Extensive field surveys in the 1990s found symptoms of ozone injury on many crops, including sweet potato, cucumber, muskmelon, spinach, potato and guava (Sheu and Liu 2003).

**9. Bean crops in the Valley of Mexico**
High concentrations of ozone and PAN close to Mexico City have been associated with characteristic visible symptoms on a range of horticultural species, and have been shown to cause significant yield losses in local bean cultivars (De Bauer 2003).

**Figure 6.16** Examples of locations where visible injury characteristic of ozone has been reported, where experimental studies in rural locations have shown reductions in crop yield, or where forest decline has been attributed to ozone (Emberson et al. 2003). The background in this global map shows the mean maximum growing season ozone concentration modelled for 1990 using the global STOCHEM model (Collins et al. 1997).

## Ozone

Epidemiological studies show that enhanced ozone levels during summer smog episodes appear to be associated with increased premature mortality and morbidity, lung function decline, airway irritation, worsening of asthma, and airway and lung tissue damage and inflammation. Many of these effects have also been found in controlled toxicological studies with human volunteers or laboratory animals. In 2000, the World Health Organization (WHO) recommended an Air Quality Guideline

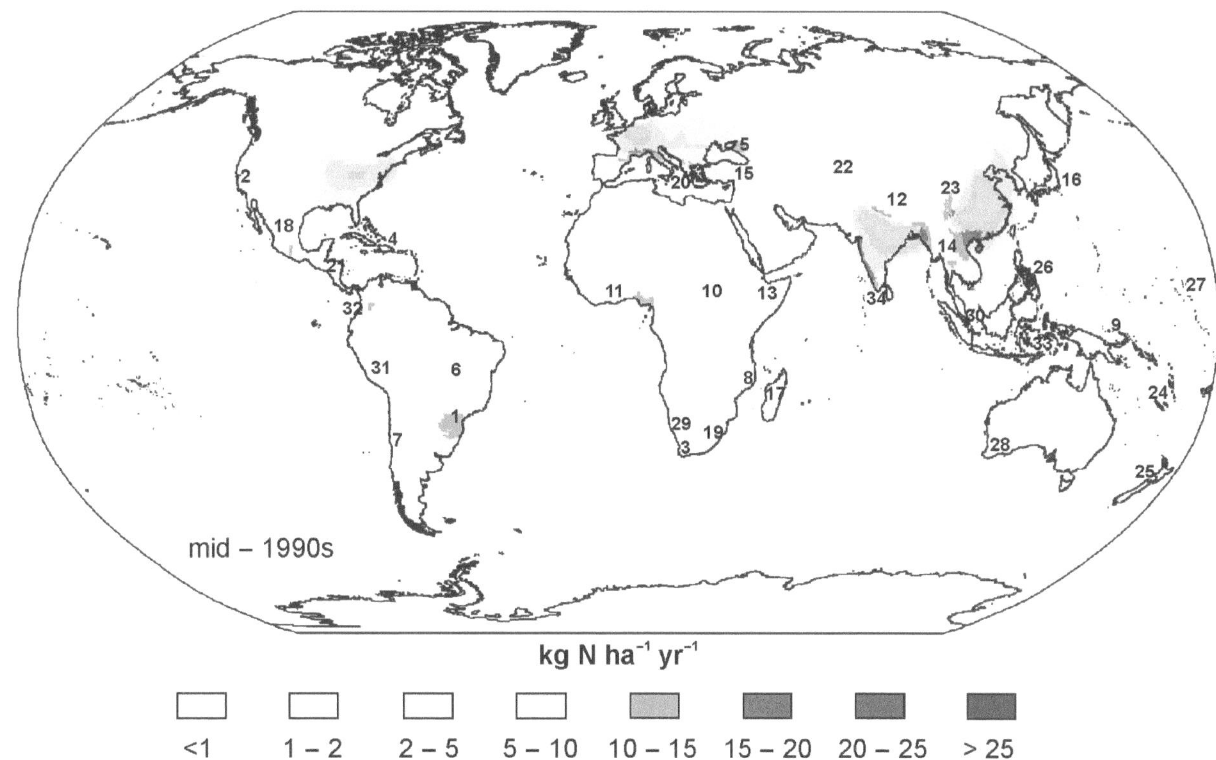


Figure 6.17 Biodiversity hotspots for conservation priorities (Myers et al. 2000; Mittermeier et al. 2005) overlaid on an estimate of the global distribution of annual N deposition (Galloway et al. 2004). The hotspot boundary map is the copyright of Conservation International and numbers indicate the 34 individual biodiversity hotspots that have been identified. To aid in identification of hotspot deposition, colouring is masked (paler) for deposition outside hotspot boundaries e.g. hotspot 34 is the Western Ghats of India and Sri Lanka and only deposition estimates for those areas are highlighted in South Asia. (Source: Phoenix et al. 2006.)

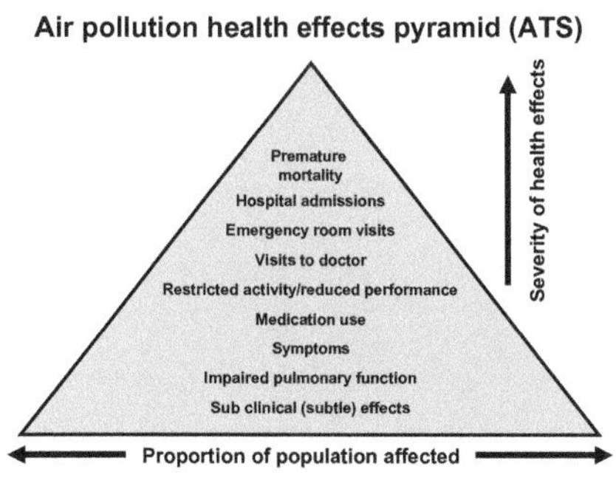


Figure 6.18 Air pollution health effects pyramid originally conceived by the American Thorasic Society (ATS).

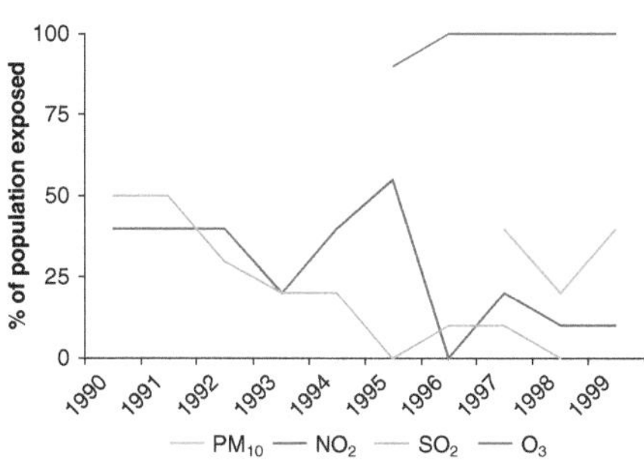


Figure 6.19 Percentage of population of Europe exposed to short-term air quality above the critical limit values for four major pollutants over the 1990s. (Source: EEA 2003a.)

for ozone (WHO 2000) of a daily maximum 8-hour mean value of 120 $\mu g/m^3$, which has been adopted by the EU as not to be exceeded on more than 25 days per year.

The current epidemiological evidence for health effects of ozone suggests that effects are seen at much lower levels. It has been recognized that the WHO 2000 guideline offers inadequate protection of public health from acute and perhaps also from multi-day, repeated, and long-term exposures. Therefore, the WHO has recently updated the Air Quality Guidelines for a number of major air pollutants (WHO, 2006). For ozone the new guideline is a daily maximum 8-hour mean value of 100 $\mu g/m^3$, assuming that this concentration will provide adequate protection of public health, even if some health effects may occur below this level. The basis for the assessment of ozone effects by WHO (2006) is summarised in Table 6.4.

Remarkably, there seems to be a continuous increase in global background concentrations of ozone, resulting in small increases in annual average concentrations of ozone (Royal Society 2008). Using the non-threshold concept, a scenario of a modest decrease in ozone peak levels, a rise in background concentrations, and an increase in the population group at risk, i.e. the elderly, suggests that the health impacts from ozone will continue or may even increase in Europe over the next decades.

The summer of 2003 in western Europe was characterised by a serious heat wave and a substantially higher mortality rate. During this heat wave, the concentration of ozone in particular broke records and frequently reached levels exceeding the EU alarm levels. It has been calculated that air pollution contributed considerably to the observed 'heat wave' mortality rate in this summer (Fischer et al. 2004; Stedman et al.

Table 6.4 Summary of basis for WHO air quality guideline and interim targets for ozone (adapted from WHO 2006).

| Level of protection | 8h mean ozone concentration ($\mu gm^{-3}$) | Likely effects |
|---|---|---|
| High level | 240 | Significant effects; substantial proportion of vulnerable populations affected |
| Interim target | 160 | Important health effects, including a 3–5% increase in daily mortality compared to a concentration of 70 $\mu gm^{-3}$ |
| Air quality guideline | 100 | Adequate protection of public health, but some health effects, including a 1–2% increase in daily mortality compared to a concentration of 70 $\mu gm^{-3}$ |

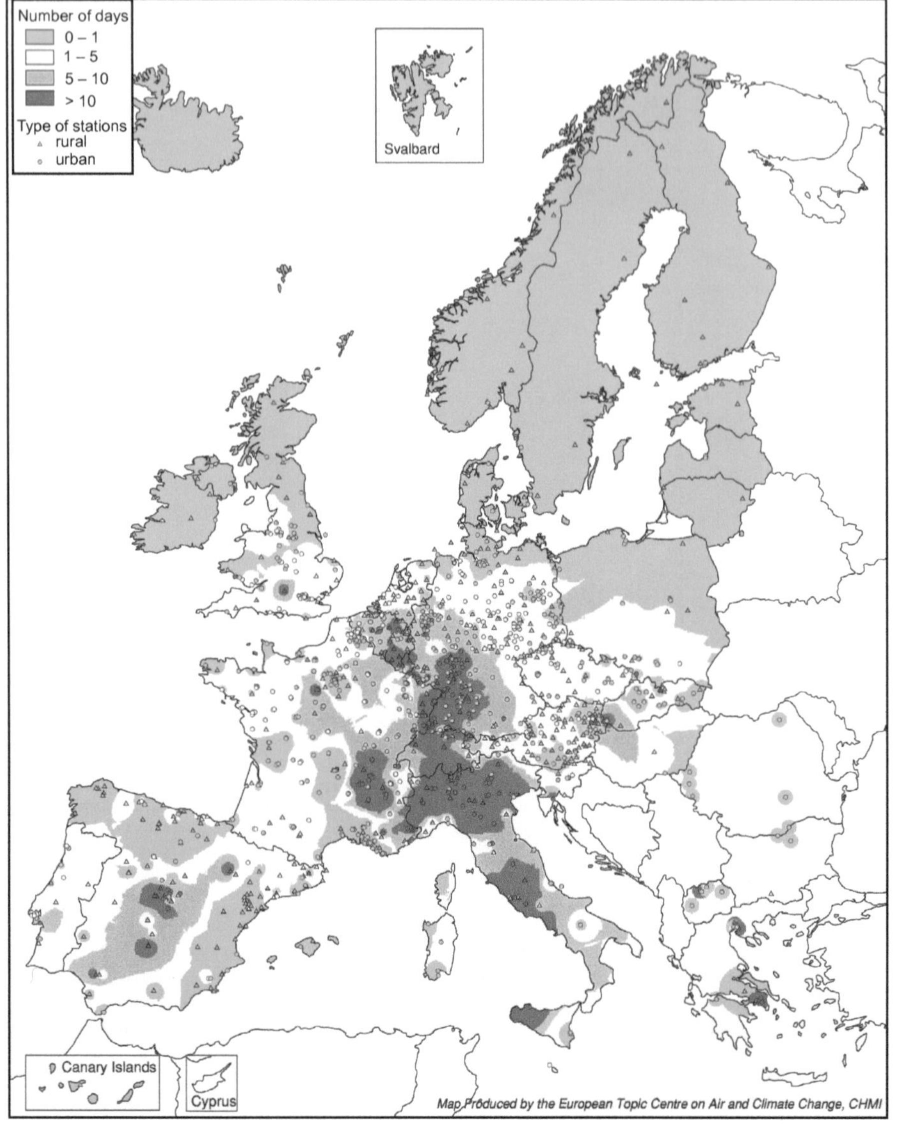


**Figure 6.20** Distribution of number of exceedances of a threshold value of 180 $\mu g/m^3$ ozone as a 1-hour mean over the summer of 2003 in Europe. (Source: EEA 2003b.)

2004). Figure 6.20 (EEA 2003b) shows the number of exceedances of the threshold value used to provide information to the general public, i.e. 180 $\mu g/m^3$ ozone as a 1-hour average, throughout Europe at various rural and urban background stations in the summer of 2003.

## Particulate Matter

Epidemiological studies have reported statistical associations between short-term, and to a limited extent also long-term, exposure to increased ambient PM concentrations ($PM_{10}$ and sometimes also $PM_{2.5}$ and ultrafine particles) and increased morbidity and premature mortality. Whether these associations are causal and which PM properties and/or mechanisms ($PM_{10}$, $PM_{2.5}$, ultrafine-mode particles, physical properties, chemical or biological components) are responsible is still unclear. However, the toxicity database on PM effects is expanding rapidly and increasingly may help to explain the health effects observed and provide a view on biologically plausible mechanisms of action. It is currently assumed that there is no threshold below which health effects are unlikely to occur. The revision of the WHO Air Quality Guidelines for PM proposed that, despite this, guidelines should be set to minimise the risk of adverse effects of both short-term and long-term effects of PM (WHO 2006). These values were set as 20 $\mu g/m^3$ for an annual mean and 50 $\mu g/m^3$ as a daily mean for $PM_{10}$, with corresponding values of 10 $\mu g/m^3$ and 25 $\mu g/m^3$ for $PM_{2.5}$, but advice is also given on exposure-effect relationships on which a health impact assessment can be based. The WHO Air Quality Guidelines for Europe (2000) estimated a number of health outcomes associated with changes in daily particulate matter ($PM_{10}$) concentrations for a population of 1 million people (with health data from epidemiological studies). The basis for the assessment of air quality guidelines for PM by WHO (2006) is summarised in Table 6.5.

**Table 6.5** Summary of basis for WHO air quality guidelines and interim targets for long-term effects of PM (adapted from WHO 2006).

| Level of protection | $PM_{10}$ ($\mu gm^{-3}$) | $PM_{2.5}$ ($\mu gm^{-3}$) | Likely effects |
|---|---|---|---|
| Interim target 2 | 50 | 25 | Risk of premature mortality increased by about 9% relative to AQG |
| Interim target 3 | 30 | 15 | Risk of premature mortality increased by about 3% relative to AQG |
| Air quality guideline (AQG) | 20 | 10 | Lowest level at which mortality is increased by long-term exposure to $PM_{2.5}$ with more than 95% confidence |

## WHO and EU Approaches for Clean and Healthy Air

The WHO Air Quality Guidelines (AQG) for Europe (WHO 2000), together with the global update for the four major air pollutants (WHO 2006), provide a basis for protecting human health from effects of air pollution and provide guidance for authorities to make risk management decisions (see also Chapter 2, section 2.7). In 2001, WHO agreed with the European Commission to provide the Clean Air For Europe (CAFE) programme with a new review of health aspects of air quality in Europe. This review focused on studies published after the second edition of the WHO AQG was elaborated, and has been influential in changing views on health-related aspects of the substances under consideration. The WHO recommended the use of fine particulate matter ($PM_{2.5}$) as an indicator for particulate pollution-induced health effects and also the need to consider the evidence for short-term ozone effects on mortality and respiratory morbidity at the ozone concentrations experienced in many areas in Europe. Based on these findings, the WHO recommended the removal of the threshold concept for ozone; and to update the exposure-response relationships for various health outcomes induced by PM or ozone.

In 1996, the EU adopted the Air Quality Framework Directive and the Air Quality Limit Value methodology. The First Daughter Directive addresses PM, $NO_2$ and lead; the Second Directive covers CO and benzene; and the Third Directive includes ozone. Air Quality Guidelines established by the WHO (1986, revised in 2006) formed the health basis for this standard setting, and the new values intend to provide increased protection to the population against a wide range of health effects. The analysis of the health benefits of different emission control strategies, based

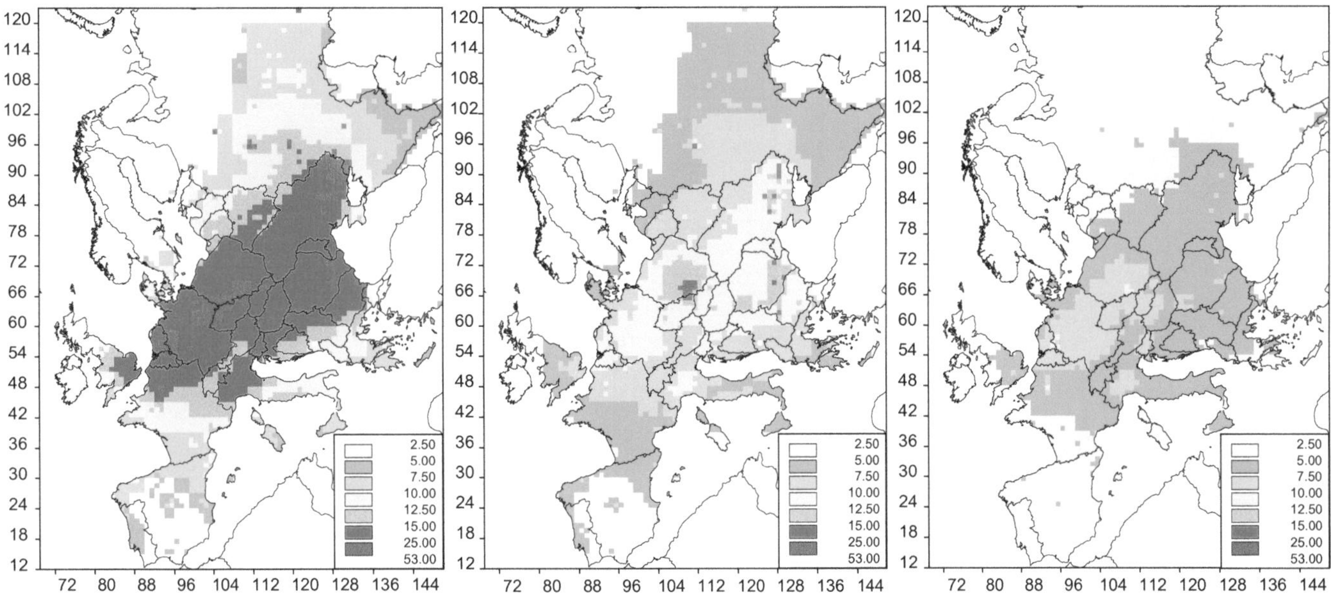


Figure 6.21 Losses in statistical life expectancy (in months) in rural areas of Europe in 1990 (left), for current legislative scenario in 2010 (central panel) and maximum technically feasible reductions (right panel).

on long-term and short-term effects of $PM_{2.5}$ and short-term effects of ozone, provided a central element in the development of a Thematic Strategy for Air Quality by the European Commission, following the CAFE programme (CEC 2005).

## Health Benefits from Air Pollution Abatement

Air pollution abatement strategies in the last decades have been successfully focused on the reduction of severe episodes and high peak levels. Substantial improvements in air quality have indeed been achieved in Europe, despite some remaining problems like ozone and PM. Nowadays it is recognized that abatement actions should also focus on reducing the total burden of health effects by reducing longer-term average exposures to healthy levels. It remains an important question to what extent there is conclusive evidence that air pollution abatement measures and emission interventions have indeed resulted in lower personal exposures and reduced health effects. Quantifying the health impact of air pollution and providing evidence that air quality regulations indeed improve public health (the 'accountability' issue) has therefore become an important component in policy and decision making and is increasingly the subject of studies. A few examples of estimating health benefits of air pollution abatement can be mentioned.

### *Ex ante predictions of decreased health effects through modelling of impacts of lower concentrations*

Although helpful, these studies also carry a large degree of uncertainty because it is unknown whether or not the causal factor(s) is reduced proportionally and whether the relative risk figures stay the same. This might lead to an over- or underestimation of possible benefits. However, valuable 'first-order' attempts have been made to show the potential power to predict possible health effects (mortality, life years lost) from $PM_{10}$ or $PM_{2.5}$ reductions (APHEIS 2002; Mechler et al. 2002). Figure 6.21 shows estimated losses of life expectancy due to particulate pollution in rural areas throughout Europe (left panel, 1990) and the projected decreases (improvement) under emission abatement scenarios (middle and right panels, 2010).

### *Ex post evaluations of the impact of implemented emission interventions on observable health*

These studies are unfortunately very rare. The few examples can be distinguished into short-term and long-term changes, depending on the duration and type of intervention:

- Traffic reduction during the Summer 1996 Olympic Games in Atlanta decreased the number of asthma acute care events (Friedman et al. 2001).
- Building a road tunnel in Oslo decreased self-reported symptoms and improved health and well-being (Bartonova et al. 1999; Clench-Aas et al. 2000).
- Lowering sulphur in fuel oil in Hong Kong reduced the adverse effects on airway functioning and premature deaths from respiratory and cardiovascular diseases (Wong et al. 1998; Hedley et al. 2002).
- Banning coal in Dublin resulted in decreased mortality rates (Clancy et al. 2002).
- Constructing a bypass to reduce congestion and heavy traffic-related air pollution in an area in North Wales (UK) resulted in improvement in respiratory health and a reduction of a number of respiratory symptoms (Burr et al. 2004).

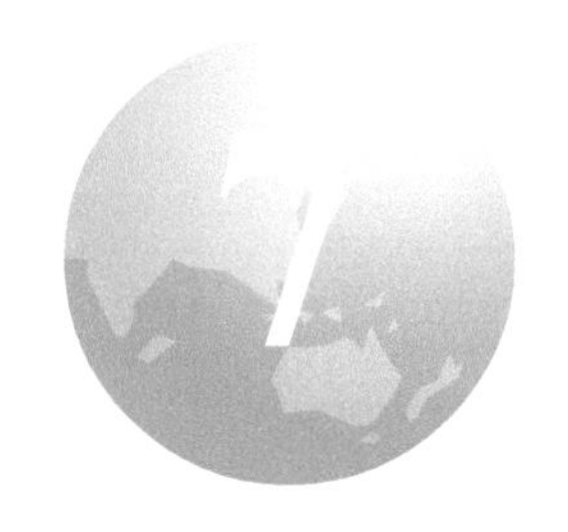

# FUTURE TRENDS IN AIR POLLUTION

*Markus Amann, Janusz Cofala, Wolfgang Schöpp and Frank Dentener*



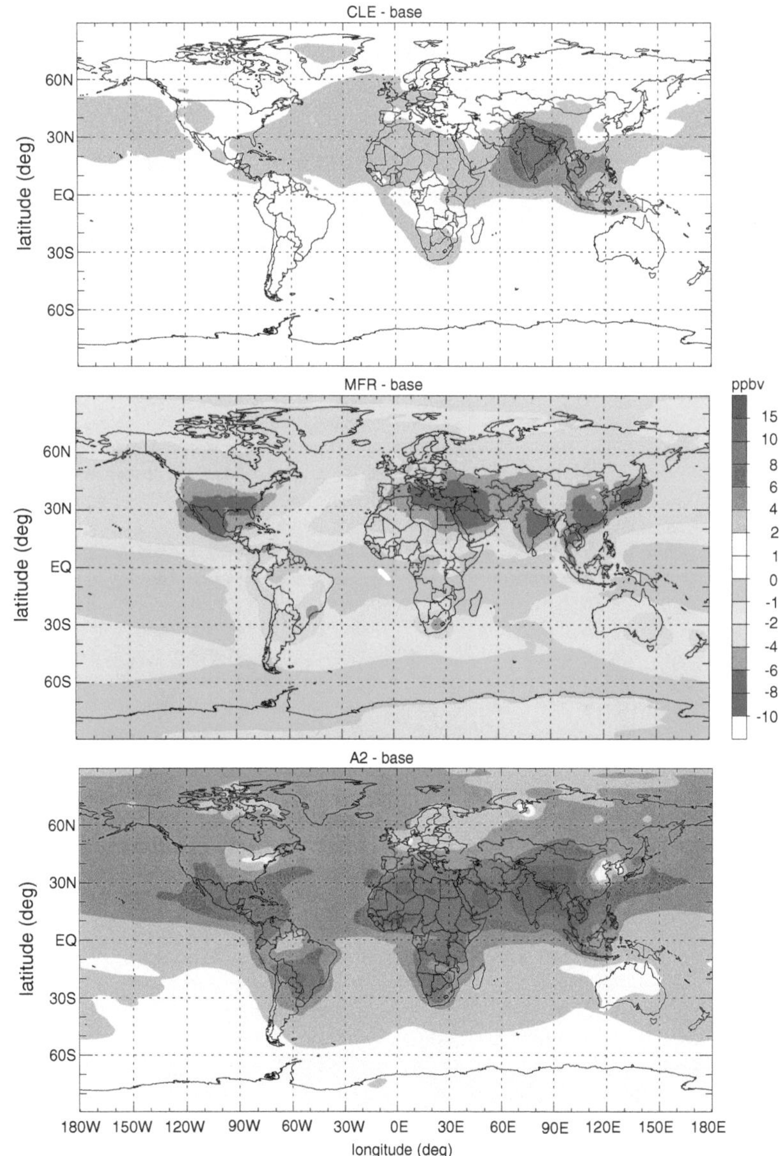


Multi-model ensemble annual average surface ozone ($O_3$) concentration differences between 2000 and the (upper) current legislation (CLE), (middle) Maximum technologically Feasible Reduction (MFR) and (lower) pessimistic IPCC SRES A2 scenarios for 2030. The CLE scenario suggests a stabilization of $O_3$ in 2030 at 2000 levels in parts of North America, Europe, and Asia but also show that $O_3$ may increase by more than 10 ppb in other areas (e.g. India). Background $O_3$ levels may also increase by 2–4 ppb in some regions (e.g. the tropical and mid-latitude northern hemisphere). In contrast if the MFR scenario is implemented by using all current technologies to abate the $O_3$ precursor emissions then a cleaner future is possible. If the more pessimistic A2 scenario is followed then annual average surface $O_3$ concentrations are projected to increases by 4 ppb worldwide and by 5–15 ppb in Latin America, Africa, and Asia. (Source: Dentener et al. 2006.)

'Prediction is very difficult, especially if it's about the future' (Niels Bohr, Nobel laureate in Physics). For instance, forecasters in Victorian London foresaw their city knee-deep in horse manure, one of the most pertinent urban environmental problems in cities at that time. A hundred years later, this prediction has not materialized and the situation has changed drastically. While traffic itself is still considered a major cause of urban air pollution, the contribution from horses has entirely disappeared and motorized vehicles are now the major source of deteriorated air quality in most modern cities.

Given the failure of simple extrapolations of present trends into the future, what can we say about air pollution in the coming decades?

To begin with, we know that population will further increase in urban areas, and we know that all societies aim to further strengthen their economic wealth. For a long time, air pollution from anthropogenic (non-natural) activities has been considered an unavoidable concomitant of economic development. Over long historic periods, we have seen air pollution levels increasing together with economic growth. Countermeasures to control air pollution have often been considered too costly to put into effect without compromising economic wealth.

Following this logic, the envisaged continued growth in global population, together with the universal target of improving prosperity, would lead to drastically worsened air quality around the globe, especially in many developing countries. On the other hand, there are a number of real-world examples showing that once a certain level of economic development has been reached, some air pollution problems ameliorate. We can identify several reasons for declining pollution: some of the most polluting economic activities (such as the production of steel and cement) decline in the course of economic development and other less polluting activities (e.g. information technology) become more important. Economic development spurs the introduction of advanced technologies – many of them are, by their nature, less polluting. Moreover, societies begin to be concerned about air pollution and find ways to actively reduce pollution levels to improve their living conditions. Thus, the relation between the levels of economic activity and air pollution is not necessarily fixed, but depends on a range of factors. At least some of these factors critically depend on the importance given by a society to what are acceptable physical and social living conditions.

What does this tell us about future air pollution in industrialized countries, where societies have already accepted and implemented costly measures to keep air pollution at acceptable levels? And what about the prospects for developing countries, especially in urban areas, where air population is expected to continue its rapid growth in the coming decades? And for societies which will reach the status of economic wealth and growth at which other countries have begun to fight air pollution?

In the following sections, we will discuss the major factors influencing the emissions of air pollutants and how they depend on the level of economic development. We will summarise the present perspectives of how these factors will develop in the coming decades in the various parts of the world, and look into the possible consequences on the emissions of various pollutants. Finally, we will sketch the implications of the anticipated changes in emissions on air quality.

Unlike in the climate field, there are relatively few research groups who are examining the future trends of air pollution on regional and global scales. Consequently, much of the results and conclusions contained in this chapter are based on a few but critical studies reported in the literature.

## 7.1 Population and Economic Development

The growth of population is a major driving force at the source of atmospheric air pollution. The number of people determines, inter alia, the demand for economic services and thus the overall amount of anthropogenic activities that give rise to air pollution. Additionally, the spatial patterns (e.g. transportation patterns) resulting from the living and working habits of people have a crucial influence on the spatial density (geographical distribution) of emissions, and thus determine the hotspot pollution areas in the world. Both the spatial patterns and densities are expected to change significantly in the future in such a way that air pollution problems should become more accentuated.

The median United Nations (UN) population projections (UN 2004) suggest a world population increase between 2000 and 2030 of 34 per cent. However, most relevant for future air pollution levels is that populations will change differently in many parts of the world. While the UN projections foresee for the more developed countries an increase of only 4 per cent up to 2030, population in less developed countries is expected to grow by 41 per cent and in the least developed countries by as much as 88 per cent. In addition, urbanisation is expected to continue throughout the world, so that by 2030 more than 60 per cent of the world population will live in urban areas, especially in developing countries. This will cause emissions to concentrate exactly at those locations where people live, thus exposing even more people to potentially harmful levels of air pollution. (Variations in population and air pollution in cities is discussed further in Chapter 2.)

This intensification of emission-generating activities that comes together with economic development aggravates the threat of air pollution to an enlarged population. All governments around the world have ambitious plans for economic growth to improve the material well-being of the population. For our analyses we have collected economic projections up to 2030 for the entire globe, as far as possible from national sources. According to these policy plans, world economic activities expressed through the gross domestic product (GDP) at world market prices are anticipated to grow by 3 per cent per year (i.e. by a factor of more than three between 1990 and 2030). Industrialized countries typically aim at a 2 per cent increase per year, while developing countries strive to reduce their gap with industrialized countries through an average growth rate of 6 per cent. This would increase the GDP per capita in developing countries by a factor of five by 2030.

## 7.2 Projections of Energy Consumption

History shows that economic development does not simply boost all anthropogenic activities uniformly. Economic development is driven primarily by technological progress. With economic progress, advanced technologies gain market shares, and old, outdated production processes become less important. This is of critical importance for air pollution. Emissions from modern production processes are different to those of traditional processes. In many cases emissions are lower, but some new technologies may actually release other types of pollution.

Traditionally, energy combustion (e.g. burning coal) was the major source of air pollution, both in industrialized and developing countries. However, energy systems are continuously transforming over time, and these changes will have a direct impact on the emissions of air pollutants. Many of these transformations occur 'autonomously' as a feature of technological progress and as a consequence of the shift of nations' economies from the focus on energy-intensive basic material industries towards new products with less material content. For instance, we can observe over the last decades a worldwide trend of decreasing energy intensity of the national economies. Over the last century energy input has declined by approximately 1 per cent per year for producing the same amount of GDP (overall economic output).

Due to the importance of energy for the overall economic and environmental performance, many quantitative economic projections specifically address the implications on energy systems. The energy projections that we have compiled for this analysis anticipate for the coming decades a continuing and even slightly accelerated decrease of energy intensity of the national economies. Up to the year 2030, the energy intensity of GDP is expected to decrease globally by approximately 1.2 per cent per year. Consequently, while global economic output would grow by 130 per cent between 2000 and 2030, world primary energy consumption would only increase by about two-thirds (i.e. by 1.65 per cent per year). This continued decoupling between economic growth and energy consumption is an important factor in determining future levels of air pollution.

Then again, perhaps even more important are the anticipated structural shifts in energy consumption, for example, from the reliance on coal towards natural gas. Consistently, all energy projections foresee less increase for the most polluting fuels and growing market shares for cleaner forms of energy. For instance, coal use is expected to increase globally by 5 per cent, while the consumption of liquid fuels is expected to grow by some 45 per cent and of natural gas by even 130 per cent up to 2030.

At the global level, these developments would lead to a 50 per cent increase in energy-related carbon dioxide ($CO_2$) emissions between 2000 and 2030, which is significantly less than the growth in total energy consumption (+66 per cent) and total economic output (+130 per cent). This global increase in anthropogenic $CO_2$ emissions is immediately relevant for climatic change. However, the evolution of air pollution is strongly determined by the spatial pattern of emissions. Most pollutants have a relatively short lifetime in the atmosphere, so they do not mix globally and thus the locations where emissions occur is critical for their impacts. Thus, air pollution will show a much more differentiated development between continents.

The projections, which we have compiled from national sources, anticipate the most rapid economic growth in Asia (+500 per cent increase in GDP) and Latin America (+200 per cent), while the developed Organization for Economic Cooperation and Development (OECD) region expects its economic output to grow by approximately 60 per cent between 2000 and 2030. Higher pace of economic development should lead to faster improvements in energy intensities, as well as to an accelerated shift away from the most polluting fuels. Thus, the fast-growing economies in Asia and Latin America would 'only' double their energy consumption, but increase their coal use by not more than 30–40 per cent.

This analysis tells us that we have to expect for the coming decades increased pressure from the main driving forces for air pollution throughout the world. For instance, population will grow further and people will concentrate even more in urban agglomerations than in the past. More people will also ask for more economic services, which cause more pollution. Nevertheless, economic development moves national economies towards cleaner production and consumption processes, so air pollution is likely to grow at a lower rate than the overall economic output.

## 7.3 Emission Control Measures

Modern societies have successfully decoupled economic growth from the consumption of natural resources, and this trend is expected to continue. Will this autonomous decoupling be sufficient to provide clean air for the world population?

Humans are not living evenly dispersed around the globe, but people tend to live together in agglomerations. At present, almost half the world population lives in urban areas which only make up a very tiny fraction of the entire earth's surface. Furthermore, the share of urban population is expected to continue to increase in the future. At the same time, most economic activities that cause pollution can be concentrated at the same places where people live, so that the most populated areas in the world are often exposed to the highest levels of air pollution.

Thus, in order to answer the questions posed above, we cannot restrict ourselves to an analysis of the global average situation, but we must look into these hotspot areas.

The preceding chapters in this book have demonstrated that present levels of air pollution impose serious threats to human health and ecosystems in many regions around the world, and in particular, in most metropolitan areas in developing countries. It is clear that, especially in these rapidly developing urban agglomerations, the pressure on the environment will further increase due to the continued urbanisation trends. The autonomous structural changes will not be sufficient to compensate for this increased pressure in the future, let alone to bring down present air pollution to levels that are not harmful to human health and the environment.

A wide range of technological measures has been developed that can prevent emissions from being emitted into the atmosphere. Many of the measures are rather efficient and advanced technologies can today eliminate up to 90–99.9 per cent of air pollutants from being released to the atmosphere. While certain costs are associated with the application of such technologies, their environmental benefits are considered high enough that they are now widely applied, especially in industrialized countries. Consequently, actual emissions in many industrialized countries are only 10–20 per cent of the volume that would normally occur without such abatement technologies. Such 'end-of-pipe' measures make it possible to reduce emissions substantially below historic levels, even with sustained economic growth, and to approach air quality conditions that will not give rise to significant damage to human health and the environment.

Thus, the extent to which such emission control measures will be applied in the future will critically determine prospective levels of air pollution.

We do not yet fully understand which factors determine whether a society finds it appropriate to apply such emission control measures. Obviously economic considerations are important. Only countries which have reached a certain stage of economic performance have issued legislation requesting such emission control measures. However, there is no obvious threshold of economic development at which emissions are actively controlled. It seems that, over time, nations have started to implement abatement measures at earlier phases of economic development. The market availability and maturation of emission control technologies are further factors. For instance, once catalytic converters for vehicles have become widely available, they also entered legislation in developing countries. Environmental pressure perceived from air pollution awareness also seems to be an additional factor to convince a society to spend money on air pollution control.

Among all the factors that influence future levels of air pollution, it is probably most difficult to predict accurately the future stringency of emission control legislation in the various countries around the world. As a conservative approach, we take stock of the emission control measures that are decided in each country at the time of writing, assume their implementation over time following the national laws and track their penetration into the future. We do not assume in this analysis that countries will take additional measures in the coming decades to further reduce their emissions, even if the environmental pressure would call for such action.

## 7.4 Emission Projections

The combined impacts of the factors that determine air quality (i.e. population growth, increase in economic wealth, technological progress and application of emission control measures) should lead to a distinct decoupling between the volume of human activities that cause emissions and, the amount of generated air pollution throughout the world. As mentioned above, although economic development will lead to increased levels of emission-generating human activities, these activities will be performed in a cleaner manner and will cause less pollution. As a result, most of the major air pollutants are expected to decline in many industrialized countries due to stringent emission control legislation. Furthermore, in developing countries, the recently adopted emission control measures will reduce the previously uncontrolled growth in air pollution. However in many cases, the currently adopted measures will not be sufficient to stop a further increase or even to reduce air pollution in the future.

For the emissions of sulphur dioxide ($SO_2$), particularly large reductions are expected in Europe due to the aggressive policies to combat acid rain and the economic restructuring in eastern Europe, see Figure 7.1. No major changes are anticipated for $SO_2$ emissions in North America, Latin America and Africa. Due to the continued reliance on coal to power the economic growth in Asian countries, we have to expect further growth of $SO_2$ emissions

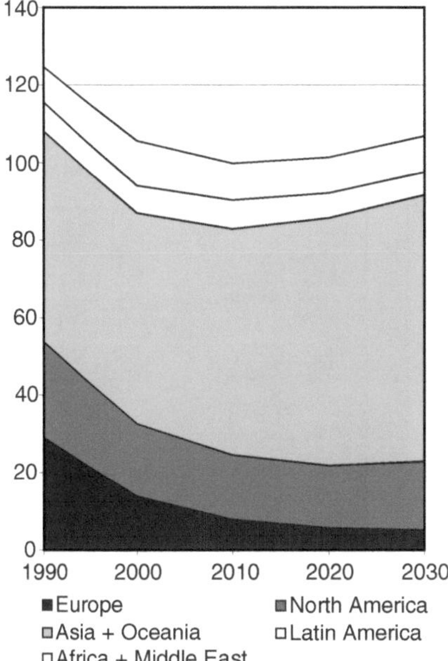


Figure 7.1 Projected development of anthropogenic emissions of sulphur dioxide ($SO_2$) by world region, assuming the implementation of all presently decided emission control legislation (million tons $SO_2$). (Note: emissions resulting from open biomass burning, international shipping and aircraft are not included in this figure.)

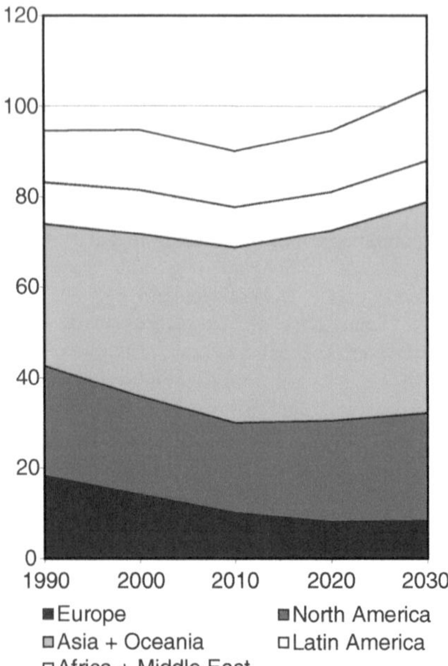


**Figure 7.2** Projected development of anthropogenic emissions of nitrogen oxides ($NO_x$) by world region, assuming the implementation of all presently decided emission control legislation (million tons $NO_2$). (Note: emissions resulting from open biomass burning, international shipping and aircraft are not included in this figure.)

in this region, despite the emission control measures that have been adopted recently in several Asian economies (Figure 7.1). Similar figures are shown for $NO_x$ (Figure 7.2) and CO (Figure 7.3) emissions. On a per capita basis, countries in North America

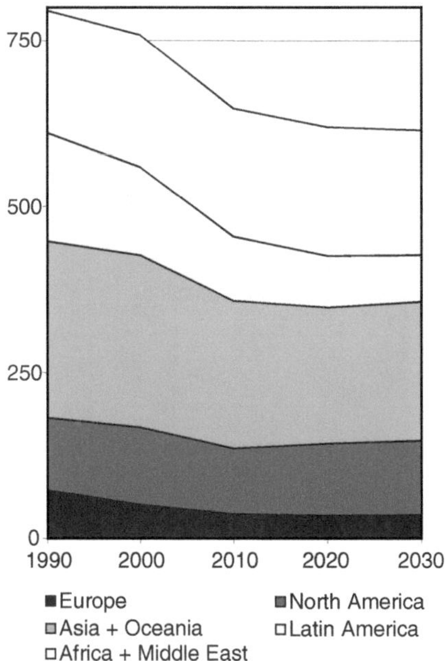


**Figure 7.3** Projected development of anthropogenic emissions of carbon monoxide (CO) by world region, assuming the implementation of all presently decided emission control legislation (million tons CO). (Note: emissions resulting from open biomass burning, international shipping and aircraft are not included in this figure.)

and developing countries in Asia show highest emission densities, both for $SO_2$ and $NO_x$. Over time, per capita emissions are expected to decline.

More recent estimates of future global emissions of $SO_2$, $NO_x$, and CO (Cofala et al. 2007, Cofala et al. 2009) are lower, which is due to revised energy forecasts and faster penetration of emission control measures in many developing countries. Recently, in preparation of the 5th IPCC report, a new set of so-called RCP scenarios has been generated (Moss et al. 2010). The emissions of air pollutant in these scenarios peak by 2030 and decline afterwards, with the lowest emissions associated with the most progressive climate policies. Thus, stronger links of climate and air pollution policies can be expected in the coming decades.

## 7.5 Projections of Future Air Quality

While a number of studies have explored future air quality for different regions in the world, we have only an incomplete global overview on the air quality implications related to the emission trends outlined in the previous section. Our global understanding is best for those pollutants that are, due to their long residence in the atmosphere, transported between continents. For instance, scientists have paid a great deal of attention to ozone pollution in the lower layers of the atmosphere, which is harmful to human health and vegetation. While peak concentrations of ground-level ozone occur around most of the world's urban areas (primarily due to local emissions), significant background ozone pollution is associated with the intercontinental transport of its precursor emissions from the entire hemisphere.

Control measures for local emissions in North America and Europe led to a decline of ozone peaks over the last decades of the twentieth and the early twenty-first centuries, but background concentrations of ozone have increased steadily throughout the northern hemisphere due to the increased emissions in the developing world. Scientists have used global-scale models of atmospheric chemistry and transport to study the potential impacts of the global increase of ozone precursor emissions on background ozone concentrations. These calculations demonstrate very clearly the importance of the long-range transport of ozone. For the year 2000, models calculate the transport of ozone from the North American continent, eastwards over the Atlantic, up to Europe, from Europe over central Asia and from East Asia over the Pacific (Figure 7.4a). The highest ozone levels are found at the east and west coasts of North America, in the Mediterranean region, over the Himalayas (due to mixing with ozone from the free troposphere) and over eastern China, Korea and Japan. Following the projected changes in pre-cursor emissions, ground-level ozone is computed to increase throughout large parts of the world, especially in China and India, but also in Africa, as shown in Figure 7.4b.

These global-scale atmospheric models focus on the intercontinental transport of ozone, but cannot tell us directly the resulting local health and vegetation impacts that are strongly influenced by locally generated ozone. In Europe and North America, local and regional emissions are computed to contribute approximately 10–20 ppb on an annual basis to the local ozone burden, while background concentrations at the time of writing typically reach 30–40 ppb in the northern hemisphere. Present understanding suggests a general increase of background ozone by approximately 5 ppb for the first few decades of the twenty-first century, which counteracts the effectiveness of the emission control measures taken in industrialized countries and enhances the expected increase from local emissions in the developing world.

## 7.6 Projections of Air Quality Impacts

### Health Impacts from Fine Particles in Europe

Recent health studies discovered a significant risk to public health from air pollution, especially from inhaled small particles (see also Chapter 6). Altogether, epidemiological studies throughout the world have found strong associations between the human exposure to particles with a diameter of less than 2.5 $\mu$m and premature mortality, notably due to an increased occurrence of cardiovascular diseases and lung cancer (Brunekreef and Holgate 2002).

These small particles can originate from many different sources. An important fraction is generated as a direct by-product of many combustion processes (e.g. diesel exhaust, coal and wood burning), while others are formed mechanically (e.g. soil dust, industrial production processes, resuspended road dust). In addition, other types of fine particles are chemically formed in the atmosphere from the precursor species, such as the gaseous air pollutants of sulphur dioxide, nitrogen oxides, ammonia and volatile organic compounds.

Recent estimates suggest that the lifetime of the European population is shortened by approximately nine months due to fine particle ($PM_{2.5}$) exposure, which is comparable to that caused by traffic accidents (CAFE 2005). In densely populated and industrialized areas with high emissions, the impacts can be significantly higher. A comparable assessment for developing countries has not yet been carried out, but there are strong indications of drastically higher impacts in the developing world. For Europe, it is estimated that the emission reductions that are imposed by the latest air quality legislation of the European Union will reduce the loss in life expectancy by 2030 to approximately six months on average (see Figures 7.5 (a) and (b)).

### Threat to Biodiversity due to Excess Nitrogen Deposition Affecting Terrestrial Ecosystems

The last decades of the twentieth century and the early years of the twenty-first experienced a marked increase in nitrogen emissions from intensified energy combustion, agricultural activities and industrial production processes (see also Chapter 6).

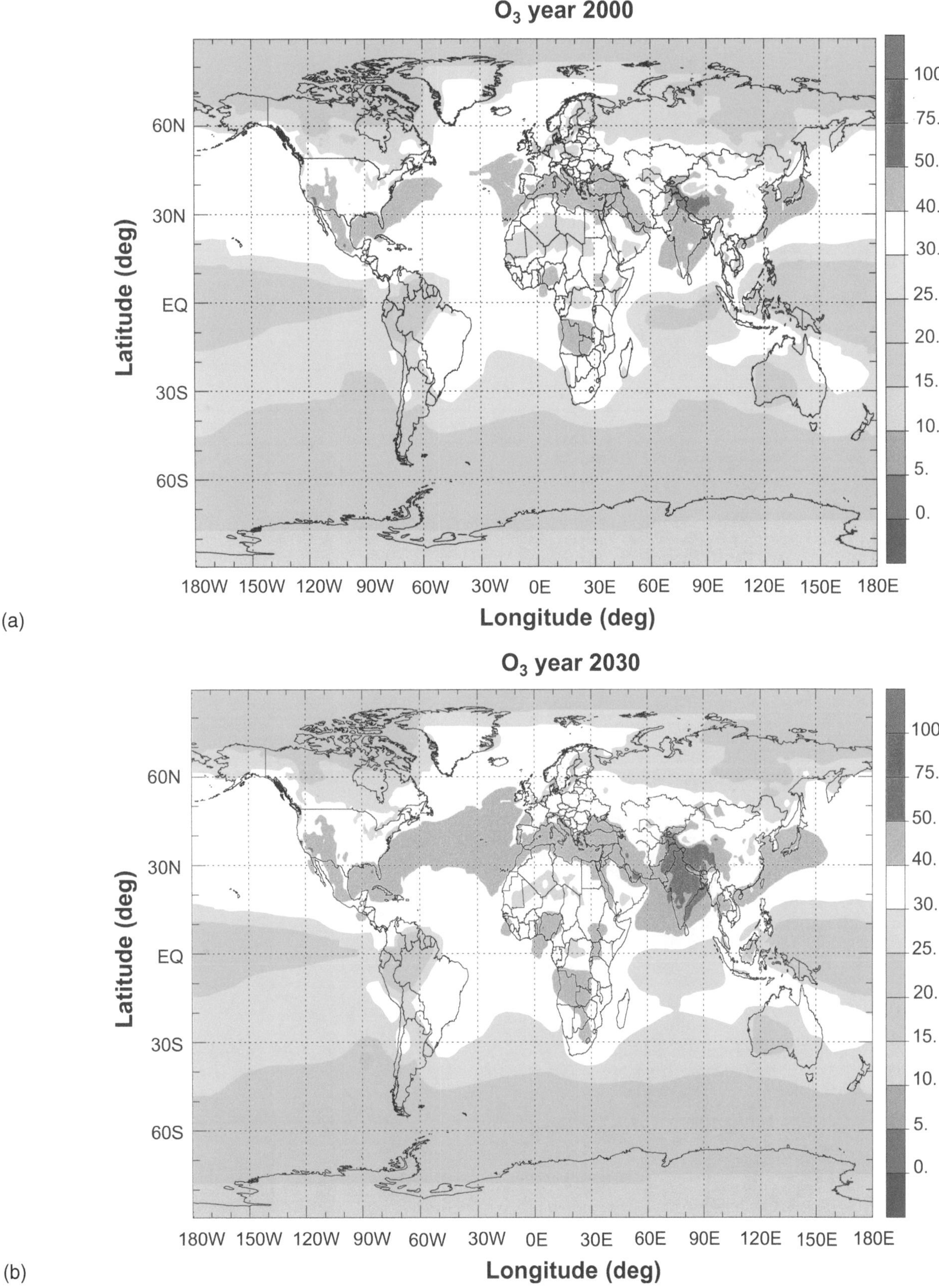


**Figure 7.4** (a) Annual mean concentrations of surface ozone (parts per billion – ppb) calculated by the global 1x1 degree TM5 model for the year 2000 and (b) annual mean concentrations of surface ozone (parts per billion – ppb) calculated by the global 1x1 degree TM5 model using a current legislation emission scenario projected for 2030 (Dentener et al. 2006).

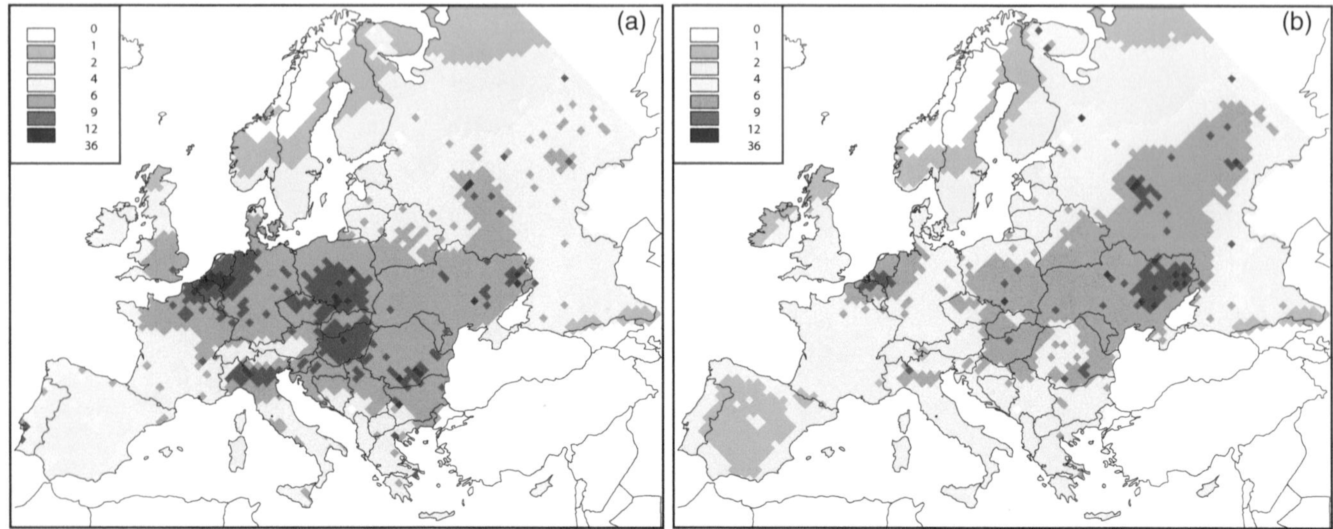


Figure 7.5 (a) and (b) Months of loss in life expectancy that can be attributed to the exposure to fine particulate matter ($PM_{2.5}$) in Europe. Situation in 2005 (a) and situation computed for 2020 (b), when all presently decided emission control measures will be implemented. (Source: CAFE 2005.)

Studies show that, as a consequence, many ecosystems throughout the world receive nitrogen input from the atmosphere above a sustainable level. These high levels of nitrogen deposition are likely to lead to negative ecological effects such as losses in biodiversity. While the nitrogen problem is now recognized at the global scale, most quantitative assessments focus on the European continent. Thus, we have only quantitative analyses of future trends available for Europe, but not for the developing countries where the situation could potentially be even worse.

In Europe, we can expect that anthropogenic emissions of nitrogen oxides will decline due to the strict emission controls that have been decided with the aim to reduce ground-level ozone. However, there is currently little effort to limit or even reduce ammonia emissions from agricultural activities, which constitute in many cases the dominant source of nitrogen to the atmosphere. Consequently, there is only little hope that harmful excess nitrogen deposition would substantially decrease in the coming decades, unless additional measures are adopted and implemented in Europe. Figures 7.6a and b show the impact of nitrogen deposition on biodiversity in Europe for 2000 and 2020.

## Acid Deposition

Acid deposition leading to damage of terrestrial and aquatic ecosystems has been recognized as one of the first serious transboundary air pollution problems. Since the 1980s a series of international environmental agreements have been established in which European and North American coun-tries commit themselves to far-reaching reduc-tions of their acidifying emissions. Model calculations have played a prominent role in these negotiations, and there is now a relatively solid understanding of the likely trends in acid deposition in Europe. This analysis compares, as a measure for the sustainability of ecosystems with respect to acidification, the amount of acid deposition from the atmosphere with site-specific thresholds (or critical loads) at which harmful effects would not occur according to current scientific understanding.

For example, Figures 7.7a and b compare the extent to which forest ecosystems received acid deposition for the year 2000 in excess of the long-term sustainable level and how the situation

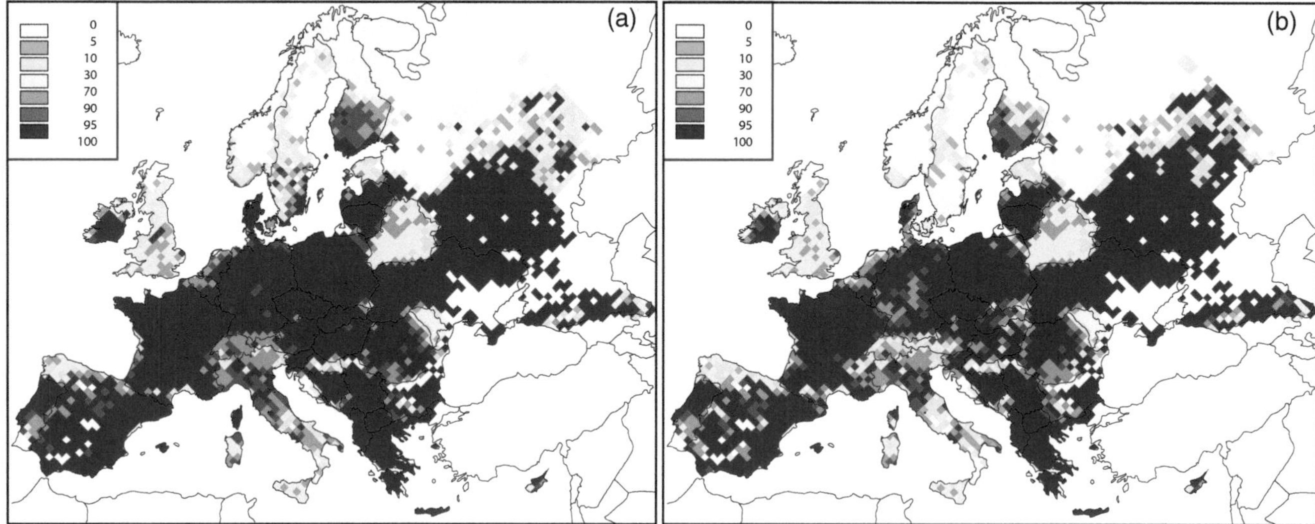


Figure 7.6 Percentage of ecosystems where biodiversity is threatened by excess nitrogen deposition from the atmosphere above the limit that can be sustained by ecosystems (i.e. ecosystems receiving nitrogen deposition above their critical loads). Situation in (a) 2000 and (b) expected situation for 2020, assuming full implementation of present emission control legislation in Europe. (Source: CAFE 2005.)

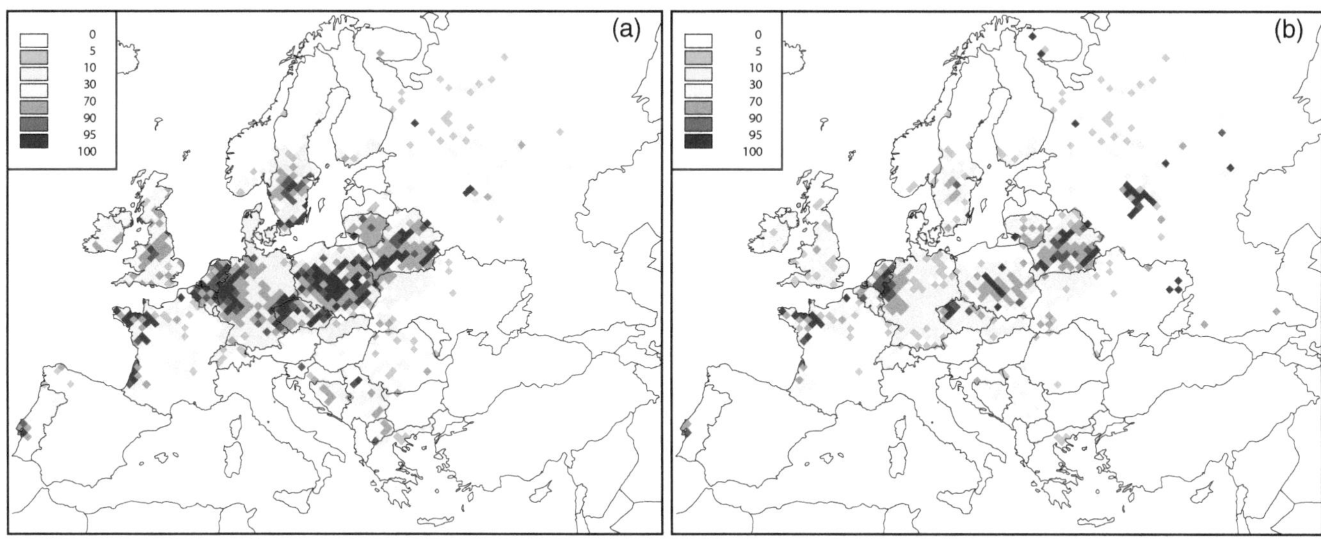


Figure 7.7 Percentage of forest ecosystems at threat from acid deposition in Europe, (a) 2000 and (b) 2020. (Source: CAFE 2005.)

is expected to improve by 2020 due to current air quality legislation in Europe. In 2000, acid deposition was a widespread threat to forest ecosystems north of the Alps. For 2020, scientists expect better conditions over large regions, with the main problem areas remaining in northern Germany, the Benelux region and some areas in eastern Europe.

A similar assessment has been carried out for Asia, highlighting the persistence of acidification problems in the eastern part of China, despite efforts to control $SO_2$ emissions in this region (Figure 7.8).

## 7.7 Conclusions

The face of air pollution will inevitably change over the course of time. In general, the important driving forces of air pollution, such as population growth, economic development, increased energy consumption and higher agricultural production, are expected to aggravate throughout the world in the coming decades. Additionally, societies have begun to be concerned about the impairment of their living conditions due to poor air quality and have started to take measures to control emissions. Thus, many of the present local and regional air quality problems will improve in the future, especially in industrialized countries.

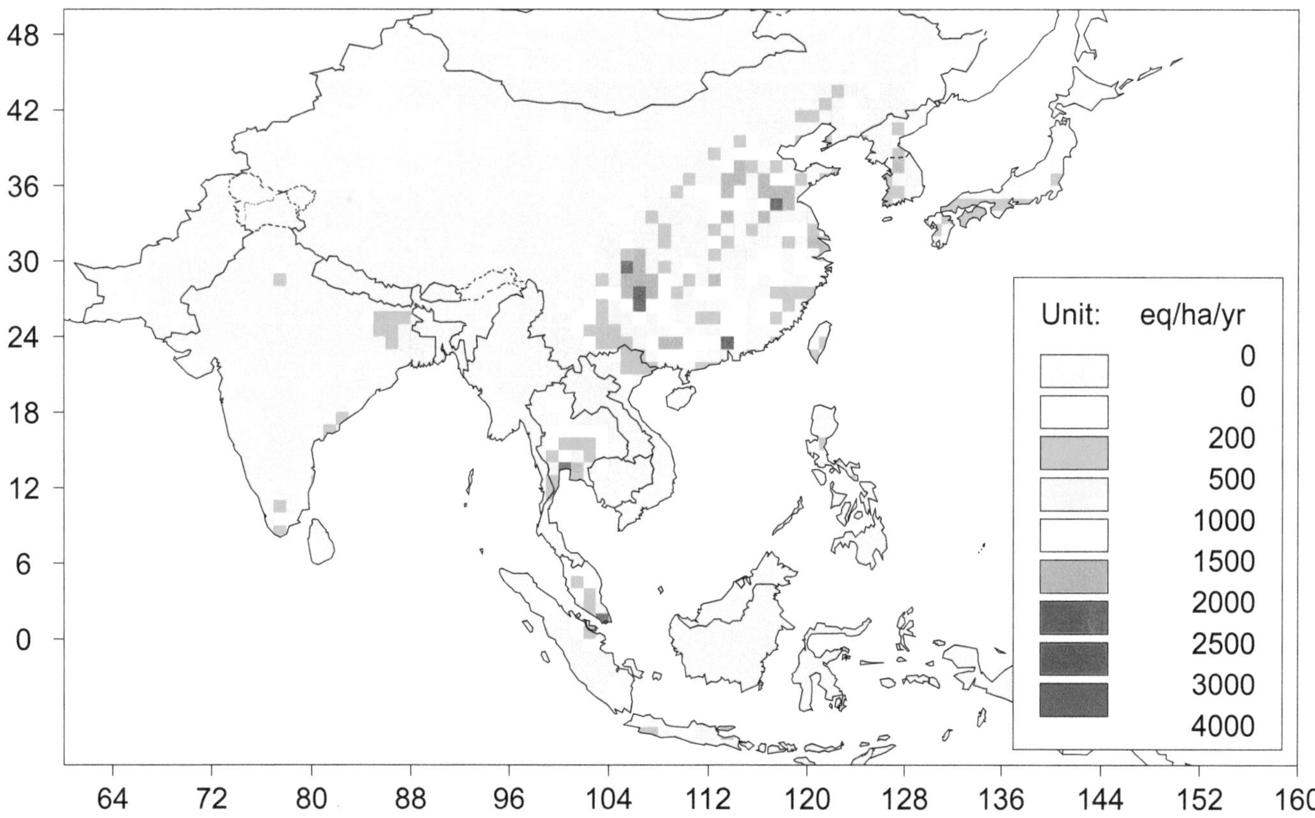


Figure 7.8 Excess sulphur deposition in Asia in 2020 above the critical loads, 'Current Legislation' scenario (acid equivalents ha-yr). (Source: Cofala et al. 2004.)

However, we have little reason to assume that these traditional air quality problems will disappear altogether.

Overall, the presently decided control measures do not appear to be sufficient to reach environmentally sustainable conditions in industrialized countries. In the developing world the combined effect of higher pollution levels, caused by the fast economic development and increased population, could lead to unprecedented levels of air pollution damage. To what extent their air quality will be kept at acceptable levels will depend on the preparedness of the societies in developing countries to allocate sufficient resources for air pollution control. Powerful technologies for controlling emissions are on the market, and many developing nations have taken the first steps to limit air pollution, at least for the worst polluted places.

While we might be modestly optimistic that local pollution hotspots will eventually be under control, there is reason for concern about the increasing levels of global background air pollution (see for example, Dentener et al. 2005 and Derwent et al. 2006). Current background concentrations alone exceed in many cases the sustainable levels, and their continuing growth counteracts the effectiveness of local and regional emission control efforts.

# REFERENCES


Abdel-Latif, N M (2003) Air pollution and vegetation in Egypt: a review. In L D Emberson, M R Ashmore, and F Murray (eds). *Air Pollution Impacts on Crops and Forests: a Global Assessment*, London: Imperial College Press, 215–35.

Agrawal, M (2003) Air pollution impacts on vegetation in India. In L D Emberson, M R Anderson, and F Murray (eds). Air *Pollution Impacts on Crops and Forests: a Global Assessment*, London: Imperial College Press, 165–87.

Ahrens, C D (2003) *Meteorology Today: An Introduction to Weather, Climate, and the Environment*, Brooks/Cole-Thomson.

AIRBASE (2005) European Topic Centre on Air and Climate Change, Topic Centre of European Environment Agency http://air-climate.eionet.eu.int/databases/airbase/airview/index_html (accessed 24 June 2006).

Anderson, E L, Albert, R E and Eisenberg, M (1999) *Risk assessment and indoor air quality*, CRC Press.

Anderson, H R, Atkinson, R W, Peacock, J L, Marston, L and Konstantinou, K (2004) *Meta-analysis of time series studies and panel studies of particulate matter (PM) and ozone ($O_3$)*. Report of a WHO task group, Copenhagen: WHO Regional Office for Europe.

Anderson, J G, Brune, W H and Proffitt, M H (1989) Ozone destruction by chlorine radicals within the Antarctic vortex-The spatial and temporal evolution of ClO-O3 anticorrelation based on in situ ER-2 data, *Journal of Geophysical Research*, 94: 11465–79.

Anderson, M R (1997) The conquest of smoke: legislation and pollution in colonial Calcutta. In D Arnold and R Guha (eds). *Nature, Culture and Imperialism*, Oxford: Oxford University Press.

Anderson, P K, Cunningham, A A, Patel, N G, Morales, F J, Epstein, P R and Daszak, P (2004) Emerging infectious diseases of plants: pathogen pollution, climate change and agrotechnology drivers. *Trends in Ecology and Evolution*, 19(10): 535–544.

Andersen, S O and Sarma, K M (2002) *Protecting the ozone layer: The United Nations History*. London: Earthscan Publications.

Anisimov, O, Vandenberghe, J, Lobanov, V and Kondratiev, A (2007) Predicting changes in alluvial channel patterns in North-European Russia under conditions of global warming *Geomorphology*, doi:10.1016/j.geomorph.2006.12.029.

APERC (2006*) APEC Energy Demand and Supply Outlook 2006*. Tokyo: Asia Pacific Energy Research Centre, vol. 1.

APHEIS (2002) *Health Impact Assessment of Air Pollution in 26 European cities*. Second-year Report 2000–2001.

APMA (2002) *Benchmarking Urban Air Quality Management and Practice in Major and Mega Cities of Asia (Stage 1)*. Air pollution in the megacities of Asia (APMA) project, Seoul: Korea Environment Institute.

AQEG (2007) *Air Quality and Climate Change: A UK Perspective*, Third report produced by the Air Quality Expert Group (AQEG) for the Department of Environment, Food and Rural Affairs (DEFRA), UK A9 Crown copyright.

Aronsson, A (1980) Frost hardiness in Scots pine. II Hardiness during winter and spring in young trees of different mineral status. *Studia Forest Suecica* 155: 1–27.

Ashmore, M R (2002) Effects of oxidants at the whole plant and community level. In J N B Bell and M Treshow (eds). *Air pollution and plant life*, 2nd edn. Chichester: John Wiley and Sons, 89–118.

Ashmore, M R (2005) Assessing the future global impacts of ozone on vegetation. *Plant Cell and Environment* 28: 949–64.

Austin, J (1983) Krakatoa sunsets. *Weather*, 38: 226–30.

Bae, H, Yang, W and Chung, M (2004) Indoor and outdoor concentrations of RSP, NO2 and selected volatile organic compounds at 32 shoe stalls located near busy roadways in Seoul, Korea, *Science of the Total Environment*, 323: 99–105.

Baede, A P M, Ahlonsou, E, Ding, Y and Schmel, D (2001) The climate system: an overview. In J T Houghton, Y Ding, D J Griggs, M Noguer, P J van der Linden, X Dai, K Maskell, and C A Johnson, (eds). *IPCC Third Assessment Report on Climate Change 2001: The Scientific Basis*, Cambridge: Cambridge University Press.

Baldasano, J M, Valera, E and Jiménez, P (2003) Air quality data from large cities. *The Science of the Total Environment*, 307: 141–65.

Baldocchi, D, Falge, E, Gu, L Olson, R, Hollinger, D, Running, S, Anthoni, P, Bernhofer, Ch., Davis, K, Evans, R, Fuentes, J, Goldstein, A, Katul, G, Law, B, Lee, X, Malhi, Y, Meyers, T, Munger, W, Oechel, W, Paw U K T, Pilegaard, K, Schmid, H P, Valentini, R, Verma, S, Vesala, T, Wilson, K and Wofsy, S (2001) FLUXNET: A new tool to study the temporal and spatial variability of ecosystem scale carbon dioxide, water vapor, and energy flux densities. *Bulletin of the American Meteorological Society*, 82: 2415–34.

Barker T, Bashmakov, I, Bernstein, L, Bogner, J E, Bosch, P R, Dave, R, Davidson, O R, Fisher, B S, Gupta, S, HalsnE6s, K, Heij, G J, Kahn Ribeiro, S, Kobayashi, S, Levine, M D, Martino, D L, Masera, O, Metz, B, Meyer, L A, Nabuurs, G-J, Najam, A, Nakicenovic, N, Rogner, H-H, Roy, J, Sathaye, J, Schock, R, Shukla, P, Sims, R E H, Smith, P, Tirpak, D A, Urge-Vorsatz, D and Zhou, D (2007) *Technical Summary. In Climate change 2007: Mitigation. Contribution of Working group III to the Fourth Assessment Report of the Intergovernmental Panel on Climate Change.* B Metz, O R Davidson, P R Bosch, R Dave, and L A Meyer (eds). Cambridge, UK and New York, USA: Cambridge University Press.

Barter, P A (1999) *An International Comparative Perspective on Urban Transport and Urban Form in Pacific Asia: The Challenge of Rapid Motorisation in Dense Cities.* Ph.D Thesis, Murdoch University, Western Australia, Perth.

Bartonova, A, Clench-Aas, J, Gram, F, Grønskei, K E, Guerreiro, C, Larssen, S, Tønnesen, D A and walker, S E (1999) Air pollution exposure monitoring and estimation: Part V: traffic exposure in adults. *Journal of Environmental Monitoring*, 1: 337–40.

Bell, J N B and Treshow M (eds) (2002) *Air Pollution and Plant Life*. Chichester: John Wiley & Sons.

Bell, M L and Davis, D L (2001) Reassessment of the lethal London fog of 1952: Novel indicators of acute and chronic consequences of acute exposure to air pollution. *Environmental Health Perspectives*, 109(3): 389–94.

Bell, M L, Davis, D L and Fletcher, T (2004) A retrospective assessment of mortality from the London smog episode of 1952: The role of influenza and pollution. *Environmental Health Perspectives*, 112(1): 6–8.

Bell, M L, Davis, D L, Gouveia, N, Borja-Aburto, V H and Cifuentes, L A (2006) The avoidable health effects of air pollution in three Latin American cities: Santiago, São Paulo, and Mexico City. *Environmental Research*, 100: 431–40.

Benedick, R E (1991) *Ozone Diplomacy: New directions in safeguarding the planet.* Cambridge, MA: Harvard University Press.

Bengtsson, M, Shen, Y and Oki, T (2006) A SRES-based gridded global populatiou dataset for 1990–2100. *Population and Environment*, 28(2): 113–131.

Berkner, L V and Marshall, L C (1965) On the origin and rise of oxygen concentration in the Earth's atmosphere. *Journal of Atmospheric Science*, 22: 225–61.

Bhohan, P and Kennedy, J (2007) Personal communication, Hadley Centre, UK Met Office.

Bickerstaff, K and Walker, G (2001) Public understandings of air pollution: the 'localisation' of environmental risk. *Global Environmental Change – Human and Policy Dimensions*, 11: 133–45.

Bindoff, N L, Willebrand, J, Artale, V, Cazenave, A, Gregory, J, Gulev, S, Hanawa, K, Le Quéré, C, Levitus, S, Nojiri, Y, Shum, C K Talley, L D and Unnikrishnan, A (2007) Observations: Oceanic Climate Change and Sea Level. In S Solomon, D Qin, M Manning, Z Chen, M Marquis, K B Averyt, M Tignor and H L Miller (eds). *Climate Change 2007: The Physical Science Basis. Contribution of Working Group I to the Fourth Assessment Report of the Intergovernmental Panel on Climate Change.* Cambridge, UK and New York, USA: Cambridge University Press.

Björn, L O (2007) Stratospheric ozone, ultraviolet radiation, and cryptogams. *Biological Conservation*, 135: 326–33.

Bobbink, R, Hornung, M and Roelofs, J G M (1998) The effects of airborne nitrogen pollutants on species diversity in natural and semi-natural European vegetation. *Journal of Ecology*, 86: 717–38.

Bobbink, R, Ashmore, M, Braun, S, Flückiger, W and van den Wyngaert, I J J (2003) Empirical nitrogen critical loads for natural and semi-natural ecosystems: 2002 update. In B Achermann and R Bobbink (eds). *Empirical critical loads for nitrogen: expert workshop, Berne, 11–13 November 2002*. Swiss Agency for the Environment, Forests and Landscape, Environmental Documentation 164, 43–170.

Bobbink, R, Hicks, K, Galloway, J, Spranger, T, Alkemade, R, Ashmore, M, Bustamante, M, Cinderby, S, Davidson, E, Dentener, F, Emmett, B, Erisman, J W, Fenn, M, Gilliam, F, Nordin, A, Pardo, L and de Vries, W (2010) Global assessment of nitrogen deposition effects on terrestrial plant diversity: a synthesis. *Ecological Applications*, 20: 30–59.

Bowler, C and Brimblecombe, P (1990) The difficulties of abating smoke in late Victorian York. *Atmospheric Environment*, 24B: 49–55.

Bowler, C and Brimblecombe, P (2000) Control of air pollution in Manchester prior to the Public Health Act, 1875. *Environment and History*, 6: 71–98.

Boxman, A W, van Dijk, H F G, Houdijk, A L F M and Roelofs, J G M (1988) Critical loads for nitrogen with special emphasis on ammonium. In J Nilsson and P Grennfelt (eds). *Critical Loads for Sulphur and Nitrogen*. Miljørapport 15, Nordic Council of Ministers, Copenhagen, Denmark: 295–322.

Boxman, A W, De Visser, P H B and Roelofs, J G M (1995) Experimental manipulations: forest ecosystem responses to changes in water, nutrients and atmospheric loads. In J-W Erisman, G J Heij and T Schneider (eds). *Acid Rain Research: Do we have enough answers?* Proceedings of a Speciality Conference, 'S-Hertogenbosch, Netherlands, 10–12 October 1994, Amsterdam: Elsevier.

Bramstedt, K, Gleason, J, Loyola, J, Thomas, D, Bracher, W, Weber, A and Burrows, J P (2003) Comparison of total ozone from the satellite instruments GOME and TOMS with measurements from the Dobson network 1996–2000. *Atmospheric Chemistry and Physics*, 3: 1409–19.

Brankov, E, Rao, S T and Porter, P S (1998) A trajectory-clustering-correlation methodology for examining the long-range transport of air pollutants. *Atmospheric Environment*, 32: 1525–34.

Brankov, E, Rao, S T and Porter, P S (1999) Identifying pollution source regions using multiply-censored data. *Environmental Science and Technology*, 33: 2273–7.

Brankov, E, Henry, R F, Civerolo, K L, Hao, W, Rao, S T, Misra, P K, Bloxam, R and Reid, N (2003) Assessing the effects of transboundary ozone pollution between Ontario, Canada and New York, USA *Environmental Pollution*, 123: 403–11.

Bravo, H A, Saavedra, M I R, Sanchez, P A, Torres, R J and Granada, L M M (2000) Chemical composition of precipitation in a Mexican Maya region. *Atmospheric Environment*, 34(8): 1197–204.

Breysse, P N, Buckley, T J, Williams, D, Beck, C M, Jo, S-J, Merriman, B, Kanchanaraksa, S, Swartz, L J, Callahan, K A, Butz, A M, Rand, C S, Diette, G B, Krishnan, J A, Moseley, A M, Curtin-Brosnan, J, Durkin, N B and Eggleston, P A (2005) Indoor exposures to air pollutants and allergens in the homes of asthmatic children in inner-city Baltimore. *Environmental Research*, 98(2): 167–76.

Brimblecombe, P (1977) Earliest Atmospheric Profile. *New Scientist*, 76(1077): 364–5.

Brimblecombe, P (1987a) *The Big Smoke*. London: Methuen.

Brimblecombe, P (1987b) The antiquity of smokeless zones. *Atmospheric Environment*, 21(11): 2485.

Brimblecombe, P (2001) Acid rain. *Water, Air and Soil Pollution*. 130: 25–30.

Brimblecombe, P (2002) The great London smog and its immediate aftermath. In T Williamson (ed). *London Smog 50$^{th}$ Anniversary*. Brighton: National Society for Clean Air: 182–95.

Brimblecombe, P (2003) Origins of smoke inspection in Britain (circa 1900). *Applied Environmental Science & Public Health*, 1: 55–62.

Brimblecombe, P (2004a) Perceptions of late Victorian air pollution. In M DePuis (ed). *Smoke and Mirrors: The Politics and Culture of Air Pollution*, New York: State University of New York Press: 15–26.

Brimblecombe, P (2004b) Acid Rain. Encyclopaedia of Environmental History.

Brimblecombe, P (2005) History of early forest fire smoke in S E Asia. In G J Sem, D Boulaud, P Brimblecombe, P Ensor, D S, Gentry, J W, Marijnissen, J C M and Preining, O (eds). *History and Reviews of Aerosol Science*. American Association for Aerosol Research: 385–96.

Brimblecombe, P (2006) The Clean Air Act after fifty years. *Weather*, 61(12): 311–4.

Brimblecombe, P, Davies, T and Tranter, M (1986) Nineteenth-Century Black Scottish Showers. *Atmospheric Environment*, 20(5): 1053–7.

Brohan, P, Kennedy, J J, Harris, I, Tett, S F B and Jones, P D (2006) Uncertainty estimates in regional and global observed temperature changes: A new dataset from 1850. *Journal of Geophysical. Research*, 111: D12106. http://dx.doi.org/10.1029/2005JD006548.

Bronmark, C and Hansson, L A (2002) Environmental issues in lakes and ponds: current state and perspectives. *Environmental Conservation*, 29: 290–307.

Bruce, N, Perez-Padilla, R and Albalak, R (2000) Indoor air pollution in developing countries: a major environmental and public health challenge. *Bulletin of the World Health Organization*, 78(9): 1078–92.

Brunekreef, B and Holgate, S T (2002) Air pollution and health. *Lancet*, 360: 1233–42.

Burr, M L, Karani, G, Davies, B, Holmes, B A and Williams K L (2004) Effects on respiratory health of a reduction in air pollution from vehicle exhaust emissions. *Occupational and Environmental Medicine*, 61: 212–18.

Bytnerowicz, A, Omasa, K and Paoletti, E (2007) Integrated effects of air pollution and climate change on forests: a northern hemisphere perspective. *Environmental Pollution*, 147, 438–45.

Byun, D and Schere, K L (2006) Review of the governing equations, computational algorithms, and other components of the Model-3 Community Multiscale Air Quality (CMAQ) Modelling system. *Applied Mechanics Reviews*, 55: 51–77.

CAFE (2005) Baseline Scenarios for the Clean Air for Europe (CAFE) Programme. Clean Air for Europe (CAFE) programme of the European Commission, http://europa.eu.int/comm/environment/air/cafe/general/pdf/cafe_lot1.pdf (accessed 24 June 2006).

Capasso, L (2000) Indoor pollution and respiratory diseases in Ancient Rome. *Lancet*. 356(9243): 1774.

Carmichael, G R, Calori, G, Hayami, H, Uno, I, Cho, S Y, Engardt, M, Kim, S B, Ichikawa, Y, Ikeda, Y, Woo, J H, Ueda, H and Amann, M (2002) The MICS-Asia Study: model intercomparison of long-range transport and sulfur deposition in East Asia. *Atmospheric Environment*, 36: 175–99.

Cazenave, A and Nerem, R S (2004) Present-day sea level change: observations and causes. *Review of Geophysics*, 42(3): RG3001, doi:10.1029/2003RG000139.

CEC (2005) *Impact Assessment of the Thematic Strategy on Air Pollution and the Directive on 'Ambient Air Quality and Cleaner Air for Europe* (Summary). http://ec.europa.eu/environment/ air/cafe/.

CfIT (2001) *European Best Practice in the Delivery of Integrated Transport, Report on stage 1: Bench marking*, http://www.cfit.gov.uk/ (accessed 24 June 2006).

Chen, C T A, Wann, J K and Luo, J Y (2001) Aeolian flux of metals in Taiwan in the past 2600 years. *Chemosphere*, 43(3): 287–94.

Christensen, O B and Christensen, J H (2004) Intensification of extreme European summer precipitation in a warmer climate. *Global and Planetary Change*, 44: 107–117.

Christophers, A J (1982) Butyl Mercaptan Poisoning in the Parnell Civil Defense Emergency – Fact or Fiction? *New Zealand Medical Journal*, 95(706): 277–8.

Church, J A and White, N J (2006) A 20th century acceleration in global sea-level rise. *Geophysical Research Letters*, 33: L01602, doi:10.1029/2005GL024826.

City of Reykjavik (2002) Latest facts and figures from Reykjavik, http://www.randburg.com/is/capital/facts-figures-about-Reykjavik.pdf (accessed 24 June 2006).

Civerolo, K, Mao, H and Rao, S T (2003) The airshed for ozone and fine particulate pollution in the eastern United States. *Pure Applied Geophysics*, 160: 81–105.

Clancy, L, Goodman, P, Sinclair, H and Dockery, D W (2002) Effect of air-pollution control on death rates in Dublin, Ireland: an intervention study. *Lancet*, 360: 1210–14.

Clench-Aas, J, Bartonova, A, Klaeboe, R and Kolbenstvedt, M (2000) Oslo traffic study, part 2: quantifying effects of traffic measures using individual exposure modelling. *Atmospheric Environment*, 34: 4737–44.

CCSP (2003) Climate Change Science Program and the Subcommittee on Global Change Research. *Strategic plan for the U S Climate Change Science Program.* A Report by the Climate Change Science Program and the Subcommittee on Global Change Research, USA.

CLRTAP (2010) *Manual on Methodologies for Modelling and Mapping Critical Loads and Levels and Air Pollution Effects, Risks and Trends. Chapter 3: Mapping Critical Levels for Vegetation. 2010 Revision.* UNECE Convention on Long-Range Transboundary Air Pollution. http://www.rivm.nl/en/themasites/icpmm/manual-and-downloads/index.html (accessed 18 April 2011).

Cofala, J, Amann, M, Gyarfas, F, Schoepp, W, Boudri, J C, Hordijk, L, Kroeze, C, Junfeng, L, Lin, D, Panwar, T S and Gupta, S (2004) Cost-effective control of $SO_2$ emissions in Asia. *Journal of Environmental Management*, 72: 149–61.

Cofala, J, Amann, M, Klimont, Z, Kupianen, K, Hoeglund, L (2007) Scenarios of global anthropogenic emissions of air pollutants and methane until 2030. *Atmospheric Environment*, 41: 8486–99.

Cofala, J, Rafaj, P, Schoepp, W, Klimont, Z, Amann, M (2009) Emissions of Air Pollutants for the World Energy Outlook 2009 Energy Scenarios. Final Report to the International Energy Agency, Paris, France, August 2009. International Institute for Applied Systems Analysis, Laxenburg, Austria. http://www.worldenergyoutlook.org/docs/Emissions_of_Air_Pollutants_for_WEO2009.pdf (accessed 18 April 2011).

Cohen, A J, Anderson, H R, Ostro, B, Pandey, K D, Krzyanowski, M, Kunzli, N, Gutschmidt, K, Pope, C A, Romieu, I, Samet, J M and Smith, K R (2005) The global burden of disease due to outdoor air pollution. *Journal of Toxicology and Environmental Health*, Part A, 68: 1–7.

Cohen, J B and Rushton, A C (1925) *Smoke: a study of town air*, London: Edward Arnold & Co.

Collins, W J, Stevenson, D S, Johnson, C E and Derwent, R G (1997) Tropospheric ozone in a global-scale three-dimensional Lagrangian model and its response to $NO_x$ emissions controls. *Journal of Atmospheric Chemistry*, 26, 223–74.

Correa, F A (2003) *The Atmospheric Prevention and Pollution Control Plan for the Metropolitan Region of Chile – Background and Perspectives.* Santiago, Chile: Cooperación Técnica Alemana (GTZ), Chile.

CPCB (2003) Polycyclic Aromatic Hydrocarbons (PAHs). In *Air and Their Effects on Human Health.* Delhi: Central Pollutant Control Board. http://www.cpcb.nic.in/ph/content1103.htm (accessed 24 June 2006), Chapter 4.

Crutzen, P J (1971) Ozone production rates in an oxygen, hydrogen, nitrogen-oxide atmosphere. *Journal of Geophysical Research*, 76: 7311–27.

Crutzen, P J and Arnold, F (1986) Nitric acid cloud formation in the cold Antarctic stratosphere: A major cause for the springtime 'ozone hole'. *Nature*, 324: 651–5.

Cyrys, J, Heinrich, J, Richter, K, Wölke, G and Wichmann, H-E (2000) Sources and concentrations of indoor nitrogen dioxide in Hamburg (West Germany) and Erfurt (East Germany). *Science of the Total Environment*, 250: 51–62.

Dai, A, Lamb, P J, Trenberth, K E, Hulme, M, Jones, P D and Xie, P (2004) The recent Sahel drought is real. *International Journal of Climatology*, 24: 1323–31.

Dasgupta, S, Huq, M, Khaliquzzaman, M, Pandey, K and Wheeler, D (2006) Indoor air quality for poor families: new evidence from Bangladesh. *Indoor Air*, 16: 426–44.

Davidson, C I (1979) Air pollution in Pittsburgh: a historical perspective. *Journal of the Air Pollution Control Association*, 29: 1035–41.

Davies, C E and Moss, D (2002) *EUNIS Habitat Classification, 2001 Work Programme.* Final Report to the European Environment Agency European Topic Centre on Nature Protection and Biodiversity, Centre for Ecology and Hydrology.

De Assunção, J V (2002) International Seminar urban air Quality management. In: *São Paulo Metropolitan Area Air Quality in Perspective*, Sao Paulo, 21–23 October 2002.

De Bauer, M (2003) Air pollution impacts on vegetation in Mexico. In L D Emberson, M R Ashmore, and F Murray (eds). *Air Pollution Impacts on Crops and Forests: a Global Assessment.* London: Imperial College Press, 263–86.

De Graaf, M C C, Bobbink, R, Verbeek, P J M and Roelofs, J G M (1997) Aluminium toxicity and tolerance in three heathland species. *Water, Air and Soil Pollution*, 98: 229–39.

De Graaf, M C C, Bobbink, R, Roelofs, J G M and Verbeek, P J M (1998) Differential effects of ammonium and nitrate on three heathland species. *Plant Ecology*, 135: 185–96.

De Leeuw, F, Moussiopoulos, N, Bartonova, A and Sahm, P (2001) *Air Quality in Larger Cities in the European Union*, Report No. 3/2001, Copenhagen: European Environment Agency, http://reports.eea.eu.int/Topic_report_No_032001/en/toprep03_2001.pdf (accessed 24 June 2006).

De Visser, P H B (1994) *Growth and nutrition of Douglas-fir, Scots pine and pedunculate oak in relation to soil acidification.* Doctoral Thesis, Agricultural University, Wageningen, Netherlands.

De Vries, W, Reinds, G J, van Kerkvoorde, M A, Hendriks, C M A, Leeters, E E J M, Gross, C P, Voogd, J C H and Vel, E M (2000) *Intensive Monitoring of Forest Ecosystems in Europe, Technical Report.* Geneva and Brussels: UNECE and EC, Forest Intensive Monitoring Coordinating Institute.

De Vries, W, Reinds, G J, van der Salm, C, Draaijers, G P J, Bleeker, A, Erisman, J W, Auee, J, Gundersen, P, Kristensen, H L, van Dobben, H, De Zwart, D, Derome, J, Voogd, J C H and Vel, E M (2001) *Intensive Monitoring of Forest Ecosystems in Europe, Technical Report.* Geneva and Brussels: UNECE and EC, Forest Intensive Monitoring Coordinating Institute.

De Vries, W, Vel, E, Reinds, G J, Deelstra, H, Klap, J M, Leeters, E E J M, Hendriks, C M A, van Kerkvoorden, M S, Landmann, G, Herkendell, J, Haussmann, T and Erisman, J W (2003a) Intensive Monitoring of Forest Ecosystems in Europe 1. Objectives, set-up and evaluation strategy. *Forest Ecology and Management*, 74(3): 77–95.

De Vries, W, Reinds, G J, van der Salm, C, van Dobben, H, Erisman, J W, De Zwart, D, Bleeker, A, Draaijers, G P J, Gundersen, P, Vel, E M and Haussman, T (2003b) Results on nitrogen impacts in the EC and UNECE ICP Forests Programme. In B Achermann and R Bobbink (eds). Empirical critical loads for nitrogen: expert workshop, Berne, 11–13 November 2002. *Swiss Agency for the Environment, Forests and Landscape, Environmental Documentation*, 164: 199–207.

De Vries, W, Solberg, S, Dobbertin, M, Sterba, H, Laubhann, D, van Oijen, M, Evans, C, Gunderson, P, Kros, J, Wamelink, G W W, Reinds, G J and Sutton, M A (2009) The impact of nitrogen deposition on carbon sequestration by European forests and heathlands. *Forest Ecology and Management*, 258: 1814–23.

DEFRA (2005) *National Air Quality Information Archive.* http://www.airquality.co.uk (accessed 24 June 2006).

DEFRA (2007) *UK Air Quality Archive.* http://www.airquality.co.uk/archive/index.php.

Del Grosso, S J, Mosier, A R, Parton, W J and Ojima, D S (2005) DAYCENT model analysis of past and contemporary soil $N_2O$ and net greenhouse gas flux for major crops in the USA *Soil Tillage Research*, 83(1): 9–24.

Denman, K L, Brasseur, G, Chidthaisong, A, Ciais, P, Cox, P M, Dickinson, R E, Hauglustaine, D, Heinze, C, Holland, E, Jacob, D, Lohmann, U, Ramachandran, S, da Silva Dias, P L, Wofsy, S C and Zhang, X (2007) Couplings between changes in the climate system and biogeochemistry. In S Solomon, D Qin, M Manning, Z Chen, M Marquis, K B Averyt, M Tignor and H L Miller (eds).*Climate Change 2007: The Physical Science Basis.* Contribution of

Working Group 1 to the Fourth Assessment Report of the Intergovernmental Panel on Climate Change Cambridge, UK and New York, USA: Cambridge University Press.

Dennis, R L (1997) Using the Regional Acid Deposition Model to Determine the Nitrogen Deposition Airshed of the Chesapeake Bay Watershed. In J E Baker (ed.) *Atmospheric Deposition to the Great Lakes and Coastal Waters*. Society of Environmental Toxicology and Chemistry, Pensacola, FL, 393–413.

Dentener, F, Stevenson, D, Cofala, J, Mechler, R, Amann, M, Bergamaschi, P, Raes, F and Derwent, R (2005) The impact of air pollutant and methane emission controls on tropospheric ozone and radiative forcing: CTM calculations for the period 1990–. *Atmospheric Chemistry and Physics*, 5: 1731–55.

Dentener, F, Stevenson, D, Ellingsen, K, van Noije, T, Schultz, M, Amann, M, Atherton, C, Bell, N, Bergmann, D, Bey, I, Bouwman, L, Butler, T, Cofala, J, Collins, B, Drevet, J, Doherty, R, Eickhout, B, Eskes, H, Fiore, A, Gauss, M, Hauglustaine, D, Horowitz, L, Isaksen, I S A, Josse, B, Lawrence, M, Krol, M, Lamarque, J F, Montanaro, V, Müller, J F, Peuch, V H, Pitari, G, Pyle, J, Rast, S, Rodriguez, J, Sanderson, M, Savage, N H, Shindell, D, Strahan, S, Szopa, S, Sudo, K, Van Dingenen, R, Wild, O and Zeng, G (2006) The Global Atmospheric Environment for the Next Generation. *Environmental Science and Technology*, 40(11): 3586–94.

Derwent, R G, Simmonds, P G, Seuring, S and Dimmer, C (2004) Observation and interpretation of the seasonal cycles in the surface concentrations of ozone and carbon monoxide at mace head, Ireland from 1990 to 1994. *Atmospheric Environment*, 32: 4769–78.

Derwent, R G, Simmonds, P G, O'Doherty, S, Stevenson, D S, Collins, W J, Sanderson, M G, Johnson, C E, Dentener, F, Cofala, J, Mechler, R and Amann, M (2006) External influences on Europe's air quality: Baseline methane, carbon monoxide and ozone from 1990 to 2030 at Mace Head, Ireland. *Atmospheric Environment*, 40: 844–55.

Dhakal, S, Kaneko, S and Imura, H (2003) $CO_2$ Emissions from Energy Use in East Asian Mega-Cities: Driving Factors, Challenges and Strategies. In *Proceedings of International Workshop on Policy Integration Towards Sustainable Urban Energy Use for Cities in Asia*, 4–5 February 2003, Honolulu, Hawaii.

Dhakal, S (2004) *Urban energy use and greenhouse gas emissions in Asian mega cities – policies for a sustainable future.* Japan: Institute for Global Environmental Strategies (IGES).

Diaz, J P, Exposito, F J, Torres, C J, Herrera, F, Prospero, J M and Romero, M C (2001) Radiative properties of aerosol in Saharan dust outbreak using ground-based and satellite data: Application to radiative forcing. *Journal of Geophysical Research*, 106D: 18403–16.

Dobson, G M B, Brewer, A W and Cwilong, B M (1946) Meteorology of the lower stratosphere. *Proceedings of the Royal Society of London A*, 185: 144–75.

Doherty, R,M, Stevenson, D S, Johnson, C E, Collins, W J and Sanderson, M G (2006) Tropospheric ozone and El Niño–Southern Oscillation: influence of atmospheric dynamics, biomass burning emissions, and future climate change. *Journal of Geophysical Research*, 111:D19304. doi:10.1029/2005JD006849.

Domingos, M and Klumpp, A (2003) Disturbances to the Atlantic rainforest in South-east Brazil. In L D Emberson, F Murray, and M R Ashmore (eds). *Air Pollution Impacts on Crops and Forests: a Global Assessment.* London: Imperial College Press, 287–308.

Draxler, R R (1992) Hybrid Single-Particle Lagrangian Integrated Trajectories (HY-SPLIT): Version 3.0 – User's guide and model description. *NOAA Technical Memorandum ERL-ARL-195*, Silver Spring, MD: Air Resources Laboratory.

Driver, G R and Miles, J C (1952) *The Babylonian laws legal commentary*, Oxford: Clarendon Press.

Dunning, J (2005) World Cities Research, Final Report, prepared by MVA for Commission for Integrated Transport, http://www.cfit.gov.uk/ (accessed 24 June 2006).

Eder, B K, Davis, J M and Bloomfield, P (1994) An automated classification scheme designed to better elucidate the dependence of ozone on meteorology. *Journal of Applied Meteorology*, 33, 1182–99.

EDGAR (2011) *Emissions Database for Global Atmospheric Research.* http://edgar.jrc.ec.europa.eu (accessed March 2011).

Edgren, B and Herschend, F (1982) Ektorp för fjärde gaongeng. *Forskning och Framsteg*, 5: 13–19.

EEA (European Environment Agency) (2001) *Air Quality in Larger Cities in the European Union: A contribution to the Auto-Oil II Programme.* Topic Report 3/2001.

EEA (European Environment Agency) (2003a) *Europe's environment: the third assessment.* Environmental assessment report no. 10.

EEA (European Environment Agency) (2003b) *Air pollution by ozone in Europe in summer 2003.* Topic report no. 3.

EEA (European Environment Agency) (2007) *Air pollution in Europe 1990–2004*, EEA Report No 2/2007. Copenhagen: EEA.

EEA (European Environment Agency) (2011) AirBase: public air quality database. http://www.eea.europa.eu/themes/air/airbase (accessed March 2011).

Ekström, M, Fowler, H J, Kilsby, C G and Jones, P D (2005) New estimates of future changes in extreme rainfall across the UK using regional climate model integrations. 2. Future estimates and use in impact studies. *Journal of Hydrology*, 300, 234–51.

Emberson, L D (2007) Personal communication, Stockholm Environment Institute, University of York, Heslington, York, YO10 5DD, UK.

Emberson, L D, Ashmore, M R and Murray, F (2003) *Air Pollution Impacts on Crops and Forests: A Global Perspective.* London: Imperial College Press.

Emberson, L D, Büker, P, Ashmore, M R, Mills, G, Jackson, L S, Agrawal, M, Atikuzzaman, M D, Cinderby, S, Engardt, M, Jamir, C, Kobayashi, K, Oanh, N T K, Quadir, Q,F and Wahid, A (2009) A comparison of North American and Asian exposure-response data for ozone effects on crop yields. *Atmospheric Environment*, 43: 1945–53.

EMEP (2006) *Transboundary acidification, eutrophication and ground-level ozone in Europe since 1990 to 2004*, Norwegian meteorological Institute, EMEP Status Report 1\ 2006

ENV ECO (2001) *Enhancing the Comparability of the Air Emission Inventories in Canada. Mexico and the United States.* Environmental Economics, Draft 9, October.

Environmental Law-Institute (2003) *Reporting on Climate Change: Understanding the Science*, 3rd edn, Washington, DC.

EOS (2003) The 2001 Asian dust events: Transport and impact on surface aerosol concentration in the US *Eos, Transactions, American Geophysical Union*, 84(46): 501–16.

Etheridge, D M, Steele, L P, Langenfelds, R L, Francey, R J, Barnola, J-M and Morgan, V I (1996) Natural and anthropogenic changes in atmospheric $CO_2$ over the last 1000 years from air in Antarctic ice and firn. *Journal of Geophysical* Research, 101(D2): 4115–28.

EUROSTAT (2006) *Statistical Office of the European Communities.* http://epp.eurostat.ec.europa.eu/ (accessed January 2006).

Fairlie, T D, Jacob, D J and Park, R J (2007) The impact of transpacific transport of mineral dust in the United States. *Atmospheric Environment*, 41, 1251–66.

Falla, J, Laval-Gilly, P, Henryon, M, Morlot, D and Ferard, J F (2000) Biological air quality monitoring; a review. *Environmental Monitoring and Assessment*, 64: 627–44.

Farman, J C, Gardiner, B Gand Shanklin, J D (1985) Large losses of ozone in Antarctica reveal seasonal ClOx/NOx interaction. *Nature*, 315: 207–10.

Feichter, J, Roeckner, E, Lohmann, U and Liepert, B (2004) Nonlinear aspects of the climate response to greenhouse gas and aerosol forcing. *Journal of Climate*, 17: 2384–98.

Fischer, P, Brunekreff, B and Lebret, E (2004) Air pollution related deaths during the 2003 heat wave in the Netherlands. *Atmospheric Environment*, 38: 1083–5.

Fishman, J and Balok, A E (1999) Calculation of daily tropospheric ozone residuals using TOMS and empirically derived SBUV measurements: application to an ozone pollution episode over the eastern United States. *Journal of Geophysical Research*, 104: 30319–40.

Flagler, R B (1998) *Recognition of Air Pollution Injury to Vegetation: A Pictorial Atlas.* Pittsburgh, PA: Air & Waste Management Association.

Flanner, M G, Zender, C S, Randerson, J T and Rasch, P J (2007) Present-day climate forcing and response from black carbon in snow, *Journal of Geophysical Research*, 112: D11202, doi:10.1029/2006JD008003.

Flückiger, W and Braun, S (1998) Nitrogen deposition in Swiss forests and its possible relevance for leaf nutrient status, parasite attacks and soil acidification. *Environmental Pollution*, 102: 69–76.

Forster, P, Ramaswamy, V, Artaxo, P, Berntsen, T, Betts, R, Fahey, D W, Haywood, J, Lean, J, Lowe, D C, Myhre, G, Nganga, J, Prinn, R, Raga, G, Schulz, M and Van Dorland, R (2007)

Changes in Atmospheric Constituents and in Radiative Forcing. In: S Solomon, D Qin, M Manning, Z Chen, M Marquis, K B Averyt, M Tignor and H L Miller (eds) *Climate Change 2007: The Physical Science Basis. Contribution of Working Group I to the Fourth Assessment Report of the Intergovernmental Panel on Climate Change*. Cambridge, UK and New York, USA: Cambridge University Press.

Freedman, B and Hutchinson, T C (1980) Long-term effects of smelter pollution at Sudbury, Ontario on forest community composition. *Canadian Journal of Botany*, 58: 2123–40.

Friedman, M S, Powell, K E, Hutwagner, L, Graham, L M and Teague, W G (2001) Impact of changes in transportation and commuting behaviours during the 1996 summer Olympic games in Atlanta on air quality and childhood asthma. *Journal of the American Medical Association*, 285: 897–905.

Fromme, H, Twardella, D, Dietrich, S, Heitmann, D, Schierl, R, Liebl, B and Rüden, H (2007) Particulate matter in the indoor air of classrooms—exploratory results from Munich and surrounding area. *Atmospheric Environment*, 41: 854–66.

Fumigalli, I, Gimeno, B S, Velissariou, D, de Temmerman, L and Mills, G (2001) Evidence of ozone-induced adverse effects on crops in the Mediterranean region. *Atmospheric Environment*, 35: 2583–7.

Gaffin, S R, Rosenzweig, C, Xing, X, and Yetman, G (2004) Downscaling and geo- spatial gridding of socio-economic projections from the IPCC Special Report on Emission Scenarios (SRES). *Global Environmental Change*, 14: 105–23.

Galloway, J N and Cowling, E B (2002) Reactive nitrogen and the world: 200 years of change. *Ambio*, 31: 64–71.

Galloway, J N, Dentener, F J, Capone, D G, Boyer, E W, Howarth, R W, Seitzinger, S P, Asner, G P, Cleveland, C C, Green, P A, Holland, E A, Karl, D M, Michaels, A F, Porter, J H, Townsend. A R and Vorosmarty, C J (2004) Nitrogen cycles: past, present and future. *Biogeochemistry*, 70: 153–226.

Gari, L (1987) Notes on air pollution in Islamic heritage. *Hamdard*, 30(3): 40–8.

Gaza, R S (1998) Mesoscale meteorology and high ozone in the Northeast United States. *Journal of Applied Meteorology*, 37, 961–7.

Gilbert, N L, Gauvin, D, Guay, M, Héroux, M-E, Dupuis, G, Legris, M, Chan, C C, Dietz, R N and Lévesque, B (2006) Housing characteristics and indoor concentrations of nitrogen dioxide and formaldehyde in Quebec City, Canada. *Environmental Research*, 102: 1–8.

GLA (2010) Clearing the air: The Mayor's Air Quality Strategy.

Godzik, S (1984) Air pollution problems in some central European countries – Czechoslovakia, the German Democratic Republic, and Poland. In M J Koziol and F R Whatley (eds). *Gaseous Air Pollutants and Plant Metabolism*, London: Butterworth, 25–34.

Gordon, J E, Haynes, V M, and Hubbard, A (2007) Recent glacier changes and climate trends on South Georgia. *Global and Planetary Change*, doi:10.1016/j.gloplacha.2006.07.037.

Goudie, A S and Middleton, N J (2006) *Desert Dust in the Global System*. Springer.

Grewe, V (2006) The origin of ozone. *Atmospheric Chemistry and Physics*, 6: 1495–511.

Grewe, V (2007) Impact of climate variability on tropospheric ozone. *Science of the Total Environment*, 374(1): 167–81.

Guderian, R, Tingey, D T and Rabe, R (1985) Effects of photochemical oxidants on plants. In R Guderian (ed.) *Air Pollution by Photochemical Oxidants. Formation, Transport, Control and Effects on Plants*, Berlin-Heidelberg: Springer-Verlag, 129–333.

Guerra, C A, Snow, R W and Hay, S I (2006) Mapping the global extent of malaria in 2005. Trends in Parasitology, 22(8), 353–8.

Gupta, I and Kumar, R (2006) Trends of particulate matter in four cities in India. *Atmospheric Environment*, 40: 2552–66.

Gurjar, B R, van Aardenne, J A, Lelieveld, J and Mohan, M (2004) Emission estimates and trends (1990–2000) for megacity Delhi and implications. *Atmospheric Environment*, 38: 5663–81.

Gurjar, B R and Lelieveld, J (2005) New directions: megacities and global change. *Atmospheric Environment*, 39: 391–3.

Guttikunda, S K, Carmichael, G R, Calori, G, Eck, C and Woo, J-H (2003) The contribution of megacities to regional sulphur pollution in Asia. *Atmospheric Environment*, 37: 11–22.

Guttikunda, S K, Tang, Y, Carmichael, G R, Kurata, G, Pan, L, Streets, D G, Woo, J-H, Thongboonchoo, N and Fried, A (2005) Impacts of Asia megacity emissions on regional air quality during spring 2001. *Journal of Geophysical Research*, 110: D20301, dio:10.1029/2004JD 004921.

Haines, A, Kovats, R S, Campbell-Lendrum, D and Corvalan, C (2006) Climate change and human health: Impacts, vulnerability and public health. *Public Health*, 120, 585–96.

Hanaoka, T, Kawase, R, Kainuma, M, Matsuoka, Y, Ishii, H and Oka, K (2006) *Greenhouse gas emissions scenarios database and regional itigation analysis*. Center for Global Environmental Research (CGER) Report, CGER D-038-2006, ISSN 1341-4356.

Hänninen, O O, Lebret, E, Ilacqua, V, Katsouyanni, K, Künzli, N, Srám, R J and Jantunen, M (2004) Infiltration of ambient $PM_{2.5}$ and levels of indoor generated non-ETS $PM_{2.5}$ in residences of four European cities. *Atmospheric Environment*, 38(37): 6411–23.

Haq, G, Han, W-J, Kim, C and Vallack, H (2002) *Benchmarking Urban Air Quality Management and Practice in Major and Mega Cities of Asia (Stage I)*. Seoul: Korea Environment Institute.

Hara, Y, Uno, I and Wang, Z (2006) Long-term variation of Asian dust and related climate factors. *Atmospheric Environment*, 40: 6730–40.

Hargreaves, J C and Annan, J D (2006) Using ensemble prediction methods to examine regional climate variation under global warming scenarios. *Ocean Modelling*, 11, 174–92.

Harris, G R, Sexton, D M H, Booth, B B B, Collins, M, Murphy, J M and Webb, M J (2006) Frequency distributions of transient regional climate change from perturbed physics ensembles of general circulation model simulations. *Climate Dynamics*, 27, 357–75.

Hartley, W N (1880) On the probable absorption of the solar ray by atmospheric ozone. *Chemical News*, 26 November: 268.

Hassan, I A, Ashmore, M R and Bell, J N B (1995) Effect of ozone on radish and turnip under Egyptian field conditions. *Environmental Pollution*, 89: 107–14.

Heck, W W (1966) The use of plants as indicators of air pollution. *Air and Water Pollution International*, 10: 99–111.

Hedley, A J, Wong, C M, Thach, T Q, Ma, S, Lam, T H and Anderson, H R (2002) Cardiorespiratory and all-cause mortality after restrictions on sulphur content of fuel in Hong-Kong: an intervention study. *Lancet*, 360: 1646–52.

Helfand, W H, Lazarus, J and Theerman, P (2001) Donora, Pennsylvania: An environmental disaster of the 20th century. *American Journal of Public Health*, 91: 553.

Hettelingh, J-P, Posch, M, De Smet, P A M and Downing, R J (1995) The use of critical loads in emission reduction agreements in Europe. *Water, Air and Soil Pollution*, 85: 2381–8.

Hettelingh, J-P, Posch, M and De Smet, P A M (2001) Multi-effect critical loads used in multi-pollutant reduction agreements in Europe. *Water, Air and Soil Pollution*, 130: 1133–8.

Hettelingh, J-P, Posch, M, Slootweg, J, Reinds, G J, Spranger, T and Tarrason, L (2007) Critical loads and dynamic modelling to assess European areas at risk of acidification and eutrophication. *Water, Air and Soil Pollution: Focus* (online) DOI: http://dx.doi.org/10.1007/s11267-006-9099-1.

Highwood, E J and Kinnersley, R P (2006) When smoke gets in our eyes: The multiple impacts of atmospheric black carbon on climate, air quality and health. *Environment International*, 32, 560–6.

Hirsch, A I, Michalak, A M, Bruhwiler, L M, Peters, W, Dlugokencky, E J and Tans, P P (2006) Inverse modelling estimates of the global nitrous oxide surface flux from 1998–2001. *Global Biogeochem.Cycles*, 20: GB1008, doi:10.1029/2004GB002443.

Hitz, S and Smith, J (2004) Estimating global impacts from climate change. *Global Environmental Change*, 14, 201–18.

Holgate, S J and Woodworth, P L (2004) Evidence for enhanced coastal sea level rise during the 1990s. *Geophysical Research Letters*, 31: L07305, doi:10.1029/2004GL019626.

Holloway, T, Levy II, H and Carmichael, G (2002) Transfer of reactive nitrogen in Asia: development and evaluation of a source-receptor model. *Atmospheric Environment*, 36: 4251–64.

Holloway, T, Fiore, A and Hastings, M G (2003) Intercontinental transport of air pollution: will emerging science lead to a new hemispheric treaty? *Environmental Science and Technology*, 37: 4535–42.

Holmes, J A, Franklin, E C and Gould, R A (1915) *The Report of the Selby Smelter Commission*. Washington, DC: Department of Interior, Bureau of Mines.

Hoogwijk, M (2005) *IPCC Expert Meeting on Emission Scenarios*. 12–14 January 2005, Washington DC IPCC Technical Support Unit Working Group III.

Houghton, J (1994) *Global Warming*, Oxford: Lion Publishing.

Houyin, Z, Longyi, S and Qiang, Y (2005) Microscopic Morphology and Size Distribution of Residential Indoor PM10 in Beijing City. *Indoor and Built Environment*, 14(6), 513–20.

Huntingford, C, Hemming, D, Gash, J H C, Gedney, N and Nuttall, P A (2007) Impact of climate change on health: what is required of climate modellers? *Transactions of the Royal Society of Tropical Medicine and Hygiene*, 101: 97–103.

Husar, R B, Tratt, D M, Schichtel, B A, Falke, S R, Li, F, Jaffe, D, Gasso, S, Gill, T, Laulainen, N S, Lu, F, Reheis, M C, Chun, Y, Westphal, D, Holben, B N, Gueymard, C, McKendry, I, Kuring, N, Feldman, G C, McClain, C, Frouin, R J, Merrill, J, DuBois, D, Vignola, F, Murayama, T, Nickovic, S, Wilson, W E, Sassen, K, Sugimoto, N and Malm, W C (2001) Asian dust events of April 1998. *Journal of Geophysical Research*, 106: 18317–30.

IPCC (1995) *Climate Change 1994: Radiative Forcing of Climate Change and an Evaluation of the IPCC IS92 Emission Scenarios*. Houghton, J T, Meira Filho, L G, Bruce, J, Hoesung Lee, B A, Callander, E, Hates, N, Harris and Maskell, K (eds), Cambridge, UK: Cambridge University Press.

IPCC (1996) *Climate Change 1995. The Science of Climate Change*. The Contribution of Working Group I to the Second Assessment Report of the Intergovernmental Panel on Climate Change. J P Houghton, L G Meira Filho, B A Callendar, A Kattenberg, and K Maskell, (eds), Cambridge, UK: Cambridge University Press.

IPCC (2000) *IPCC Special Report: Land Use, Land Use Change and Forestry*, Cambridge, UK: Cambridge University Press.

IPCC (2001a) *Climate Change 2001: Impacts, Adaptation, and Vulnerability*. J J McCarthy, O F Canziani, N A Leary, D J Doknen and K S White (eds). Cambridge, UK and New York, USA: Cambridge University Press.

IPCC (2001b) *IPCC Third Assessment Report on Climate Change 2001: The Scientific Basis*. Houghton, J T, Ding, Y, Griggs, D J, Noguer, M, van der Linden, P J, Dai, X, Maskell, K and Johnson, C A (eds). Cambridge, UK and New York, USA: Cambridge University Press.

IPCC (2007) *Climate Change 2007: The Physical Science Basis. Contribution of Working Group I to the Fourth Assessment Report of the Intergovernmental Panel on Climate Change*. S Solomon, D Qin, M Manning, Z Chen, M Marquis, K B Averyt, M Tignor and H L Miller (eds). Cambridge, UK and New York, USA: Cambridge University Press, 996.

IPCC WGII (2007) *Climate Change 2007: Impacts, Adaptation and Vulnerability Working Group II Contribution to the Intergovernmental Panel on Climate Change Fourth Assessment Report. Summary for Policymaker.*

Jacob, D J, Logan, J A and Murti, P P (1999) Effect of rising Asian emissions on surface ozone in the United States. *Geophysical Research Letters*, 26: 2175–8.

Jacobson, M Z (2001) Strong radiative heating due to the mixing state of black carbon in atmospheric aerosols. *Nature*, 409: 695–7.

Jacobson, M Z (2002) *Atmospheric Pollution – History, Science and Regulation*. Cambridge: Cambridge University Press.

Jansen, E, Overpeck, J, Briffa, K R, Duplessy, J-C, Joos, F, Masson-Delmotte, V, Olago, D, Otto-Bliesner, B, Peltier, W R, Rahmstorf, S, Ramesh, R, Raynaud, D, Rind, D, Solomina, O, Villalba, R and Zhang, D (2007) Palaeoclimate. In S Solomon, D Qin, M Manning, Z Chen, M Marquis, K B Averyt, M Tignor and H L Miller, (eds). *Climate Change 2007: The Physical Science Basis. Contribution of Working Group I to the Fourth Assessment Report of the Intergovernmental Panel on Climate Change*. Cambridge, UK and New York, NY, USA: Cambridge University Press.

Janssen, N A H, Lanki, T, Hoek, G, Vallius, M, de Hartog, J J, Van Grieken, R, Pekkanen, J and Brunekreef, B (2005) Associations between ambient, personal, and indoor exposure to fine particulate matter constituents in Dutch and Finnish panels of cardiovascular patients. *Occupational and Environmental Medicine*, 62: 868–77. http://oem.bmj.com/cgi/content/full/62/12/868.

Jones, H G (1992) *Plants and Microclimate: a Quantitative Approach to Environmental Plant Physiology*, 2nd edn, Cambridge: Cambridge University Press.

Johnston, H (1971) Reduction of stratospheric ozone by nitrogen oxide catalysts from supersonic transport exhaust. *Science*, 173: 517–22.

Kallos, G, Kotroni, V, Lagouvardos, K, Varinou, M, Papadopoulos, A, Kakaliagou, O, Luria, M, Peleg, M, Wanger, A and Sharf, G (1997a) Ozone production and transport in the eastern Mediterranean. *Proceedings of the Technical workshop on Tropospheric Ozone Pollution in Southern Europe*, 4–7 March, Valencia, Spain.

Kallos, G, Kotroni, V, Lagouvardos, K, Papadopoulos, A, Varinou, M, Kakaliagou, O, Luria, M, Peleg, M, Wanger, A and Uliasz, M (1997b) Temporal and spatial scales for transport and transformation processes in the Mediterranean. *Proceedings of the 22nd NATO/CCMS International Technical Meeting on Air Pollution Modelling and Its Application*, 2–6 June, Clermont Ferrand, France, ed. Gryning, S-E and Chaumerliac, N, New York: Plenum Press, vol. 20.

Kallos, G, Kotroni, V, Lagouvardos, K, Varinou, M and Papadopoulos, A (1998) The role of the Black Sea on the long-range transport from southeastern Europe towards middle east during summer. *Proceedings of the 23rd NATO/CCMS International Technical Meeting on Air Pollution Modelling and its Application*, 6–10 October, Sofia, Bulgaria.

Kashulina, G, Reimann, C and Banks, D (2002) Sulphur in the Arctic environment (3): environmental impact. *Environmental Pollution*, 124: 151–71.

Keeling, C D and Whorf, T P (2005) Atmospheric $CO_2$ records from sites in the SiO air sampling network. *Trends: A Compendium of Data on Global change*. Carbon Dioxide Information Analysis Center, Oak Ridge National Laboratory, US Department of Energy, Oak Ridge, TN http://cdiac.esd.ornl.gov/trends/co2/sio-keel-flask/sio-keel-flask.html.

Kenworthy, J R (1995) Automobile Dependence in Bangkok: An International Comparison with Implications for Planning Policies. *World Transport Policy & Practice*, 1(3): 31–41.

Kenworthy, J R and Laube, F (1999) Patterns of automobile dependence in cities: an international overview of key physical and economic dimensions with some implications for urban policy. *Transportation Research part A: Policy and Practice*, 33(11): 691–723.

Kenworthy, J R (2003) Transport energy use and greenhouse gases in urban passenger transport systems: a study of 84 global cities, in Proceedings of the second meeting of the academic forum of regional government for sustainable development, Fremantle, Western Australia, http://www.sustainability.dpc.wa.gov.au/conferences/refereed%20papers/Kenworthy,J%20-%20paper.pdf (accessed 24 June 2006).

Khasnis, A A and Nettleman, M D (2005) Global Warming and Infectious Disease. *Archives of Medical Research*, 36, 689–96.

Khokhar, M F, Frankenberg, C, van Roozendael, M, Beirle, S, Kühl, S, Richter, A, Platt, U and Wagner, T (2005) Satellite observations of atmospheric SO2 from volcanic eruptions during the time-period of 1996–2002. *Advances in Space Research*, 36: 879–87.

Kiehl, J T and Trenberth, K E (1997) Earth's Annual Global Mean Energy Budget. *Bulletin of the American Meteorological Society*, 78(2): 197–208.

Kiester, E (1999) A darkness in Donora – When smog killed 20 people in a Pennsylvania mill town in 1948, the Clean Air Movement got its start. *Smithsonian*, 30(8): 22–4.

Koerner, R M and Fisher, D A (2002) Ice-core evidence for widespread Arctic glacier retreat in the Last Interglacial and the early Holocene. *Annals of Glaciology*, 35(1): 19–24(6).

Krishnan, P, Swain, D K, Chandra Bhaskar, B, Nayak, S K and Dash, R N (2007) Impact of elevated $CO_2$ and temperature on rice yield and methods of adaptation as evaluated by crop simulation studies. *Agriculture, Ecosystems & Environment*, 122(2), 233–42.

Krupa, S V and Manning, W J (1988) Atmospheric ozone: formation and effects on vegetation. *Environmental Pollution*, 50: 101–37.

Kukkonen, J, Pohjola, M, Sokhi, R S, Luhana, L, Kitwiroon, N, Fragkou, L, Rantamäki, M, Berge, E, Ødegaard, V, Slørdal, L H, Denby, B and Finardi, S (2005) Analysis and evaluation of selected local-scale PM10 air pollution episode in four European cities: Helsinki, London, Milan and Oslo. *Atmospheric Environment*, 39: 2759–73.

Kunkel, K E and Liang, X-Z (2005) GCM Simulations of the Climate in the Central

United States. *Journal of Climate*, 18(7): 1016–1031.

Kuylenstierna, J and Hicks, K (2002) *Air pollution in Asia and Africa: The approach of RAPIDC programme*. SEI (Stockholm Environment Institute).

LAEI (2003) *London Atmospheric Emissions Inventory 2001*. London: Greater London Authority.

LAEI (2005) *London Atmospheric Emissions Inventory 2002*. London: Greater London Authority.

LAEI (2010) *London Atmospheric Emissions Inventory 2010*. London: Greater London Authority. http://data.london.gov.uk/laei-2008 (accessed 18 April 2011).

Lai, H K Kendall, M, Ferrier, H, Lindup, I, Alm, S, Hänninen, O, Jantunen, M, Mathys, P, Colvile, R, Ashmore, M R, Cullinan, P and Nieuwenhuijsen, M J (2004) Personal exposures and microenvironment concentrations of $PM_{2.5}$, VOC, $NO_2$ and CO in Oxford, UK *Atmospheric Environment*, 38(37): 6399–410.

Lai, H K, Bayer-Oglesby, L, Colvile, R, Götschi, T, Jantunen, M J, Künzli, N, Kulinskaya, E, Schweizer, C and Nieuwenhuijsen, M J (2006) Determinants of indoor air concentrations of $PM_{2.5}$, black smoke and $NO_2$ in six European cities (EXPOLIS study). *Atmospheric Environment*, 40(7): 1299–313.

Lamb, A J and Klaussner, E (1998) Response of the fynbos shrubs *Protea repens* and *Erica plukenetii* to low levels of nitrogen and phosphorus applications. *South African Journal of Botany*, 54: 558–64.

Langner, J, Bergström, R and Foltescu, V (2005) Impact of climate change on surface ozone and deposition of sulphur and nitrogen in Europe. *Atmospheric Environment*, 39, 1129–41.

Larssen, T, Cosby, B J, Lund, E and Wright, RF (2010) Modeling future acidification and fish populations in Norwegian surface waters. *Environmental Science and Technology*, 44: 5345–51.

Lawrence, A J, Masih, A and Taneja, A (2004) Indoor/outdoor relationships of carbon monoxide and oxides of nitrogen in domestic homes with roadside, urban and rural locations in a central Indian region. *Indoor Air*, 15: 76–82.

Lazaridis, M, Aleksandropoulou, V, SmolEDk, J, Hansen, J E, Glytsos, T, Kalogerakis, N and Dahlin, E (2006) Physico-chemical characterization of indoor/outdoor particulate matter in two residential houses in Oslo, Norway: measurements overview and physical properties – URBAN-AEROSOL Project, *Indoor Air*, 16: 282–95.

Lee, E H, Tingey, D T, Hogsett, W E and Laurence, J A (2003) History of tropospheric ozone for the San Bernardino Mountains of Southern California, 1963–1999. *Atmospheric Environment*, 37(19): 2705–17.

Lee, J-T, Son, J-Y and Cho, Y-S (2007) A comparison of mortality related to urban air particles between periods with Asian dust days and without Asian dust days in Seoul, Korea, 2000–2004. *Environmental Research*, doi:10.1016/j.envres.2007.06.004

Lefohn, A S, Husar, J D and Husar, J D (1999) Estimating historical anthropogenic global sulfur emission patterns for the period 1850–1990. *Atmospheric Environment*, 33(21): 3435–44.

Leggett, J, Pepper, W J, Swart, R J, Edmonds, J, Meira Filho, L G, Mintzer, I, Wang, M X and Watson, J (1992) Emissions Scenarios for the IPCC: an Update. *Climate Change 1992: The Supplementary Report to The IPCC Scientific Assessment*, Cambridge, UK: Cambridge University Press, 68–95.

Lemke, P, Ren, J, Alley, R B, Allison, I, Carrasco, J, Flato, G, Fujii, Y, Kaser, G, Mote, P, Thomas, R H and Zhang, T (2007) Observations: Changes in Snow, Ice and Frozen Ground. In S Solomon, D Qin, M Manning, Z Chen, M Marquis, K B Averyt, M Tignor and H L Miller (eds). *Climate Change 2007: The Physical Science Basis. Contribution of Working Group I to the Fourth Assessment Report of the Intergovernmental Panel on Climate Change.* Cambridge, UK and New York, USA: Cambridge University Press.

Le Treut, H, Somerville, R, Cubasch, U, Ding, Y, Mauritzen, C, Mokssit, A, Peterson T and Prather, M (2007) Historical Overview of Climate Change. In S Solomon, D Qin, M Manning, Z Chen, M Marquis, K B Averyt, M Tignor and H L Miller (eds). *Climate Change 2007: The Physical Science Basis. Contribution of Working Group I to the Fourth Assessment Report of the Intergovernmental Panel on Climate Change.* Cambridge, UK and New York, USA Cambridge University Press.

Leuliette, E W, Nerem, R S and Mitchum, G T (2004) Calibration of TOPEX/Poseidon and Jason altimeter data to construct a continuous record of mean sea level change. *Marine Geodesy*, 27(1): 79–94.

Li, Q, Jacob, D J, Bey, I, Palmer, P I, Duncan, B N, Field, B D, Martin, R V, Fiore, A M, Yantosca, R M, Parrish, D D, Simmonds, P G and Oltmans, S J (2002) Transatlantic transport of pollution and its effects on surface ozone in Europe and North America. *Journal of Geophysical Research*, doi 10.1029/2001JD001422.

Liao, H and Seinfeld, J H (2005) Global impacts of gas-phase chemistry-aerosol interactions on direct radiative forcing by anthropogenic aerosols and ozone. *Journal of Geophysical Research*, 110: D18208, doi:10.1029/2005JD005907.

Lichtheim, M (1980) *Ancient Egyptian Literature*, Berkeley, University of California Press.

Lohmann, U (2002) A glaciation indirect aerosol effect caused by soot aerosols. *Geophysical Research Letters*, 29: 10.1029/2001GL014357.

Lorenz, M, Mues, V, Becher, G, Müller-Edzards, C, Luyssaert, S, Raitio, H, Fürst, A and Langouche, D (2003) *Forest Condition in Europe*, Technical Report. Hamburg, Germany: Federal Research Centre for Forestry and Forest Products.

Loso, M G, Anderson, R S, Anderson, S P and Reimer, P J (2006) A 1500-year record of temperature and glacial response inferred from varved Iceberg Lake, southcentral Alaska. *Quaternary Research*, 66(1): 12–24.

Luria, M, Peleg, M, Sharf, G, Alper Siman-Tov, D, Shpitz, N, Ben-Ami, Y, Yitzchaki, A and Seter, I (1996) Atmospheric sulfur over the east Mediterranean region. *Journal of Geophysical Research*, 101: 25917–25.

Mabaso, M L H, Kleinschmidt, I, Sharp, B and Smith, T (2007) El Niño Southern Oscillation (ENSO) and annual malaria incidence in Southern Africa. *Transactions of the Royal Society of Tropical Medicine and Hygiene*, 101: 326–30.

MacFarling Meure, C, Etheridge, D, Trudinger, C, Steele, P, Langenfelds, R, Van Ommen, T, Smith, A and Elkins, J (2006) The Law Dome CO2, CH4 and N2O ice core records extended to 2000 years BP *Geophysical Research Letters*, 33: L14810. http://dx.doi:10.1029/2006GL026152.

MacLeod, R M (1965) The Alkali Acts administration. *Victorian Studies*, 9, 86–112.

Mamane, Y (1987) Air-pollution control in Israel during the 1st and 2nd century. *Atmospheric Environment*, 21(8): 1861–3.

Manning, W J and Feder, W A (1980) *Biomonitoring Air Pollutants with Plants*. Dordrecht, Netherlands: Kluwer Academic Publishers.

Mao, H and Talbot, R (2004) Role of meteoro logical processes in two New England ozone episodes during summer 2001. *Journal of Geophysical Research*. 109, D20305. http://dx.doi:1029/2004JD004850.

Marchenko, S S, Gorbunov, A P and Romanovsky, V E (2007) Permafrost warming in the Tien Shan Mountains, Central Asia. *Global and Planetary Change*, 56, 311–27.

Märker, M, Angeli, L, Bottai, L, Costantini, R, Ferrari, R, Innocenti, L and Siciliano, G (2007) Assessment of land degradation susceptibility by scenario analysis: A case study in Southern Tuscany, Italy. *Geomorphology*, doi:10.1016/ j.geomorph.2006.12.020.

Martens, P (1998) *Health and Climate Change: Modelling the Impacts of Global Warming and Ozone Depletion*. London: Earthscan Publications.

Martens, P (1999) How will climate change affect human health? *American Scientist*, 87: November–December, 534–41.

Martens, P (2004) Personal communication, International Centre for Integrated Assessment and Sustainable Development (ICIS), University Maastricht, The Netherlands.

Martens, P, Kovats, R S, Nijhof, S, de Vries, P, Livermore, M T J, Bradley, D J, Cox, J and McMichael, A J (1999) Climate change and future populations at risk of malaria. *Global Environmental Change*, S9: 89–107.

Masoli, M, Fabian, D, Holt, S and Beasley, R (2004) *Global Burden of Asthma*, reproduced from the WHO Population Statistics 2001. *Allergy*, 59(5): 469–78.

McElroy, M B, Salawitch, R J, Wofsy, S C and Logan, J A (1986) Reductions of Antarctic ozone due to synergistic interactions of chlorine and bromine. *Nature*, 321, 759–62.

McInnes, H, Laupsa, H and Larssen, S (2005) Private communication, Norwegian Institute for

Air Research (NILU), http://www.nilu.no/ (accessed 24 June 2006).

McKendry, I G, Hacker, J P, Stull, R, Sakiyam, S, Mignacca, D and Reid, K (2001) Long-range transport of Asian dust to the Lower Fraser Valley, British Columbia, Canada. *Journal of Geophysical Research*, 106, 18361–70.

McMichael, A J, Haines, A, Slooff, R and Kovats, S (1996) *Climate Change and Human Health: an Assessment Prepared by a Task Group on Behalf of the World Health Organization, the World Meteorological Organization and the United Nations Environment Programme*. Geneva: World Health Organization.

Mechler, R, Amann, M and Schöpp, W (2002) *A Methodology to Estimate Changes in Statistical Life Expectancy Due to the Control of Particulate Matter Air Pollution*, IR-02-035, Laxenburg, Austria: International Institute for Applied Systems Analysis.

Medina, S, Boldo, E, Saklad, M, Niciu, E M, Krzyzanowski, M, Frank, F, Cambra, K, Muecke, H G, Zorilla, B, Atkinson, R, Le Tertre, A, Forsberg, B and the contribution members of the APHEIS group (2005). *APHEIS Health Impact Assessment of Air Pollution and Communications Strategy*, Third-year report. Institut de Veille Sanitaire, Saint-Maurice and European Commission.

Meehl, G A, Stocker, T F, Collins, W D, Friedlingstein, P, Gaye, A T, Gregory, J M, Kitoh, A, Knutti, R, Murphy, J M, Noda, A, Raper, S C B, Watterson, I G, Weaver, A J and Zhao, Z-C (2007) Global climate projections. In S Solomon, D Qin, M Manning, Z Chen, M Marquis, K B Averyt, M Tignor and H L Miller (eds). *Climate Change 2007: The Physical Science Basis. Contribution of Working Group I to the Fourth Assessment Report of the Intergovernmental Panel on Climate Change.* Cambridge, UK and New York, USA Cambridge University Press.

Meywerk, J and Ramanathan, V (2002) Influence of anthropogenic aerosols on the total and spectral irradiance at the sea surface during the Indian Ocean experiment (INDOEX) 1999. *Journal of Geophysical Research*, 107: No.D19, 8018, doi:10.1029/2000JD000022.

Meteorological Service of Canada (2003) *2001 in Review: an Assessment of New Research Developments Relevant to the Science of Climate Change, $CO_2$/Climate Report, Summer.*

Mieck, I (1990) Reflections on a typology of historical pollution: complementary conceptions. In P Brimblecombe and C Pfister (eds). *The Silent Countdown*. Berlin: Springer-Verlag, 73–80.

Mikami, M, Shi, G Y, Uno, I, Yabuki, S, Iwasaka, Y, Yasui, M, Aoki, T, Tanaka, T Y, Kurosaki, Y, Masuda, K, Uchiyama, A, Matsuki, A, Sakai, T, Takemi, T, Nakawo, M, Seino, N, Ishizuka, M, Satake, S, Fujita, K, Hara, Y, Kai, K, Kanayama, S, Hayashi, M, Du, M, Kanai, Y, Yamada, Y, Zhang, X Y, Shen, Z, Zhou, H, Abe, O, Nagai, T, Tsutsumi, Y, Chiba, M and Suzuki, J (2006) Aeolian dust experiment on climate impact: An overview of Japan–China joint project ADEC *Global and Planetary Change*, 52, 142–72.

Miller, P and McBride, J (1999) *Oxidant Air Pollution Impacts in the Montane Forests of Southern California: The San Bernadino Mountain Case Study*. New York: Springer-Verlag.

Miller, R L and Tegen, I (1998) Climate response to soil dust aerosols. *Journal of Climate*, 11: 3247–67.

Mills, G, Hayes, F, Simpson, D, Emberson, L, Norris, D, Harmens, H and Büker, P (2011) Evidence of widespread effects of ozone on crops and (semi)-natural vegetation in Europe in relation to AOT40- and flux-based risk maps. *Global Change Biology*, 17: 592–613.

Mirasgedis, S, Sarafidis, Y, Georgopoulou, E, Kotroni, V, Lagouvardos, K and Lalas, D P (2007) Modelling framework for estimating impacts of climate change on electricity demand at regional level: Case of Greece. *Energy Conversion and Management*, 48, 1737–750.

Mittermeier, R A, Robles Gil, P, Hoffmann, M, Pilgrim, J, Brooks, T, Goettsch Mittermeier, C, Lamoreux, J and Da Fonseca, G A B (2005) *Hotspots Revisited: Earth's Biologically Richest and Most Threatened Terrestrial Ecoregions*. Mexico: CEMEX.

Mocarelli, P (2001) Seveso: a teaching story. *Chemosphere*, 43(4–7): 391–402.

Molina, L T and Molina, M J (1987) Production of Cl2O2 from the self-reaction of the ClO radical. *Journal of Physical Chemistry*, 91: 433–6.

Molina, M J and Molina, L T (2002) *Air Quality in the Mexico Megacity: an Integrated Assessment*. Dordrecht, Netherlands: Kluwer Academic Publishers.

Molina, M J and Molina, L T (2004) Megacities and atmospheric pollution. *Journal of Air & Waste Management Association*, 54: 644–80.

Molina, M J and Rowland, F S (1974) Stratospheric sink for chlorofluoromethanes: chlorine atom catalyzed destruction of ozone. *Nature*, 249: 810–4.

Monnin, E, Indermühle, A, Dällenbach, A, Flückiger, J, Stauffer, B, Stocker, T F, Raynaud, D and Barnola, J-M (2001) Atmospheric $CO_2$ concentrations over the last glacial termination. *Science*, 291(5501): 112–4.

Monnin, E, Steig, E J, Siegenthaler, U, Kawamura, K, Schwander, J, Stauffer, B, Stocker, T F, Morse, D L, Barnola, J-M, Bellier, B, Raynaud, D and Fischer, H (2004) Evidence for substantial accumulation rate variability in Antarctica during the Holocene, through synchronization of $CO_2$ in the Taylor Dome, Dome C and DML ice cores. *Earth Planetary Science Letters*, 224(1–2): 45–54.

Montero, J (2004) Market-based policies for the control of air pollution: the case of Santiago-Chile. In *Integrated Program on Urban, Regional and Global Air Pollution Seventh Workshop on Mexico Air Quality*, 8–21 January, Mexico City.

Mosley, S (2001) *The Chimney of the World*, Cambridge: The White Horse Press.

Moss, R H, Edmonds, J A, Hibbard, K A, Manning, M R, Rose, S K, van Vuuren, D P, Carter, T R, Emori, S, Kainuma, M, Kram, T, Meehl, G A, Mitchell, J F B, Nakicenovic, N, Riahi, K, Smith, S J, Stouffer, R J, Thomson, A M, Weyant, J P, Wilbanks, T, J (2010) The next generation of scenarios for climate change research and assessment. *Nature*, 463: 747–56 (11 February 2010) doi:10.1038/nature08823.

Myers, N, Mittermeier, R A, Mittermeier, C G, da Fonesca, G A B and Kent, J (2000) Biodiversity hotspots for conservation priorities. *Nature*, 403: 853–8.

NAEI (2011) *National Atmospheric Emissions Inventory* 2008. http://www.naei.org.uk/ (accessed March 2011).

Nakicenovic, N, Alcamo, J, Davis, G, de Vries, B, Fenhann, J, Gaffin, S, Gregory, K, Grübler, A, Jung, T Y, Kram, T, La Rovere, E L, Michaelis, L, Mori, S, Morita, T, Pepper, W, Pitcher, H, Price, L, Raihi, K, Roehrl, A, Rogner, H-H, Sankovski, A, Schlesinger, M, Shukla, P, Smith, S, Swart, R, van Rooijen, S, Victor, N and Dadi, Z (2000) *Emissions Scenarios. A Special Report of Working Group III of the Intergovernmental Panel on Climate Change.* Cambridge, UK and New York, USA Cambridge University Press, 599.

NASA (1997) *NASA Satellite Tracks Hazardous Smoke and Smog Partnership*, http://visibleearth.nasa.gov/cgi-bin/viewrecord?7613 (accessed 24 June 2006).

NASA (2000) *First Global Carbon Monoxide (Air Pollution) Measurements*, http://visibleearth.nasa.gov/cgi-bin/viewrecord?8086 (accessed 24 June 2006).

NASA (2001a) *The 'Perfect Dust Storm' of April 2001*, http://jwocky.gsfc.nasa.gov/aerosols/today_plus/yr2001/asia_dust.html (accessed 24 June 2006).

NASA (2001b) *The Pacific Dust Express*, http://science.nasa.gov/headlines/y2001/ast17may_1.htm (accessed 24 June 2006).

Neill, C, Steudler, P A, Garcia-Montiel, D C, Melillo, J M, Feigl, B J, Piccolo, M C and Cerri, C C (2005) Rates and controls of nitrous oxide and nitric oxide emissions following conversion of forest to pasture in Rondônia. *Nutrient Cycling in Agroecosystems*, 71: 1–15.

Nemery, B, Hoet, P H M and Nemmar, A (2001) The Meuse Valley fog of 1930: an air pollution disaster. *Lancet*, 357(9257): 704–8.

Nerem, R S, Leuliette, E and Cazenave, A (2006) Present-day sea-level change: a review. *Comptes Rendus Geoscience*, 338(14–15), 1077–83.

Ní Riain, C M, Mark, D, Davies, M, Harrison, R M and Byrne, M A (2003) Averaging periods for indoor–outdoor ratios of pollution in naturally ventilated non-domestic buildings near a busy road. *Atmospheric Environment*, 37: 4121–32.

Nicholson, W (1907/8) Practical smoke abatement. *Sanitary Inspector's Journal*, 13: 89–97.

Nilsson, J and Grennfelt, P (1988) *Critical Loads for Sulphur and Nitrogen*. Milj-rapport 15, Copenhagen, Denmark: Nordic Council of Ministers.

Nylander, W (1866) Les lichens du Jardin de Luxembourg. *Bulletin de la Société Botanique de France*, 13 : 364–71.

OECD (2002) *OECD Environmental Data 2002*, Environmental Performance and Information Division, OECD Environment Directorate.

Olivier, J G J, van Aardenne, J A, Dentener, F, Pagliari, V, Ganzeveld, L N and Peters, J A H W

(2005) Recent trends in global greenhouse gas emissions: regional trends 1970–2000 and spatial distribution of key sources in 2000. *Enviromental Science*, 2(2–3) : 81–99. DOI: 10.1080/15693430500400345. http://www.mnp.nl/edgar/global_overview/

Olivier, J G J, Pulles, T and van Aardenne, J A (2006) Part III: Greenhouse gas emissions: 1. 45 Shares and trends in greenhouse gas emissions; 2. Sources and Methods; Greenhouse gas emissions for 1990, 1995 and 2000. In *CO2 emissions from fuel combustion 1971–2004*, 2006 Edition, pp. III.1–III.41. International Energy Agency (IEA), Paris. ISBN 92-64-10891-2 (paper) 92-64-02766-1 (CD ROM).

Oltmans, S J, Lefohn, A S, Harris, J M, Galbally, I, Scheel, H E, Bodeker, G, Brunke, E, Claude, H, Tarasick, D, Johnson, B J, Simmonds, P, Shadwick, D, Anlauf, K, Hayden, K, Schmidlin, F, Fujimoto, T, Akagi, K, Meyer, C, Nichol, S, Davies, J, Redondas, A and Cuevas, E (2006) Long-term changes in tropospheric ozone. *Atmospheric Environment*, 40: 3156–73.

Osterkamp, T E (2005) The recent warming of permafrost in Alaska. *Global and Planetary Change*, 49: 187– 202.

Park, S-U, Chang, L-S and Lee, E-H (2005) Direct radiative forcing due to aerosols in East Asia during a Hwangsa (Asian dust) event observed on 19–23 March 2002 in Korea. *Atmospheric Environment*, 39, 2593–606.

Pellizzari, E D, Clayton, C A, Rodes, C E, Mason, R E, Piper, L L, Fort, B, Pfeifer, G and Lynam, D (1999) Particulate matter and manganese exposures in Toronto, Canada. *Atmospheric Environment*, 33(5): 721–34.

Phoenix, G K, Hicks, W H, Cinderby, S, Kuylenstierna, J C I, Stock, W D, Dentener, F J, Giller, K E, Austin, A T, Lefroy, R D B, Gimeno, B D, Ashmore, M R and Ineson, P (2006) Atmospheric nitrogen deposition in world biodiversity hotspots: the need for a greater global perspective in assessing N deposition impacts. *Global Change Biology*, 12: 1–7.

Piechocki-Minguy, A, Plaisance, H, Schadkowski, C, Sagnier, I, Saison, J Y, Galloo, J C and Guillermo, R (2006) A case study of personal exposure to nitrogen dioxide using a new high sensitive diffusive sampler. *Science of the Total Environment*, 366: 55–64.

Piringer, M and Joffre, S (2005) *The Urban Surface Energy Budget and Mixing Height in European Cities: Data, Models and Challenges for Urban Meteorology and Air Quality.* Final report of Working Group 2 of COST 715 Action on Urban Meteorology Applied to Air Pollution Problems, COST Office, European Science Foundation, Brussels: Demetra Publishers.

Pluschke, P (2004) *Indoor Air Pollution.* Springer.

Pope, A C, Burnett, R T, Thun, M J, Calle, E E, Krewski, D, Ito, K and Thurston, G D (2002) Lung cancer, cardiopulmonary mortality, and long-term exposure to fine particulate air pollution. *Journal of the American Medical Association*, 287: 1132–41.

Posch, M, Hettelingh, J-P and De Smet, P A M (2001) Characterization of critical load exceedances in Europe. *Water, Air and Soil Pollution*, 130: 1139–44.

Posch, M, Hettelingh, J-P and Slootweg, J (2003a) *Manual for Dynamic Modelling of Soil Response to Atmospheric Deposition.* Coordination Centre for Effects, RIVM Report 259101012, Bilthoven, Netherlands, 71 pp. http://www.mnp.nl/cce.

Posch, M, Hettelingh, J-P, Slootweg, J and Downing R J (2003b) *Modelling and Mapping of Critical Thresholds in Europe: Status Report 2003*, RIVM Report 259101013, Bilthoven, Netherlands: Coordination Center for Effects.

Posthumus, A C (1982) Biological indicators of air pollution. In M H Unsworth and D P Ormrod (eds). *Effects of Gaseous Air Pollution in Agriculture and Horticulture.* London: Butterworth Scientific, 27–42.

Potter, L (2001) Drought, fire and haze in the historical records of Malaysia. In P Eaton and M Radojevic. *Forest Fires and Regional Haze in Southeast Asia*, Huntington, NY: Nova, 23–40.

Ramanathan, V, Crutzen, P J, Lelieveld, J, Mitra, A P, Althausen, D, Anderson, J, Andreae, M O, Cantrell, W, Cass, G R, Chung, C E, Clarke, A D, Coakley, J A, Collins, W,D, Conant, W C, Dulac, F, Heintzenberg, J, Heymsfield, A J, Holben, B, Howell, S, Hudson, J, Jayaraman, A, Kiehl, J T, Krishnamurti, T N, Lubin, D, McFarquhar, G, Novakov, T, Ogren, J A, Podgorny, I A, Prather, K, Priestley, K, Prospero, J M, Quinn, P K, Rajeev, K, Rasch, P, Rupert, S, Sadourny, R, Satheesh, S K, Shaw, G E, Sheridan, P and Valero, F P J (2001) Indian ocean experiment: an integrated analysis of the climate forcing and effects of the great Indo-Asian haze. *Journal of Geophysical Research*, 106(D22): 28371–98.

Ramanathan, V and Crutzen, P J (2003) New directions: Atmospheric brown 'clouds'. *Atmospheric Environment*, 37(28): 4033–5.

Rao, S T, Zalewsky, E, Zurbenko, I G, Porter, P S, Sistla, G, Hao, W, Zhou, N, Ku, J-Y, Kallos, G and Hansen, D A (1998) Integrating observations and modelling in ozone management efforts. In S-E Gryning and N Chaumerliac (eds). *Air Pollution Modelling and Its Applications XII*, New York: Plenum Press, 115–24.

Rao, S T, Ku, J-Y, Berman, S, Zhang, K and Mao, H (2003) Summertime characteristics of the atmospheric boundary-layer and relationships to ozone levels over the eastern United States. *Pure Applied Geophysics*, 160: 21–55.

Reinds, G J, Posch, M and de Vries, W (2009) Modelling the long-term soil response to atmospheric deposition at intensively monitored forest plots in Europe. *Environmental Pollution*, 157: 1258–69.

Richards, B L, Middleton, J T and Hewitt, W B (1958) Air pollution with relation to agronomic crops. V Oxidant stipple to grape. *Agronomy Journal*, 50: 559–61.

Roche, A E, Kumer, J B, Mergenthaler, J L, Nightingale, R W, Uplinger, W,G, Ely, G A, Potter, J F, Wuebbles, D J, Connell, P S and Kinnison, D E (1994) Observations of lower-stratospheric ClONO2, HNO3, and aerosol by the UARS CLAES experiment between January 1992 and April 1993. *Journal of Atmospheric Science*, 52: 2877–902.

Roeckner, E, Bengtson, L, Feichter, J, Lelieveld, J and Rodhe, H (1999) Transient climate change simulations with a coupled atmosphere–ocean GCM including the tropospheric sulphur cycle. *Journal of Climate*, 12: 3004–32.

Roelofs, J G M, Kempers, A J, Houdijk, A L F M and Jansen, J (1985) The effect of air-borne ammonium sulphate on Pinus nigra var. maritima in the Netherlands. *Plant and Soil*, 84: 45–56.

Roelofs, J G M, Bobbink, R, Brouwer, E and De Graaf, M C C (1996) Restoration ecology of aquatic and terrestrial vegetation on non- calcareous sandy soils in The Netherlands. *Acta Botanica Neerlandica*, 45: 517–41.

Rojas-Bracho, L, Suh, H H and Oyola, P (2002). Measurements of children's exposures to particles and nitrogen dioxide in Santiago, Chile, *The Science of the Total Environment*, 287: 249–264.

Rostayn, L D and Lohmann, U (2002) Tropical rainfall trends and the indirect aerosol effect. *Journal of Climate*, 15: 2103–16.

Rounsevell, M D A, Reginster, I, Araújo, M B, Carter, T R, Dendoncker, N, Ewert, F, House, J I, Kankaanpää, S, Leemans, R, Metzger, M J, Schmit, C, Smith, P and Tuck, G (2006) A coherent set of future land use change scenarios for Europe. *Agriculture, Ecosystems and Environment*, 114: 57–68.

Royal Society (2008) *Ground-level ozone in the 21st century: future trends, impacts and policy implications*. Science Policy Report 15/08, The Royal Society, London.

SAI (Systems Applications International). (1995) *User's Guide to the Variable Grid Urban Airshed Model (UAM-V)*, San Rafael, CA: Systems Applications International (available from Systems Applications International, 101 Lucas Valley Road. San Rafael, CA 94903).

Sakai, R, Siegmann, H C, Sato, H and Voorhees, A S (2002) Particulate matter and particle-attached polycyclic aromatic hydrocarbons in the indoor and outdoor air of Tokyo measured with personal monitors. *Environmental Research*, 89: (1), 66–71.

Sánchez-Ccoyllo, O P, Ynoue, R Y, Martins, L D and Andrade, M F (2006) Impacts of ozone precursor limitation and meteorological variables on ozone concentration in São Paulo, Brazil. *Atmospheric Environment*, 40: 552–562.

Santee, M L, Manney, G L, Waters, J W and Livesey, N J (2003) Variations and climatology of ClO in the polar lower stratosphere from UARS Microwave Limb Sounder measurements. *Journal of Geophysical Research*,108: Art. No. 4454.

Sarofim, M C, Forest, C E, Reiner, D M and Reilly, J M (2005) Stabilization and global climate policy. *Global and Planetary Change*, 47: 266–272.

Satake, K (2001) New eyes for looking back to the past and thinking of the future. *Water Air and Soil Pollution*, 130(1–4): 31–42.

Satheesh, S K and Moorthy, K K (2005) Radiative effects of natural aerosols: A review. *Atmospheric Environment*, 39, 2089–110.

Sawa, T (1997) *Japan's Experience in the Battle against Air Pollution.* Tokyo: The Pollution Related Health Damage Compensation and Prevention Association.

Schichtel, B A and Husar, R B (1996) Summary of ozone transport, Report of the Ad Hoc Air Trajectory Workgroup, http://capita.wustl.edu/otag/Reports/AQATransport/Transport.html (accessed 24 June 2006).

Schichtel, B A and Husar, R B (2001) Eastern North American transport climatology during high- and low-ozone days. *Atmospheric Environment*, 35: 1029–38.

Schimel, D, D Alves, I Enting, M Heimann, F Joos, D Raynaud, T Wigley, M Prather, R Derwent, D Ehhalt, P Fraser, E Sanhueza, X Zhou, P Jonas, R Charlson, H Rodhe, S Sadasivan, K P Shine, Y Fouquart, V Ramaswamy, S Solomon, J Srinivasan, D Albritton, I Isaksen, M Lal, and D Wuebbles, 1995: Radiative Forcing of Climate Change. In *Climate Change 1995 - The Science of Climate Change*, IPCC, Cambridge University Press, Cambridge, 65–131.

Schönherr, J and Riederer, M (1989) Foliar penetration and accumulation of organic chemicals in plant cuticles. *Reviews of Environmental Contamination and Toxicology*, 108: 1–70.

Schöpp, W, Amann, M, Cofala, J, Heyes, C and Klimont, Z (1999) Integrated assessment of European emission control strategies. *Environmental Modelling & Software*, 14: 1–9.

Schramm, E (1990) Experts in the smelter smoke debate. In P Brimblecombe and C Pfister. *The Silent Countdown*, Berlin: Springer-Verlag, 196–209.

Schwela, D, Haq, G, Huizenga, C, Han, W-J, Fabian, H and Ajero M (2006) *Urban Air Pollution in Asian Cities*. UK: Earthscan.

SEI (Stockholm Environment Institute) (2004) *A Strategic Framework for Air Quality Management in Asia*, Stockholm Environment Institute, Korea Environment Institute and Ministry of Environment – Korea.

Seinfeld, J H and Pandis, S N (1998) *Atmospheric Chemistry and Physics – From Air Pollution to Climate Change*. Chichester: John Wiley and Sons.

Sem, G J, Boulaud, D, Brimblecombe, P, Ensor, D S, Gentry, J W, Marijnissen, J C M and Preining, O (2005) *History and Reviews of Aerosol Science*. Mount Laurel, NJ: American Association for Aerosol Research, pp. 385–96.

Shao, Y and Dong, C H (2006) A review on East Asian dust storm climate, modelling and monitoring. *Global and Planetary Change*, 52: 1–22.

Sharma, C, Dasgupta, A and Mitra, A P (2002) Inventory of GHGs and other urban pollutants from transport sector in Delhi and Calcutta. *Proceeding of IGES/APN Mega-cities Project*, 23–25 January, Kitakyushu, Japan.

Shepard Krech III, J R McNeill, and Merchant, C (2003) *Encyclopedia of World Environmental History*, New York: Routledge.

Sheu, B H and Liu, C P (2003) Air pollution impacts on vegetation in Taiwan. In L D Emberson, M R Ashmore, and F Murray (eds), *Air Pollution Impacts on Crops and Forests: a Global Perspective*. London: Imperial College Press, 145–64.

Siegenthaler, U, Monnin, E, Kawamura, K, Spahni, R, Schwander, J, Stauffer, B, Stocker, T F, Barnola, J-M and Fischer, H (2005) Supporting evidence from the EPICA Dronning Maud Land ice core for atmospheric CO2 changes during the past millennium. *Tellus*, 57B(1): 51–57.

Simpson, D, Ashmore, M R, Emberson, L and Tuovinen, J-P (2007) A comparison of different approaches for mapping potential ozone damage to vegetation. A model study. *Environmental Pollution*, 146, 715–25.

Skov, H, Christensen, C S, Fenger, J, Essenbaek, M, Larsen, D and Sorensen, L (2000) Exposure to indoor air pollution in a reconstructed house from the Danish Iron Age. *Atmospheric Environment*, 34(22): 3801–4.

Slootweg, J, Posch, M and Hettelingh, J-P (2003) Summary of national data. In M Posch, J-P Hettelingh, J Slootweg, and R J Downing (eds). *Modelling and Mapping of Critical Thresholds in Europe: Status Report 2003, RIVM Report 259101013/2003*, Bilthoven, Netherlands: Coordination Center for Effects, 11–27.

Smith, K A, and Conen, F (2004) Impacts of land management on fluxes of trace greenhouse gases. *Soil Use Management*, 20: 255–63.

Smith, K R (2002) Indoor air pollution in developing countries: recommendations for research. *Indoor Air*, 12(3): 198–207.

Smith, S J, Pitcher, H and Wigley, T M L (2001) Global and regional anthropogenic sulfur dioxide emissions. *Global and Planetary Change*, 29: 99–119.

Smith, S J, Pitcher, H and Wigley, T M L (2005) Future Sulfur Dioxide Emissions. *Climatic Change*, 73(3): 267–318.

Smith, T M and Reynolds, R W (2005) A global merged land and sea surface temperature reconstruction based on historical observations (1880–1997). *Journal of Climate*, 18, 2021–36.

Sokhi, R S (2005) Urban air quality special issue, *Atmospheric Environment*, 39: 2695–817.

Sokhi, R S (2006) Urban air quality modelling. *Environmental Modelling and Software*, 21: 430–599.

Sokhi, R S and Bartzis, J G (2002) *Urban Air Quality – Recent Advances*, Dordrecht, Netherlands: Kluwer Academic Publishers.

Solomon, S, Garcia, R R, Rowland, F S and Wuebbles, D J (1986) On the depletion of Antarctic ozone. *Nature*, 321: 755–8.

Solomon, S, Qin, D, Manning, M, Alley, R B, Berntsen, T, Bindoff, N L, Chen, Z, Chidthaisong, A, Gregory, J M, Hegerl, G C, Heimann, M, Hewitson, B, Hoskins, B J, Joos, F, Jouzel, J, Kattsov, V, Lohmann, U, Matsuno, T, Molina, M, Nicholls, N, Overpeck, J, Raga, G, Ramaswamy, V, Ren, J, Rusticucci, M, Somerville, R, Stocker, T F, Whetton, P, Wood, R A and Wratt, D (2007) Technical Summary. In S Solomon, D Qin, M Manning, Z Chen, M Marquis, K B Averyt, M Tignor and H L Miller (eds). *Climate Change 2007: The Physical Science Basis. Contribution of Working Group I to the Fourth Assessment Report of the Intergovernmental Panel on Climate Change*. Cambridge, United Kingdom and New York, USA: Cambridge University Press.

Sparrow, C J (1968) Some geographic aspects of air pollution in Auckland. *Clean Air*, 2/4. December: 3–10.

Sparrow, C J, Skam, A W and Thom, N G (1969) The growth and work of Auckland air pollution research committee. *Clean Air*, 3/1. March, 3–12.

Spiecker, H, Mielikäinen, K, Köhl, M and Skovsgaard, J P (1996) *Growth Trends in European Forests*, EFI Research Report 5. Berlin: Springer-Verlag.

Srivastava, A and Jain, V K (2007) A study to characterize the suspended particulate matter in an indoor environment in Delhi, India. *Building and Environment*, 42: 2046–52.

Stedman, J (2004) The predicted number of air pollution related deaths in the UK during the August 2003 heatwave. *Atmospheric Environment*, 38: 1087–90.

Steele, H M, Hamill, P, McCormick, M P and Swissler, T J (1983) The formation of polar stratospheric clouds. *Journal of Atmospheric Science*, 40: 2055–67.

Stephens, E R, Darley, E F, Taylor, O C and Scott, W E (1961) Photochemical reaction products in air pollution. *International Journal of Air and Water Pollution*, 4: 79–100.

Stern, D I (2006). Reversal of the trend in global anthropogenic sulphur emissions. *Global Environmental Change*, 16: 207–20.

Stier, P, Feichter, J, Kloster, S, Vignati, E, and Wilson, J (2006) Emission-induced nonlinearities in the global aeorsol system: Results from the ECHAM5-HAM aerosol-climate model. *Journal of Climate*, 19(16), 3845–62.

Stoddard, J L, Jeffries, D S, Lükewille, A, Clair, T A, Dillon, P J, Driscoll, C T, Forsius, M, Johannessen, M, Kahl, J S, Kellog, J H, Kemp, A, Mannio, J, Monteith, D, Murdoch, P S, Patrick, S, Rebsdorf, A, Skjelkvåle, B L, Stainton, M P, Traaen, T S, van Dam, H, Webster, K E, Wieting, J and Wilander, A (1999) Regional trends in aquatic recovery from acidification in North America and Europe 1980–95. *Nature*, 401: 575–8.

Stohl, A, Eckhard, S, Forster, C, James, P and Spichtinger, N (2002) On the pathways and timescales of intercontinental air pollution transport. *Journal of Geophysical Research*, 107, DOI: 10.1029/2001JD001396.

Stolarski, R S, Krueger, A J, Schoeberl, M R, McPeters, R D, Newman, P A and Alpert, J A (1986) Nimbus 7 Satellite Measurements of the Springtime Antarctic Ozone Decrease. *Nature*, 332: 808–11.

Stolarski, R S, McPeters, R D, Herman, J R and Bloomfield, P (1991) Total Ozone Trends Deduced from Nimbus 7 TOMS Data. *Geophysical Research Letters*, 18: 1015–18. This paper showed the first global ozone trends based on calibrated satellite data.

Stull, R B (1998) *An Introduction to Boundary Layer Meteorology*, Dordrecht, Netherlands: Kluwer Academic Publishers.

Sud, Y C and Lee, D (2007) Parameterization of aerosol indirect effect to complement McRAS cloud scheme and its evaluation with the 3-year ARM-SGP analyzed data for single column models. *Atmospheric Research*, doi:10.1016/j.atmosres.2007.03.007.

Sundt, N (2007) Personal communication, US Global Change Research Program/ Climate Change Science Program, 1717 Pennsylvania

Ave., NW, Suite 250, Washington, DC 20006, USA.

Sverdrup, H and Warfvinge, P (1993) The effect of soil acidification on the growth of trees, grass and herbs as expressed by the (Ca+Mg+K)/Al ratio. *Reports in Ecology and Environmental Engineering 2*, Lund University, Department of Chemical Engineering II.

Svirejeva-Hopkins, A, Schellnhuber, H J and Pomaz, V L (2004) Urbanised territories as a specific component of the global carbon cycle. *Ecological Modelling*, 173, 295–312.

Swanger, K M and Marchant, D R (2007) Sensitivity of ice-cemented Antarctic soils to greenhouse-induced thawing: Are terrestrial archives at risk? *Earth and Planetary Science Letters*, 259, 347–59.

Takemura, T, Uno, I, Nakamura, T, Hignrashi, A and Sano, I (2002) Modelling study of long-range transport of Asian dust and anthropogenic aerosols from East Asia. *Geophysical Research Letters*, 29 (24): 10.1029/2002 GL016251.

Takemura T, Nozawa, T, Emori, S, Nakajima, T Y and Nakajima, T (2005) Simulation of climate response to aerosol direct and indirect effects with aerosol transport-radiation model. *Journal of Geophysical Research*, 110: D02202, doi:10.1029/2004JD005029.

Tanaka, T (Lead Author) Howard, H (Topic Editor) (2007) Global dust budget. In C J Cleveland (ed). *Encyclopedia of Earth*. Washington, D C: Environmental Information Coalition, National Council for Science and the Environment. (Published April 30, 2007; Retrieved July 8, 2007) http://www.eoearth.org/article/Global_dust_budget.

Tans, P P and Conway, T J (2005) Monthly atmospheric CO2 mixing ratios from the NOAA CMDL Carbon Cycle Cooperative Global Air Sampling Network, 1968–2002. In *Trends: A Compendium of Data on Global Change*. Carbon Dioxide Information Analysis Center, Oak Ridge National Laboratory, U S Department of Energy, Oak Ridge, TN.

Taylor, H J, Ashmore, M R and Bell, J N B (1988) *Air Pollution Injury to Vegetation*. London: Institution of Environmental Health Officers.

Tingey, D T, Olsyk, D M, Herstrom, A A and Lee, E H (1993) Effects of ozone on crops. In D J McKee (ed.) *Tropospheric Ozone: Human Health and Agricultural Impacts*. Boca Raton: Lewis Publishers, 175–206.

Tol, R S J (2007) Carbon dioxide emission scenarios for the USA *Energy Policy*, doi:10.1016/j.enpol.2006.01.039.

Tonneijck, A E G and van Dijk, C J (2002) Assessing effects of ambient ozone on injury and yield of bean with ethylenediurea (EDU): Three years of plant monitoring at four sites in The Netherlands. *Environmental Monitoring and Assessment*, 77: 1–10.

Tonneijck, A E G, ten Berge, W F and Jansen, B P (2003) Monitoring the effects of atmospheric ethylene near polyethylene manufacturing plants with two sensitive plant species. *Environmental Pollution*, 123: 275–9.

Trenberth, K E, Jones, P D, Ambenje, P, Bojariu, R, Easterling, D, Klein Tank, A, Parker, D, Rahimzadeh, F, Renwick, J A, Rusticucci, M, Soden, B and Zhai, P (2007) Observations: Surface and Atmospheric Climate Change. In S Solomon, D Qin, M Manning, Z Chen, M Marquis, K B Averyt, M Tignor and H L Miller (eds). *Climate Change 2007: The Physical Science Basis*. Contribution of Working Group I to the Fourth Assessment Report of the Intergovernmental Panel on Climate Change. Cambridge, UK and New York, USA: Cambridge University Press.

Trepte, C R, Veiga, R E and McCormick, M P (1993) The poleward dispersal of Mount-Pinatubo volcanic aerosol. *Journal of Geophysical Research*, 98: 18563–73.

Tripathi, S N, Pattnaik, A and Dey, S (2007) Aerosol indirect effect over Indo-Gangetic plain, *Atmospheric Environment*, doi:10.1016/j.atmosenv.2007.05.007.

Turco, R P, Toon, O B, Park, C, Whitten, R C, Pollack, J B and Noerdlinger, P (1981) Tunguska Meteor Fall of 1908 – Effects On Stratospheric Ozone. *Science*, 214(4516): 19–23.

UBA (2004) *Manual on Methodologies and Criteria for Mapping Critical Levels/Loads and Geographical Areas Where They Are Exceeded*. UNECE Convention on Long- range Transboundary Air Pollution. Berlin: Umweltbundesamt.

UDI PRAHA (2002) *The Yearbook of Transportation in Cities 2001*, Institute of Transportation Engineering of the City of Prague, http:// www.udi-praha.cz/rocenky/Rocenka01vm/uk01vm.htm (accessed 24 June 2006).

UK Ministry of Health (1954) Mortality and Morbidity during the London Fog of December 1952. *Report on Public Health and Medical Subjects* 95, London: Ministry of Health.

UN (2001) *State of the World's Cities 2001*, Nairobi: United Nations Human Settlements Program (UN-HABITAT).

UN (2002) *World Urbanization Prospects: The 2001 Revision*, New York: United Nations, Report # ESA/P/WP.173.

UN (2004) *World Urbanization Prospects: The 2003 Revision*, New York: United Nations.

UN (2005) *World Population Prospects: The 2004 Revision*, New York: United Nations, http://esa.un.org/unpp (accessed 24 June 2006).

UN (2006) United Nations, Department of Economics and Social Affairs, Population Division (2006). *World Urbanization Prospects: The2005 Revision*. Work Paper No ESA/P/WP/200. http://www.un.org/esa/population/publications/WUP2005/2005 wup.htm.

UN (2009) World Urbanization Prospects: The 2009 Revision, United Nation, POP/DB/WUP/Rev.2009/2/F12

UNECE (1988) *Workshop on critical levels for direct effects of air pollution on forests, crops and materials (report)*. Bad Harzburg, FRG, March.

UNECE (2002) *Damage to Vegetation by Ozone Pollution*. Brochure produced by the ICP Vegetation and the ICP Forests, ed. Mills G, Sanz M J and Fischer R.

UNECE (2005) *Trends in Europe and North America: The Statistical Yearbook of the Economic Commission for Europe 2005*. Geneva: UNECE.

UNEP (2000) Sustainable mobility. *Industry and Environment* 23(4): ISSN 0378–9993.

UNEP/DEWA/GRID (2005) *Global Environment Outlook (Geo Data Portal)*, Geneva: United Nations Environment Programme, http://geodata.grid.unep.ch/ (accessed 24 June 2006).

UNEP/WMO (1996) *The science of climate change*, contribution to working group 1 to the second assessment report of the International Panel on Climate Change, UNEP and WMO, Cambridge University Press, UK.

USEPA (2007) Images were provided by J Clark J (personal communication) on Assignment with the Air and Waste Management Association & the International Union of Air Pollution Prevention and Environmental Protection Associations, Brighton, UK.

Valaoras, G, Huntzicker, J J and White, W H (1988) On the Contribution of Motor Vehicles to the Athenian Nephos – an Application of Factor Signatures. *Atmospheric Environment*, 22(5): 965–71.

Vallejo, M, Lerma, C, Infante, O, Hermosillo, A G, Riojas-Rodriguez, H and Cárdenas, M (2004) Personal exposure to particulate matter less than 2.5 $\mu m$ in Mexico City: a pilot study. *Journal of Exposure Analysis and Environmental Epidemiology*, 14: 323–29.

Vandvik, V, Aarestad, P A, Muller, S and Dise, N B (2010) Nitrogen deposition threatens species richness of grasslands across Europe. *Environmental Pollution*, 158: 2940–45.

van Dijk, H F G, van der Gaag, M, Perik, P J M and Roelofs, J G M (1992) Nutrient availability in Corsican pine stands in The Netherlands and the occurrence of Sphaeropsis sapinea: a field study. *Canadian Journal of Botany*, 70: 870–5.

van Dingenen, R, Dentener, F J, Raes, F, Krol, M C, Emberson, L and Cofala, J (2009) The global impact of ozone on agricultural crop yields under current and future air quality legislation. *Atmospheric Environment*, 43: 604–18.

van Dobben, H F (1996) Decline and recovery of epiphytic lichens in an agricultural area in The Netherlands (1900–1988). *Nova Hedwigia*, 62: 477–85.

van Dobben, H F and Ter Braak, C J F (1999) Ranking of epiphytic lichen sensitivity to air pollution using survey data: a comparison of indicator scales. *Lichenologist*, 31: 27–39.

van Dobben, H F, van Hinsberg, A, Kros, J, Schouwenberg, E P A G, de Vries, W, Jansen, M, Mol-Dijkstra, J P and Wieggers, H J J (2006) Simulation of critical load for nitrogen for terrestrial plant communities in The Netherlands. *Ecosystems*, 9: 32–45.

van Herk, C M (2001) Bark pH and susceptibility to toxic air pollutants as independent causes of changes in epiphytic lichen composition in space and time. *Lichenologist*, 33: 419–41.

van Herk, C M, Aptroot, A and van Dobben, H F (2002) Long-term monitoring in the Netherlands suggests that lichens respond to global warming. *Lichenologist*, 34: 141–54.

van Lieshout, M, Kovats, R S, Livermore, M T J and Martens, P (2004) Climate change and malaria: analysis of the SRES climate and socio-economic scenarios. *Global Environmental Change*, 14: 87–99.

van Noije, T P C, Eskes, H J, Van Weele, M and van Velthoven, P F J (2004) Implications of enhanced Brewer-Dobson circulation in European Centre for Medium-Range Weather Forecasts reanalysis for the stratosphere-troposphere exchange of ozone in global chemistry transport models. *Journal of Geophysical Research*, 109: D19308, doi:10.1029/2004JD004586.

van Noije, T P C, Eskes, H J, Dentener, F J, Stevenson, D S, Ellingsen, K, Schultz, M G, Wild, O, Amann, M, Atherton, C S, Bergmann, D J, Bey, I, Boersma, K F, Butler, T, Cofala, J, Drevet, J, Fiore, A M, Gauss, M, Hauglustaine, D A, Horowitz, L W, Isaksen, I S A, Krol, M C, Lamarque, J-F, Lawrence, M G, Martin, R V, Montanaro, V, Müller, J-F, Pitari, G, Prather, M J, Pyle, J A, Richter, A, Rodriguez, J M, Savage, N H, Strahan, S E, Sudo, K, Szopa, S and van Roozendael, M (2006) Multi-model ensemble simulations of tropospheric NO2 compared with GOME retrievals for the year 2000. *Atmospheric Chemistry and Physics*, 6: 2943–79.

van Roosbroeck, S, Jacobs, J, Janssen, N A H, Oldenwening, M, Hoek, G and Brunekreef, B (2007) Long-term personal exposure to $PM_{2.5}$, soot and $NO_x$ in children attending schools located near busy roads, a validation study. *Atmospheric Environment*, 41(16): 3381–94.

van Vuuren, D P and O'Neill, B C (2006) The consistency of IPCC's SRES scenarios to recent literature and recent projections. *Climatic Change*, 75: 9–46.

van Vuuren, D P, Lucas, P L and Hilderink, H (2007) Downscaling drivers of global environmental change: Enabling use of global SRES scenarios at the national and grid levels. *Global Environmental Change*, 17, 114–30.

Vasil'ev, N V and Fast, N P (1973) New material on the 'light nights' of summer 1908. In J Ikaunieks (ed.) *The Physics of Mesopheric (Noctiluscent) Clouds*, Proceedings of the Conference on Mesospheric Clouds, Riga, 20–23 November 1968, Jerusalem: Israel Program for Scientific Translation, 80–5.

Vaughan, J K, Claiborn, C and Finn, D (2001) April 1998 Asian dust event over the Columbia Plateau. *Journal of Geophysical Research*, 106: 18381–402.

Verma, S, Boucherb, O, Upadhyaya, H C and Sharma, O P (2006) Sulfate aerosols forcing: An estimate using a three-dimensional interactive chemistry scheme. *Atmospheric Environment*, 40: 7953–62.

Vitousek, P M and Howarth, R W (1991) Nitrogen limitation on land and in the sea – how can it occur? *Biogeochemistry*, 13: 87–115.

Vukovich, F M (1995) Regional-scale boundary-layer ozone variations in the eastern United States and their association with meteorological variations. *Atmospheric Environment*, 29: 2259–73.

Wahid, A (2003) Air pollution impacts on vegetation in Pakistan. In L D Emberson, M Ashmore, F Murray (eds). *Air Pollution Impacts on Crops and Forests: a Global Perspective*, London: Imperial College Press, 189–213.

Wahid, A, Maggs, R, Shamsi, S R A, Bell, J N B and Ashmore, M R (1995) Effects of air pollution on rice yield in the Pakistan Punjab. *Environmental Pollution*, 90: 323–9.

Wallace, L (1996) Indoor particles: a review, *Journal of Air Waste and Management Association*, 46: 98–126.

Wallace, L (2000) Correlations of personal exposure to particles with outdoor air measurements: a review of recent studies, *Aerosol Science and Technology*, 32: 15–25.

Wang, X and Mauzerall, D L (2004) Characterising distributions of surface ozone and its impacts on grain production in China, Japan and South Korea. *Atmospheric Environment*, 38: 4383–402.

Wang, X, Bi, X, Sheng, G and Fu, J (2006) Hospital indoor PM10/PM2.5 and associated trace elements in Guangzhou, China. *Science of the Total Environment*, 366: 124– 35.

Warwick, H and Doig, A (2004) *Smoke – the Killer in the Kitchen, Indoor Air pollution in Developing Countries*, London: ITDG Publishing.

Weschler, C J (2004) Chemical reactions among indoor pollutants: what we've learned in the new millennium. *Indoor Air*, 14, 184–94.

WHO (World Health Organization) (2000 [1986]) *WHO Air Quality Guidelines for Europe*, 2nd edn, Copenhagen: World Health Organization.

WHO (World Health Organization) (2003) *Health Aspects of Air Pollution with Particulate Matter, Ozone and Nitrogen Dioxide*, Report on a WHO working group, Bonn, Germany, 13–15 January.

WHO (World Health Organization) (2005) *WHO air quality guidelines global update 2005*, Report on a working group meeting, Bonn, Germany, 18–20 October.

WHO (World Health Organization) (2006) *WHO Air quality guidelines for particulate matter, ozone, nitrogen dioxide and sulfur dioxide: Global update 2005. Summary of Risk Assessment.* World Health Organization, Geneva.

WHO/UNEP (World Health Organization / United Nations Environment Programme) (1992) *Urban Air Pollution in Megacities of the World*. Oxford: Blackwell.

Williamson, T (2002) *London Smog 50th Anniversary*, Brighton: National Society for Clean Air.

Wilson, S R, Solomon, K R and Tang, X (2007) Changes in tropospheric composition and air quality due to stratospheric ozone depletion and climate change, *Photochemical & Photobiological Sciences*, 6, 301–310, DOI: 10.1039/b700022g.

Wirahadikusumah, K (2002) *Air quality management in Jakarta: trends and challenges, Better Air Quality Workshop*, Hong Kong, 16–18 December.

World Bank. (2003) *The World Bank Clean Air Initiative in Sub-Saharan African Cities*. 1998–2002 Progress Report. World Bank African Region.

WMO (World Meteorological Organization) (2003a) *The Global Climate System Review, June 1996 – December 2001*. Geneva: WMO, WMO-No. 950.

WMO (World Meteorological Organization) (2003b) *Scientific assessment of ozone depletion: 2002*. Global Ozone Research and Monitoring Project, Rep., vol. 47.

Wong, C M, Lam, T H, Peters, J, Hedley, A J, Ong, S G, Tam, A Y, Liu, J and Spiegelhalter, D J (1998) Comparison between two districts of the effects of an air pollution intervention on bronchial responsiveness in primary school children in Hong Kong. *Journal of Epidemiology and Community Health*, 52, 571–8.

WRI (World Resources Institute) (1996) *World Resources Institute 1996–1997*, New York: Oxford University Press.

WRI (World Resources Institute) (1998) *World Resources 1998–99: Environmental Change and Human Health*, Washington, DC: the World Resources Institute, the United Nations Environment Programme, the United Nations Development Programme and the World Bank, http://population.wri.org/pubs_content.cfm?PubID=2889 (accessed 24 June 2006).

Xu Y, Zhao, Z-C, Luo, Y and Gao, X (2005) Climate change projections for the 21st century by the NCC/IAP T63 with SRES scenarios. *Acta Meteorologica Sinica*, 19, 407–17.

Yttri, K E and Tørseth, K (2005) Transboundary particulate matter in Europe Status report 2005. EMEP Report 4/2005.

Yttri, K E and Aas, W (2006) Transboundary particulate matter in Europe Status report 2006. EMEP Report 4/2006.

Yttri, K E, Aas, W, Tørseth, K, Stebel, K, Vik, A F, Fjæraa, A M, Hirdman, D, Tsyro, S, Simpson, D, Marečková, K, Wankmüller, R, Klimont, Z, Kupiainen, K, Amann, M, Bergström, R, Nemitz, E, Querol, X, Alastuey, A, Pey, J, Gehrig R (2010) Transboundary particulate matter in Europe Status report 2010. EMEP Report 4/2010.

Yunus, M and Iqbal, M (1996) *Plant Response to Air Pollution*, Chichester: John Wiley and Sons.

Zender, C, Miller, R and Tegen, I (2004) Quantifying Mineral Dust Mass Budgets: Terminology, Constraints, and Current Estimates. *EOS, Transactions American Geophysical Union*, 85(48): 509. 10.1029/2004EO480002.

Zhang, J, Rao, S T and Daggupaty, S M (1998) Meteorological processes and ozone exceedances in the northeastern United States during the 12–16 July 1995 episode. *Journal of Applied Meteorology*, 37, 776–89.

Zhang, J and Rao, S T (1999) The role of vertical mixing in the temporal evolution of ground-level ozone concentrations. *Journal of Applied Meteorology*, 38: 1674–91.

Zheng, Y and Shimizu, H (2003) Air pollution impacts on vegetation in China. In L D Emberson, M Ashmore, F Murray (eds). *Air Pollution Impacts on Crops and Forests: a Global Assessment*, London: Imperial College Press, 123–44.

Ziemke, J R, Chandra, S, Duncan, B N, Froidevaux, L, Bhartia, P K, Levelt, P F and Waters, J W (2006) Tropospheric ozone determined from Aura OMI and MLS: Evaluation of measurements and comparison with the Global Modelling Initiative's Chemical Transport Model. Journal of Geophysical Research, 111: D19303 http://dx.doi.org/ 10.1029/2006JD007089.

# LIST OF USEFUL READING MATERIAL


Andrews, D G (2000) *An Introduction to Atmospheric Physics*, Cambridge: Cambridge University Press.

Bell, J N B and Treshow, M (eds) (2002) *Air pollution and plant life*, 2nd edn, Chichester: John Wiley and Sons.

Botkin, D B and Keller, E A (2011) *Environmental Science: Earth as a Living Planet*, Hoboken, NJ: John Wiley and Sons.

Gego, E Hogrefe, C, Rao, S T and Porter, P S (2003) Probabilistic assessment of regional-scale ozone pollution in the eastern United States, in Melas, D and Syrakov, D (eds), *Air Pollution Processes in Regional Scale*, Dordrecht, Netherlands: Kluwer Academic Publishers.

Hill, M K (1997) *Understanding Environmental Pollution*, Cambridge: Cambridge University Press.

Houghton, J (2004) *Global Warming: The Complete Briefing*, 3rd edn, Cambridge: Cambridge University Press.

IPCC (2001) *IPCC Third Assessment Report on Climate Chance 2001: The Scientific Basis*, ed. Houghton, J T, Ding, Y, Griggs, D J, Noguer, M, van der Linden, P J, Dai, X, Maskell, K and Johnson, C A, Cambridge: Cambridge University Press.

Jacobson, M Z (2002) *Atmospheric Pollution – History, Science and Regulation*, Cambridge: Cambridge University Press.

Kallenborn R (ed.) (2011) *Long-range Transport of Man-made Contamination into the Arctic and Antarctica (From Pole to Pole)*, New York: Springer.

Larssen, S, Barrett, K J, Fiala, J, Goodwin, J, Hagan, L O, Henriksen, J F, de Leeuw, F and Tarrason, (2002) *Air Quality in Europe; State and Trends 1990–99*, Copenhagen, Denmark: European Environment Agency.

Maslin, M (2007) *Global Warming – Causes, Effects and the Future, St Paul*, MN: MBI Publishing Company and Voyageur Press.

McGranahan, G and Murray, F (eds) (2003) *Air Pollution and Health in Rapidly Developing Countries*, London: Earthscan.

Metcalfe, S and Derwent, D (2005) *Atmospheric Pollution and Environmental Change*, London: Hodder Arnold.

Schneider, S H, Rosencranz, A, Mastrandrea, M D, Kuntz-Duriseti, K (eds) (2010) *Climate Change – Science and Policy*, Washington DC: Island Press.

Schwela, D, Haq, G, Huizenga, C, Han, W-J, Fabian, H and Ajero M (2006) *Urban Air Pollution in Asian Cities*. Earthscan, UK.

Scorer, R S (1997) *Dynamics of Meteorology and Climate*, Chichester: John Wiley and Sons.

Seinfeld, J H and Pandis, S N (1998) *Atmospheric Chemistry and Physics – From Air Pollution to Climate Change*, Chichester: John Wiley and Sons.

Solomon, S, Qin, D, Manning, M, Chen, Z, Marquis, M, Averyt, K B, Tignor, M and Miller, H L (eds) (2007) *Contribution of Working Group I to the Fourth Assessment Report of the Intergovernmental Panel on Climate Change*, Cambridge: Cambridge University Press.

*State of the World 2007: Our Urban Future*. A Worldwatch Institute report on the progress towards a sustainable society. Earthscan, UK.

Tiwary, A and Colls, J (2010) *Air Pollution – Measurement, Modelling and Mitigation*, Abingdon: Routledge.

Tsonis, A A (2002) *An Introduction to Atmospheric Thermodynamics*, Cambridge: Cambridge University Press.

Vallero, D (2007) *Fundamentals of Air Pollution*, London: Academic Press.

Wallace, J M and Hobbs, P V (2006) *Atmospheric Science – An Introductory Survey*, London: Academic Press.

Wright, R T and Boorse D (2010) *Environmental Science: Toward a Sustainable Future*, Boston, MA: Addison Wesley.

# INDEX

Printed in the USA
CPSIA information can be obtained
at www.ICGtesting.com
JSHW041426221024
72172JS00001B/1

* 9 7 8 1 8 4 3 3 1 8 9 1 0 *

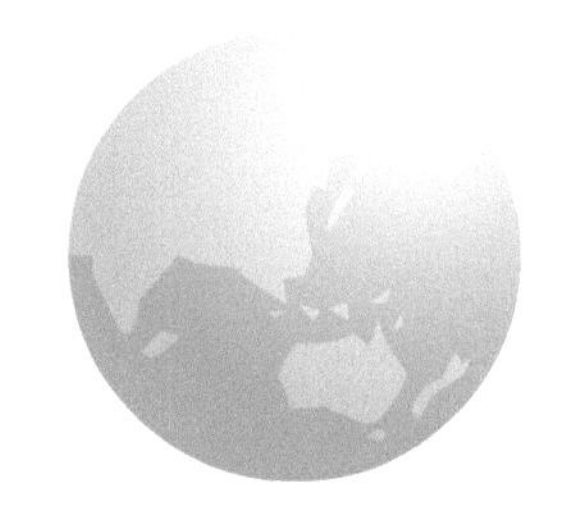

Manchester anticipated the sanitary reforms that became more general in Britain, Europe and North America by the mid-nineteenth century. In Manchester inspectors were employed to examine industrial boilers and chimneys and tried to encourage good practice. The problems of the early nineteenth century were significant because population had shifted from the countryside to the polluted cities. Early attempts to control pollutants were largely overwhelmed by a lack of effective technology, industrial pressure and insufficient political will (Bowler and Brimblecombe 1990; Mosley 2001).

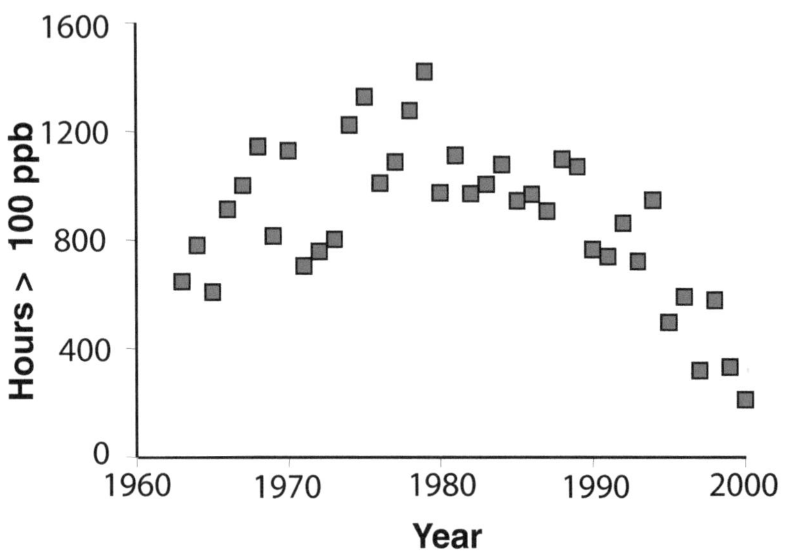


Figure 1.3d Variation of ozone concentrations from 1950 to 2000 measured in Crestline, California.

***Pittsburgh***

An abundance of coal in Pittsburgh led to its early use and the growth of powerful industry. As early as 1804 General Presley Neville complained of the smoke, and although an air pollution ordinance appeared by 1888, it was ineffective. For a short time in the 1880s, natural gas was cheap and readily available, which lowered the amount of smoke. The situation was so bad again by 1912 that the Mellon Institute undertook an in-depth study of the smoke problem. The early twentieth century saw faltering local steps to control the smoke, which finally took effect with the major changes in fuel use of the 1950s. Improvements continued, although increasingly driven by national legislation (Davidson 1979).

## 1.4 Air Pollution Disasters and Episodes

Air pollution episodes and disasters may be brief, but their social and legislative impacts can be exceedingly important. Some major air pollution disasters are described below and highlighted in Figure 1.4.

**1948, Donora**
Particulate matter and sulphur dioxide from the zinc works in Donora, Pennsylvania, became associated with stagnant and foggy air in October 1948. What had for years been a commonplace nuisance became a tragedy as firemen had to take oxygen to residents, especially the elderly, struggling to breathe.

**1952, London**
In the winters of the late nineteenth and twentieth centuries, stationary high pressure systems settled over western Europe, wind speeds fell and temperature inversions formed. In the winter of 1952, a serious fog developed in a calm winter week (5–9 December). Some 4000 to 12000 Londoners died in December from smog-related respiratory illnesses.

**1930, Meuse Valley, Belgium**
In the first week of December 1930, Belgium was engulfed by fog and along the Meuse River there were large numbers of factories whose chimneys were just below the inversion layer (approximately 80 m). Under the still conditions and increasing air pollution concentrations, some 60 people died, while thousands may have become ill.

**1910, Selby Smelter**
In the late nineteenth century there was increasing concern about widespread damage to vegetation from industrial emissions. The US Bureau of Mines' extensive 1915 study of Selby, California, recognized the importance of sulphur dioxide.

**1984, Bhopal**
A catastrophic accident at the Union Carbide factory in Bhopal, India, in 1984 released 50000 gallons of methyl isocyanate (MIC), which caused more than 2500 deaths.

**1950, Poza Rica, Mexico**
A little-known industrial accident occurred in the oil-refining town of Poza Rica, when a dense cloud of hydrogen sulphide from natural gas processing passed through the town for almost half an hour. Illnesses soon appeared and 22 people died.

**1976, Seveso**
In July 1976, an explosion at a chemical plant in Seveso, Italy, released a range of chemicals such as ethylene glycol, trichlorophenol and sodium hydroxide. However, a harmful chemical called 2,3,7,8 tetrachlorodibenzodioxin was also released in the explosion.

**1973, Auckland**
The case that occurred in Parnell, a suburb of Auckland, New Zealand, in 1973 is truly paradigmatic. This incident started after merphos, a pesticide that contained two organophosphorus compounds, leaked from some barrels.

45°
45°
0°
-45°
-45°
-180° 0° 180°

Figure 1.4 World map of air pollution disasters and episodes.

# LONG-RANGE TRANSPORT OF ATMOSPHERIC POLLUTANTS AND TRANSBOUNDARY POLLUTION

*S Trivikrama Rao, Christian Hogrefe, Tracey Holloway and George Kallos*

3.1 Regional Air Pollution Transport

- Transport of ozone in the eastern United States
- Transport of sulphur compounds in the Mediterranean

3.2 Hemispheric Air Pollution Transport

3.3 Methods for Analysing Long-Range Transport of Air Pollution

- Satellite observations
- Statistical analysis of measurements
- Trajectory analysis
- Dynamic air quality models

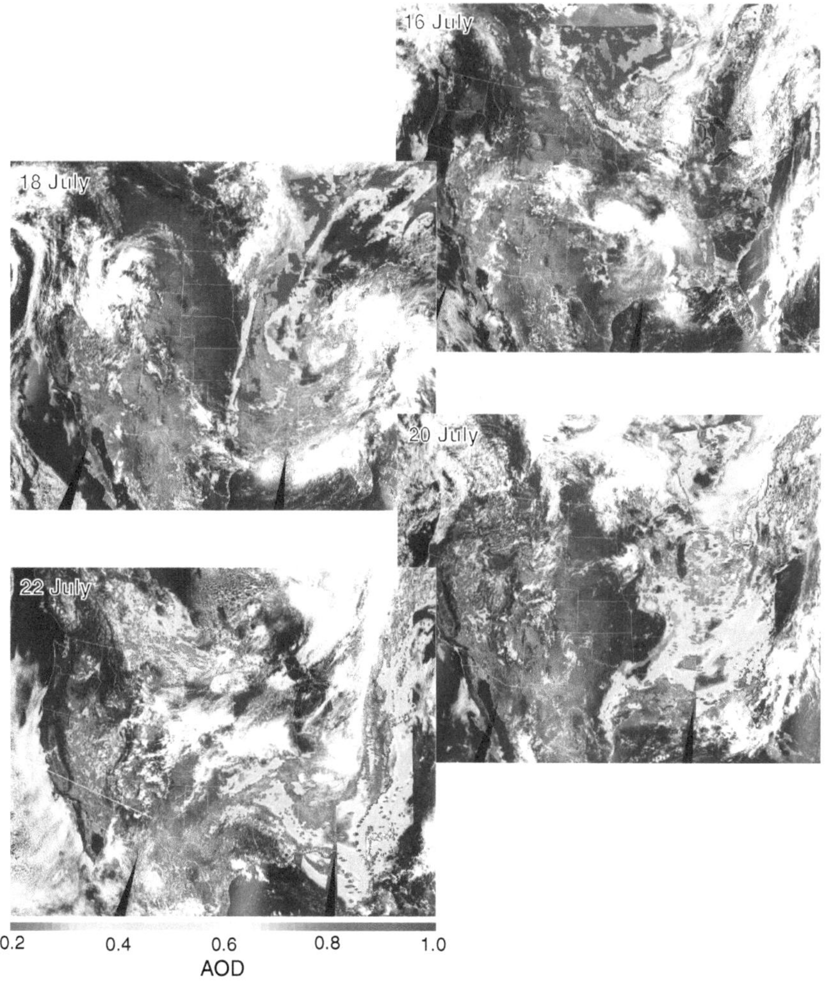


During the 2004 summer, the largest Alaskan wild fire event on record occurred in late June-July and consumed 2.72 million hectares of boreal forest. The Figure shows the aerosol optical depth (AOD) data from the MODIS instrument aboard the Terra satellite for a series of days in July 2004. The MODIS AOD is plotted over the MODIS Terra true color image for each day. These series of days show high aerosol loading associated with long-range transport of the Alaskan wild fire plume as it crosses over the northern border of the United States on July 16. This aerosol plume was advected south-eastward behind the cold front (evident in the clouds captured in the MODIS true color) over the following days, eventually affecting surface $PM_{2.5}$ levels along the Eastern United States.

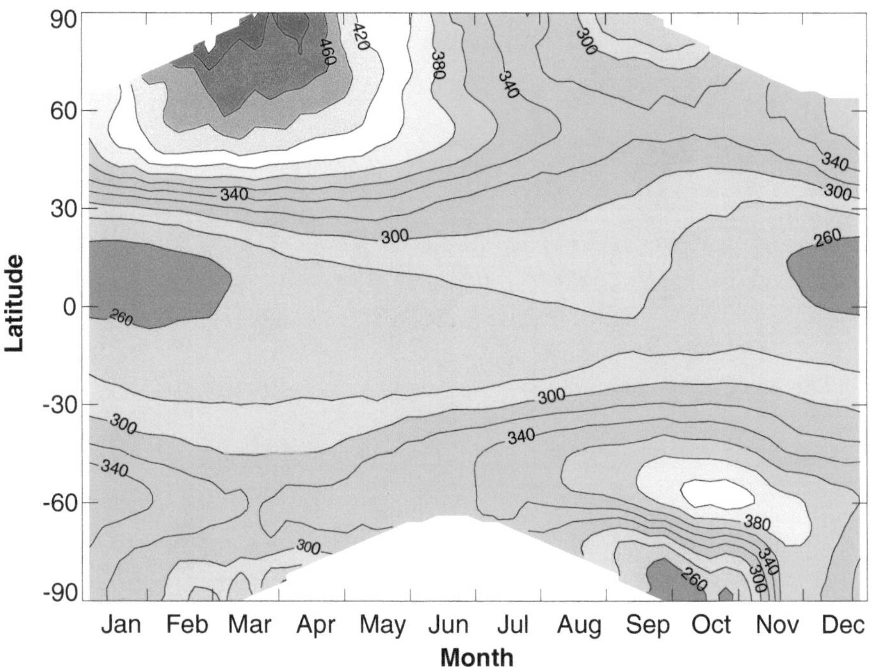


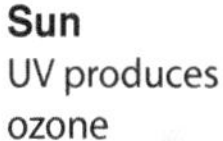


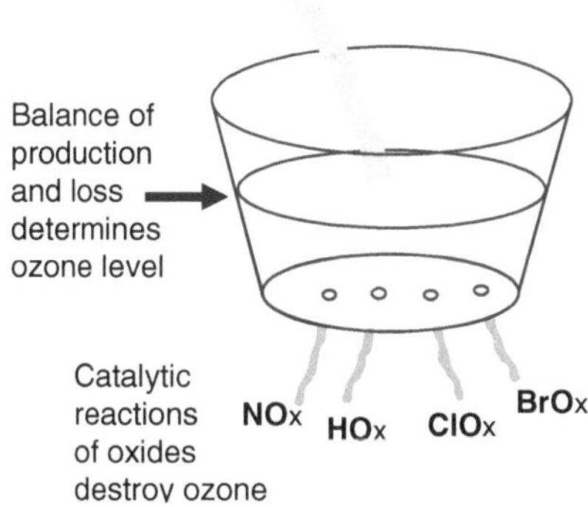


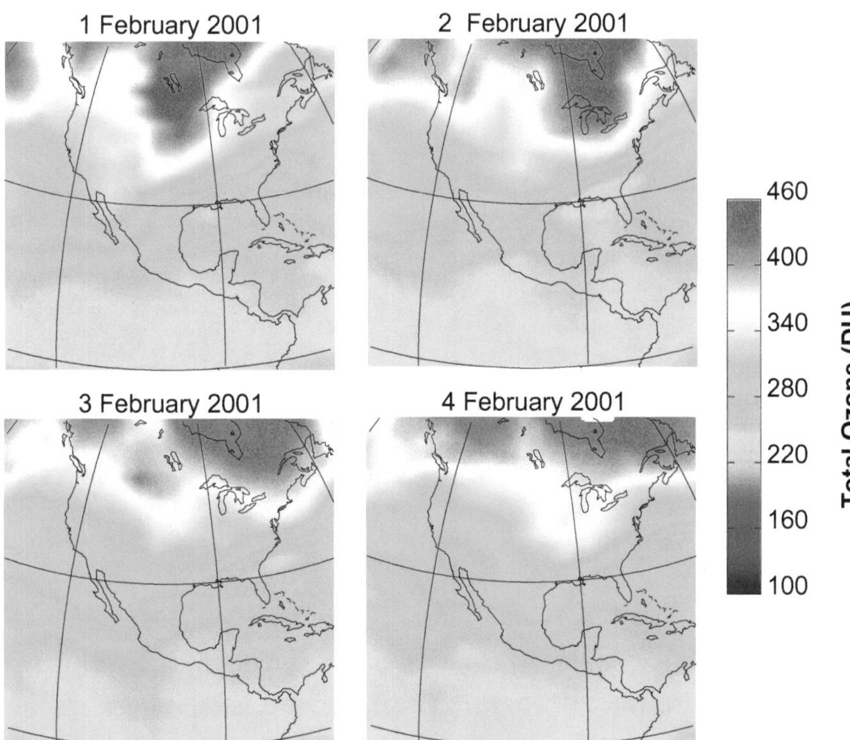

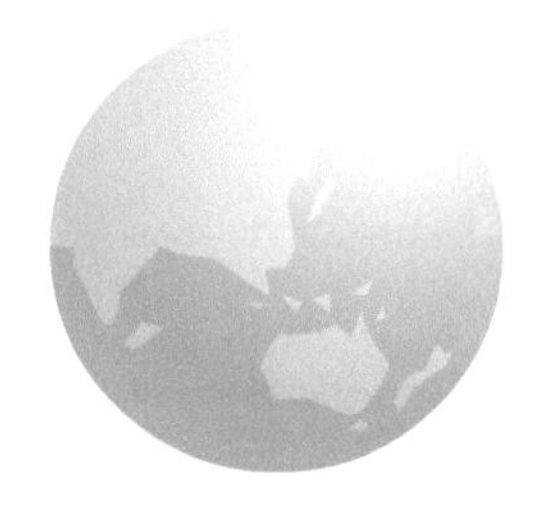